University of Michigan Studies

SCIENTIFIC SERIES

VOLUME V

REPORTS OF THE GREENLAND EXPEDITIONS OF THE UNIVERSITY OF MICHIGAN

Vol. V = Part I
Vol. VI = Part II
Vol. VII = Part III
Vol. VIII = Part IV

THE volumes of the University of Michigan Studies and Publications are published by authority of the Board of Regents under the direction of the Executive Board of the Graduate School. The Contributors are chiefly, but not exclusively, members of the faculties or graduates of the University. The expense is borne in part by gifts, in part by appropriations of the Board of Regents. A list of volumes already published or in preparation is given at the end of this volume. The volumes of this series are listed on page v of the advertising.

PLATE I

A. View of the summit of Mount Evans and the aërological station (photograph by Fred Herz)

B. The theodolite anchorage and other meteorological equipment (photograph by David M. Potter)

REPORTS

OF THE

GREENLAND EXPEDITIONS

OF THE

UNIVERSITY OF MICHIGAN

(1926–31)

WILLIAM HERBERT HOBBS, Director

PART I

AËROLOGY

EXPEDITIONS OF 1926 AND 1927-29

S. P. FERGUSSON, Editor

ANN ARBOR
UNIVERSITY OF MICHIGAN PRESS
1931

Set up and printed,
April, 1931

PRINTED IN THE UNITED STATES OF AMERICA
BY THE PLIMPTON PRESS · NORWOOD · MASS.

Paperback ISBN: 978-0-472-75135-8

ACKNOWLEDGMENTS

THE Greenland Expeditions of the University of Michigan (1926–29) were financed entirely by subscriptions received from alumni and other friends of the enterprise. These subscriptions amounted to $55,405.66 and included a grant of $10,000 from the Daniel Guggenheim Foundation for the Promotion of Aeronautics. The largest individual benefactors were the late James B. Ford of New York City, who contributed $5,696.60; Mr. Roy D. Chapin of Detroit, who contributed $5,375; and the late Charles F. Brush of Cleveland, whose contributions amounted to $5,000. The expeditions have been placed under obligation to Dr. C. F. Marvin, the head of the United States Weather Bureau, for permission granted to members of its staff to join them, and for the loan of much meteorological equipment. A full list of the benefactors of the Expeditions has been published in the Appendix to a narrative account of the Expeditions under the title, *Exploring about the North Pole of the Winds* (G. P. Putnam's Sons, New York, 1930).

The reduction of the observations and their preparation for publication has been made possible through a grant from the Faculty Research Fund of the University. The printing and publication of the report are due to action of the Executive Board of the Graduate School of the University, to which Board indebtedness is hereby gratefully acknowledged. Other reports of the Expeditions are now in preparation and their scope is indicated by a list which is published at the close of this volume.

THE DIRECTOR

February 20, 1931

CONTENTS

ILLUSTRATIONS

PLATES

FIGURES IN THE TEXT

I. INTRODUCTION

BY

WILLIAM HERBERT HOBBS

Object of the expeditions. — The Greenland Expeditions of the University of Michigan (1926–29) have had for their prime object the study of the meteorological conditions over and about the inland-ice dome which so nearly covers the continent (Fig. 1). Earlier studies based on the observations made by explorers have clearly indicated that over this northern ice-mass is located a fixed permanent and powerful anticyclone (glacial anticyclone), which as a reversing mechanism of the general circulation may be regarded as a northern wind pole of the earth in the same sense that the similar mechanism over the Antarctic continent may be looked upon as a southern wind pole.[1] The studies projected have depended upon pilot-balloons, supplemented in 1926 by registering balloons — *ballons sondes.*

Earlier studies had indicated that the domination of atmospheric conditions by the circulation above the ice is much modified as regards the surface winds at distances a few tens of miles only outside the ice margin, and so success in our studies would depend very largely upon the location of the observing station. It was clear from the start that the meteorological records, particularly the direction and force of the wind made regularly at a few coastal settlements (generally partly sheltered within the mouths of fjords), would afford quite misleading indications. In part on this account that portion of Greenland was selected for our studies where the land ribbon surrounding the inland-ice is the widest, viz., the Holstensborg district in the southwest, where this land stretches out to a width of close upon a hundred miles (Fig. 2).

The Holstensborg district already enjoyed a Greenland reputation for relative dryness, which had been confirmed by Professor Otto Nordenskjöld when he visited the district in 1909.[2]

[1] William Herbert Hobbs, *The Glacial Anticyclones; The Poles of the Atmospheric Circulation*, with an Introduction by Hugh Robert Mill. University of Michigan Studies, Scientific Series, Vol. 4, 1926, pp. xxiv+198, 3 pls. and 53 text-figures.

[2] Otto Nordenskjöld, "Einige Züge der physischen Geographie und der Entwickelungsgeschichte Süd-Grönlands," *Geogr. Zeitschr.*, Vol. 20, 1914, pp. 425–441, 505–524, 628–641.

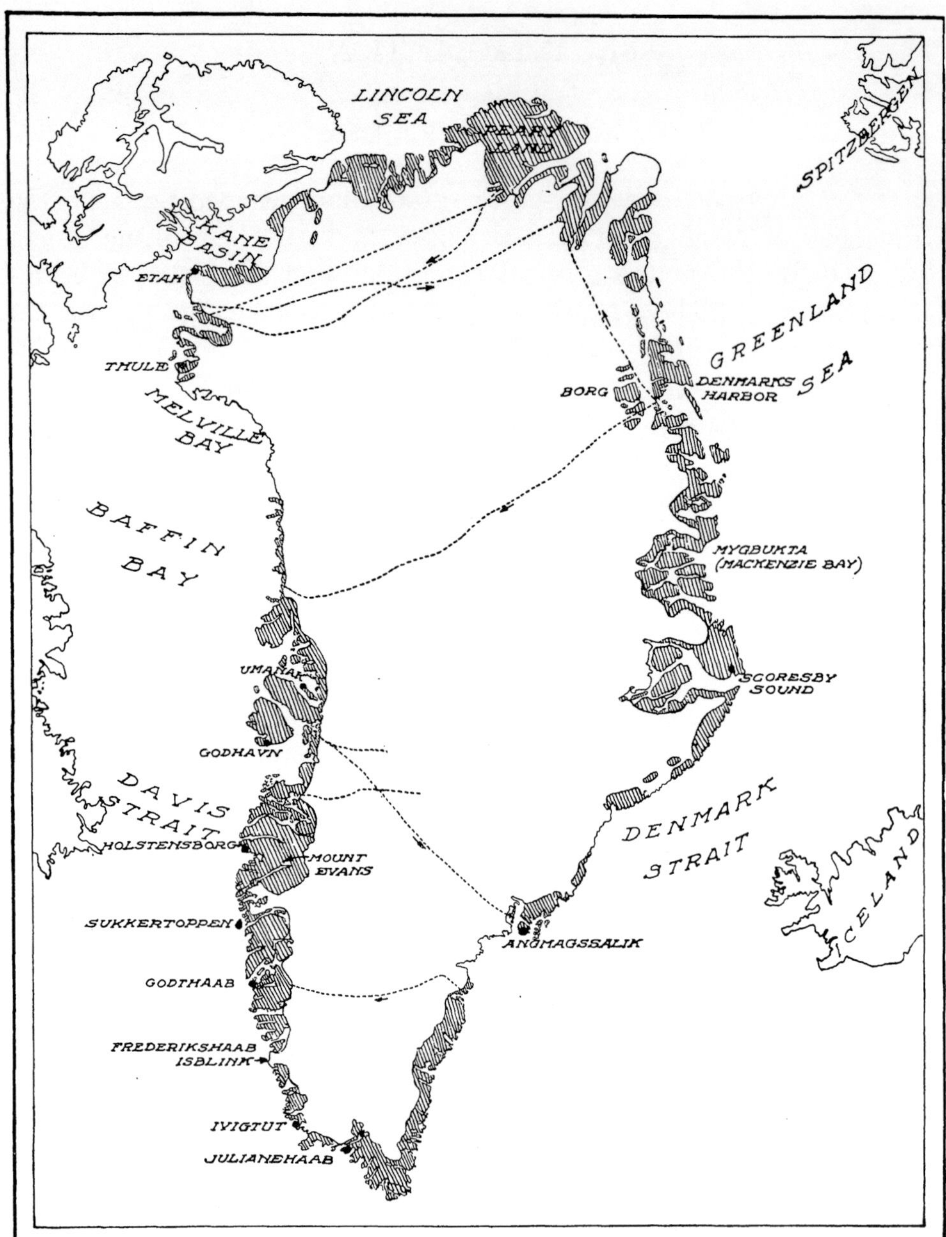

FIG. 1. Map of Greenland

Personnel of the expeditions. — The members of the provisional expedition of 1926, which was based on the Maligiakfjord (Camp Michigan) were, besides the director: Dr. Laurence M. Gould, assistant professor of geology at the University, second-in-command, geologist and photographer; S. P. Fergusson, associate meteorologist of the United States Weather Bureau, aërologist; Professor James E. Church, Jr., of the University of Nevada, meteorologist; Ralph L. Belknap, instructor in geology at the University, surveyor

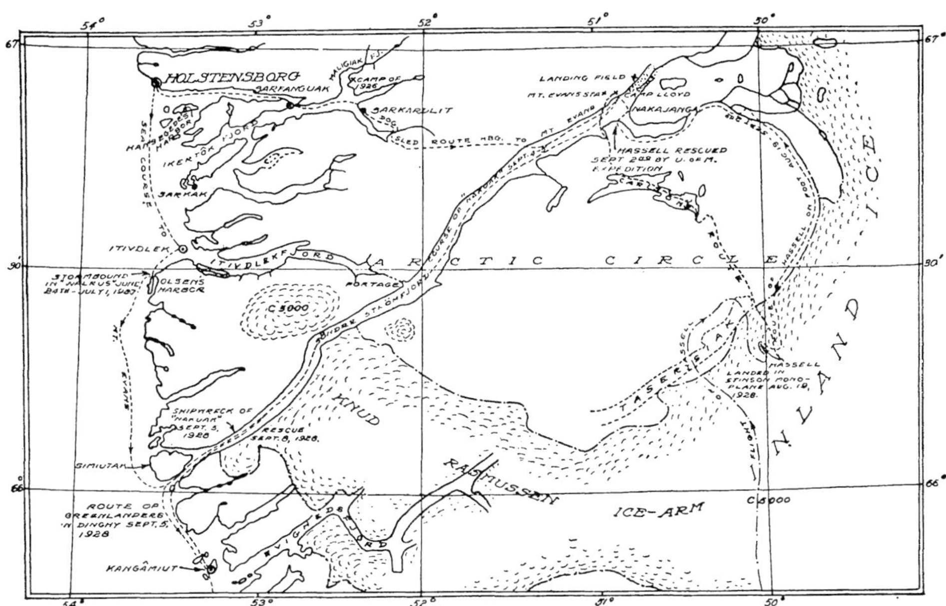

Fig. 2. Outline map of the Holstensborg district of southwest Greenland

and geologist; P. C. Oscanyan, Jr., radio operator and assistant to Mr. Fergusson in the balloon ascensions.

In the Second Expedition of 1927, which was based on the head of the Söndre Strömfjord at Camp Lloyd and Mount Evans, there were, in addition to the director: Ralph L. Belknap, second-in-command, surveyor and geologist; Professor J. E. Church, Jr., meteorologist; Clarence R. Kallquist of the United States Weather Bureau, aërologist; Carl O. Erlanson, instructor in botany at the University of Michigan, botanist; Fred Herz, mechanic; and P. C. Oscanyan, Jr., radio operator. Of this group, Church, Kallquist and Oscanyan, with Helge Bangsted, the well-known Greenland explorer of Copenhagen, composed the winter staff of the station. Bangsted, in charge, Church, and the Eskimo Marius composed the winter icecap party of that year. William S. Carlson, assistant in the department of geology at the university, assistant aërologist, joined the Expedition in April, 1928, and was a member of the station's staff until the spring of 1929. From May, 1928, until July 10, 1929, he was in charge of the station. He also led the dog-sled expedition to the inland-ice during the winter of 1928–29.

The Third Expedition of 1928, based like the Second Expedition on Mount Evans, comprised, besides the director: R. L. Belknap, second-in-command, surveyor and geologist; Clarence R. Schneider, graduate student at Clark University, aërologist; William S. Carlson, assistant aërologist, May, 1928, to May, 1929, and Evans S. Schmeling, assistant aërologist, May to July, 1929; Elmer Etes, mechanic; Francis M. Baer, radio operator, July to October, 1928; Karl Hansen, radio operator, November, 1928, to July, 1929; Helge Bangsted in charge of Eskimo helpers, July to late August, 1928. The winter party consisted of Schneider, in charge, Carlson, Hansen, and, later, in place of Carlson, Schmeling.

Acknowledgments. — The expeditions have been greatly indebted to Dr. C. F. Marvin, chief of the U. S. Weather Bureau for the loan of much valuable equipment supplied to all expeditions, as well as for the loan of Mr. Fergusson. In preparing the observations for pub-

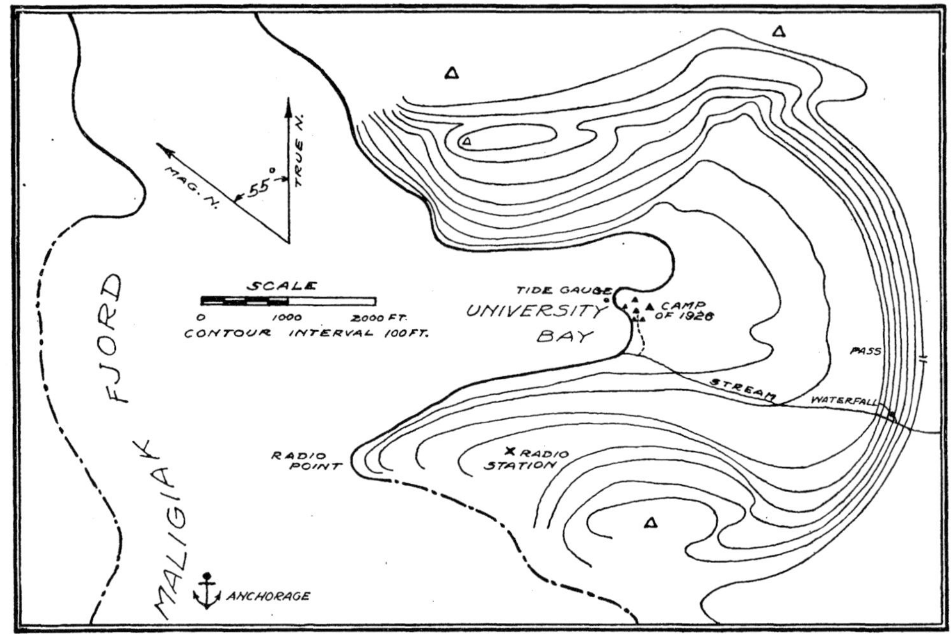

FIG. 3. Map of the base of the Expedition in 1926

lication Mr. S. P. Fergusson has been assisted by William S. Carlson, Evans S. Schmeling and Mary Elizabeth Cooley. The graphs are the work of William Cristanelli.

The base of the First Expedition. — The first expedition of 1926 was based at Camp Michigan on the head of the Maligiakfjord, thirty miles due east of the coast settlement of Holstensborg in latitude 66°55′. The site was the bottom of an amphitheatre about a half mile in diameter close to sea-level and surrounded by precipitous walls of rock approximately one thousand feet in height (see Fig. 3 and Pl. VI A, B). This site, while undesirable for our balloon studies, was admirably adapted for a camping place because of a supply of good water and a fair landing place near the head of motor-sloop navigation. Search in the neighborhood developed no better site that was accessible, and so the observations of the first season were made within this amphitheatre.

As was to be expected, the ground observations on the wind force and direction had little value, since they were determined by the local topography. In fact, within all narrower fjords of southwest Greenland winds blow almost invariably along the fjord either in one direction or the other, whereas upon the sea close to the shore, owing to the precipitous rock walls, winds blow almost invariably along the shore, at least during the season of navigation, which is from April to November or even December. It is in part for these reasons that synoptic charts prepared from the daily ground observations made at the coast settlements of Greenland are generally misleading.

Observations on the ice-cap expedition of 1926. — On July 26, 1926, an expedition set out for the inland-ice. This expedition lasted twenty-one days and was made by *umiak*, canoe and on foot. It was participated in by four of the party: Dr. Laurence M. Gould, Professor James E. Church, Jr., Mr. Ralph L. Belknap, and the Director, together with four Eskimo helpers. Professor Church acted as meteorologist and throughout the expedition at three-hour intervals carried out observations of air pressure, temperature, humidity, wind force and direction, to be used for comparison with the ground observations being simultaneously taken at the camp on the Maligiakfjord.

A temporary camp, which was named Camp Mortimer E. Cooley, was set up for a few days beside the Otto Nordenskjöld ice-tongue, and here on August 5 pilot-balloons were sent up and followed to altitudes of 4,000 and 5,400 meters. On August 7 two pilot-balloons were sent up from the surface of the Otto Nordenskjöld Glacier, which balloons were both lost at about 2,000 meters altitude. Those sent up from the ice surface at first moved toward the inland-ice, and at an elevation of only 1,000 meters reversed direction and moved off to the northwest under the influence of slope winds, and they continued in this direction until lost. A balloon sent up on the same day from the camp on the Maligiakfjord nearly seventy-five miles farther to the west continued to be under the influence of the southeast winds up to an altitude of 3,500 meters (Pl. III A and B).

Balloon ascensions on the Maligiakfjord. — At the camp on the Maligiakfjord more than ninety pilot-balloon runs were obtained between July 22 and September 1. Owing to the protection afforded by the local amphitheatre from which these balloons were launched, they rose to the height of the rim of the amphitheatre on steep angles (generally 75 degrees or more) that could not be read on the theodolite until a special sighting attachment had been made use of. Fortunately, as the balloons rose higher they came under the influence of winds from easterly quadrants, and since the opening to the amphitheatre was on the western side they could thus be followed without difficulty. At higher levels they generally reversed direction, and since they were by that time so high in the air as to be within the field afforded by the amphitheatre, the site chosen proved to be much less unsatisfactory than might have been supposed. These balloon observations are to be found in Chapter III of this report, followed

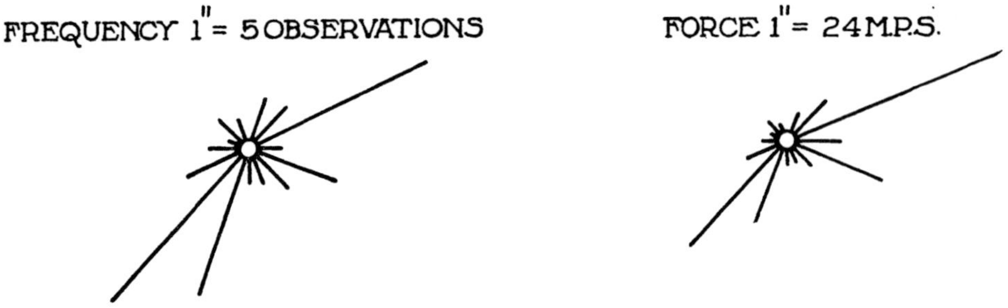

FREQUENCY 1" = 5 OBSERVATIONS

FORCE 1" = 24 M.P.S.

FIG. 4. Wind roses for the 500-meter level at Camp Michigan in 1926

in Chapter IV by graphs showing progressive changes at all heights reached. The wind roses for the 500-meter level, which is near the average level of the country, are given in Figure 4.

Observations with ballons sondes. — Mr. Fergusson was successful in applying the Rossby device for deflating sounding balloons at any desired height by use of a fuse. In these experiments he employed the light type of meteorograph designed for the United States Weather Bureau, but owing to the nearness of the fjord it was necessary to fit a light raft made from three six-inch balloons, which had been inflated to about ten inches diameter, beneath the meteorograph. Two successful ascents were made to heights of 400 and 1,700 meters, and a third balloon was sent up for 3,000 meters. It was lost, but afterwards recovered intact after reaching an altitude of 5,300 meters. A fourth attempt resulted in a permanent loss of the equipment. The results are discussed in Chapter VII (see Pl. VI B).

Secondary meteorological station at Holstensborg. — At the end of the season of 1926 a ground meteorological station was established at Holstensborg with instruments provided by the United States Weather Bureau, and the governor's assistant was at first engaged to read these instruments at the usual intervals. Later, these duties were assumed by others, but the records were kept up until the closing of the Mount Evans observatory late in July, 1929. In the report on meteorology comparative study of the observations made at Holstensborg with those regularly taken at Mount Evans and on special occasions, both on the inland-ice itself and within the hinterland, will be fully discussed by Mr. Fergusson.

The Mount Evans base. — The base on the Maligiakfjord for the season of 1926 failed to meet the conditions which it had been hoped would be secured in Greenland. In the following year it was found possible to establish a base at the head of the Söndre Strömfjord (*Kangerdlugssuak*) in latitude 66°51′ north and longitude 50° 55′ west (determination by our surveyor, Mr. Belknap). This is in approximately the same latitude as both Holstensborg and the camp on the Maligiakfjord, though situated in the hinterland only about twenty-five miles distant from the margin of the inland-ice and more than three times that distance inland from the coast. Nowhere else has it been possible to get so far removed from coastal conditions and so near to the inland-ice, and still be accessible by water navigation from the coast. It was thought, moreover, that the great arm of ice which at a high level is thrust out westward toward the coast from the main mass of the inland-ice, the Knud Rasmussen ice-arm, which is a unique feature for Greenland (Fig. 2), would make of the area to the north a relatively arid region. This it proved to be, for the annual precipitation at the new base on the north side of the fjord proved to be between five and six inches of water for each of the two years that the station was open.

The observatory was set up on a rounded summit 1,294 feet above tide, as determined by theodolite readings from a base line near the fjord (Fig. 5). On the summit of this elevation, which was named Mount Evans, an observatory combining radio station and living quarters was erected. Later an additional storehouse and also a balloon-inflating shelter were provided, though the latter

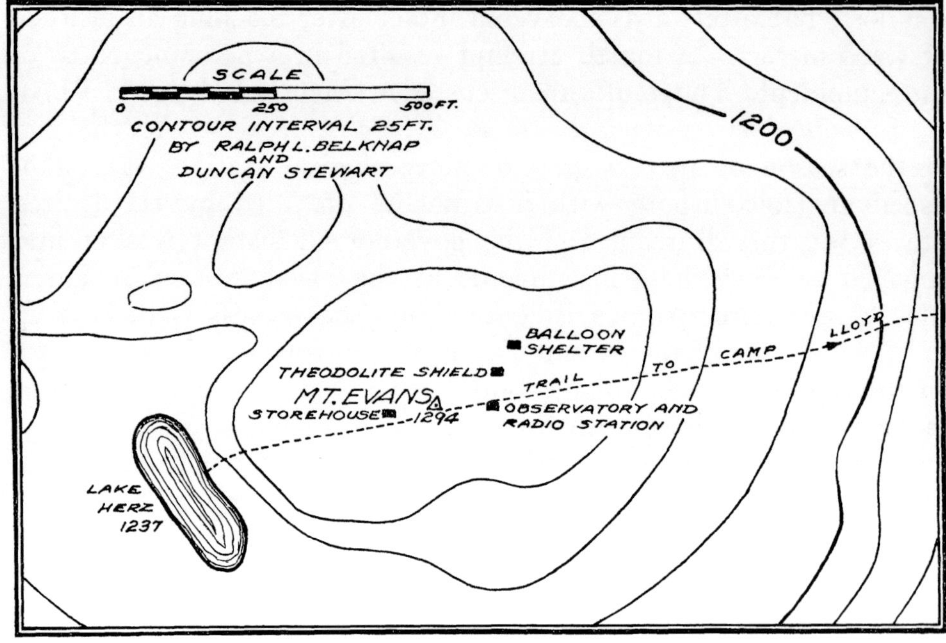

FIG. 5. Map of the summit of Mount Evans

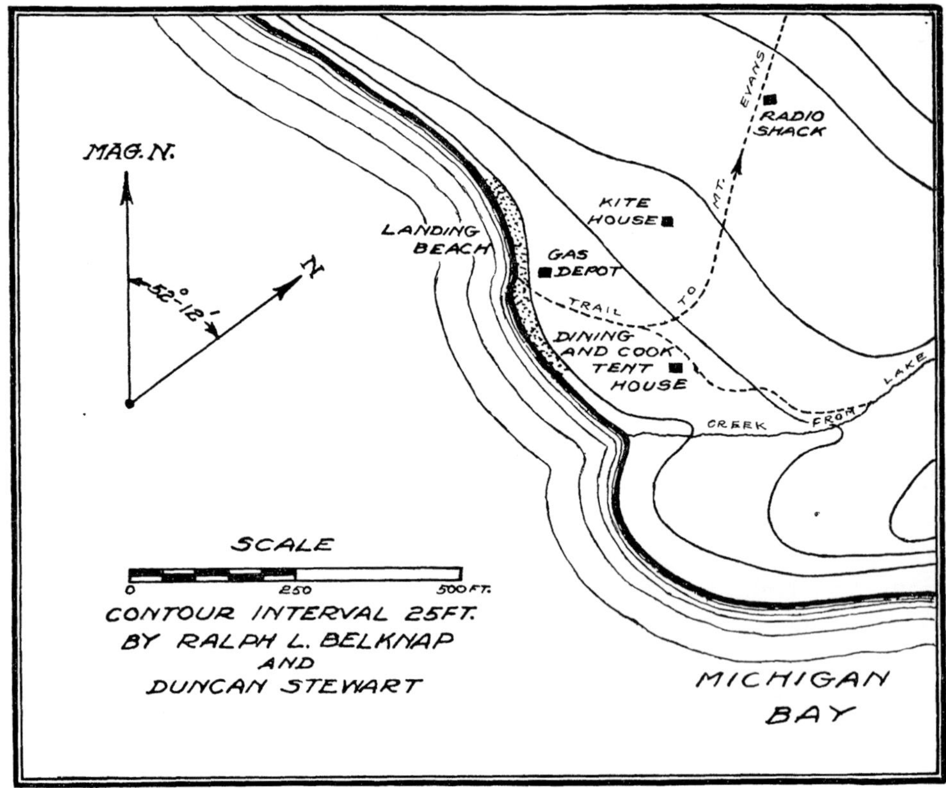

FIG. 6. Map of Camp Lloyd and vicinity

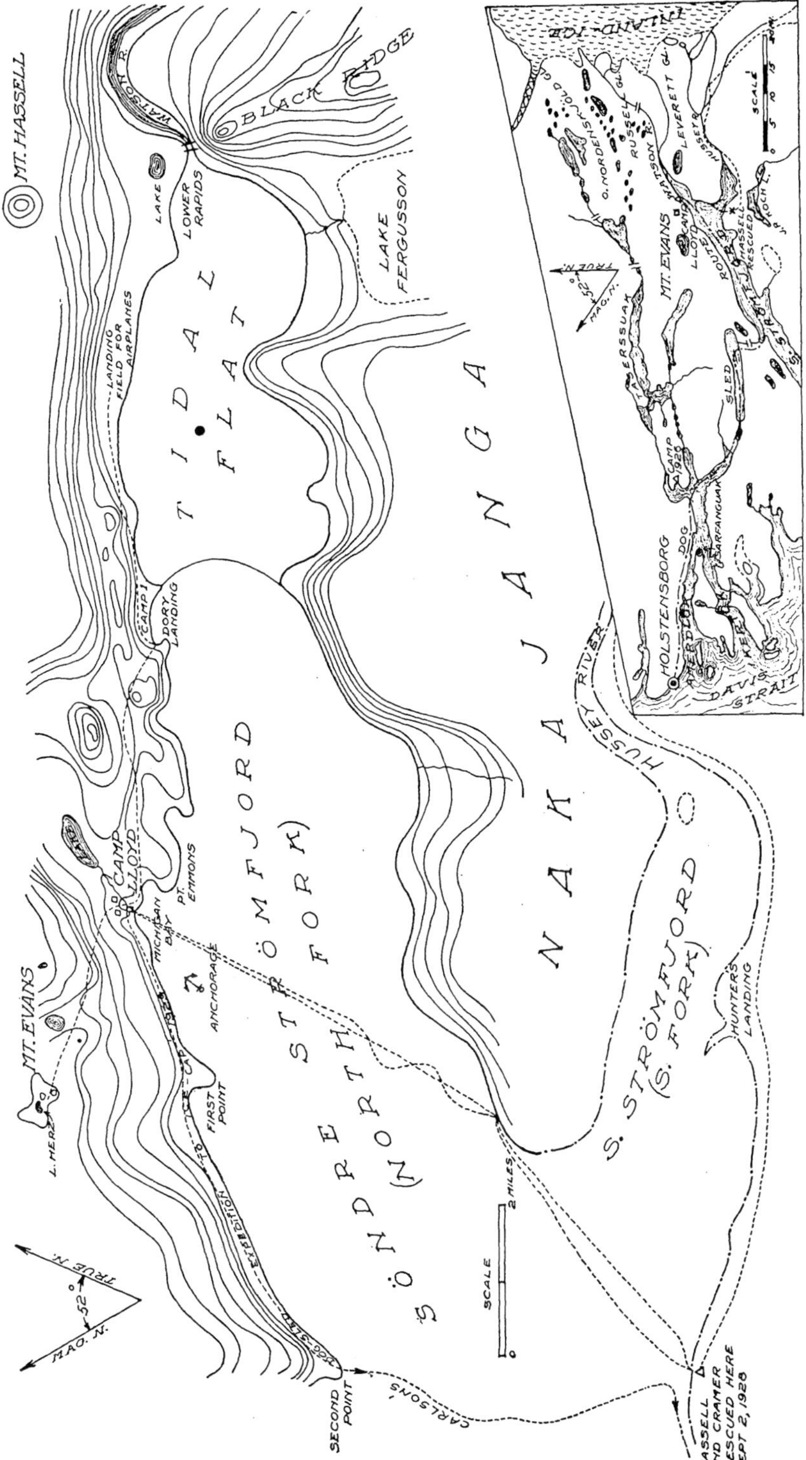

Fig. 7. Map of the area about the head of the Söndre Strömfjord

(Insert shows the Holstensborg district. Maps by Ralph Belknap)

could be used only during the summer months (Fig. 5 and Pl. II B). The district surrounding Mount Evans and Camp Lloyd is shown in Figures 6–7.

The tripod of the Buff and Buff theodolite used in following the balloons it was necessary to anchor in position by heavy rocks (see

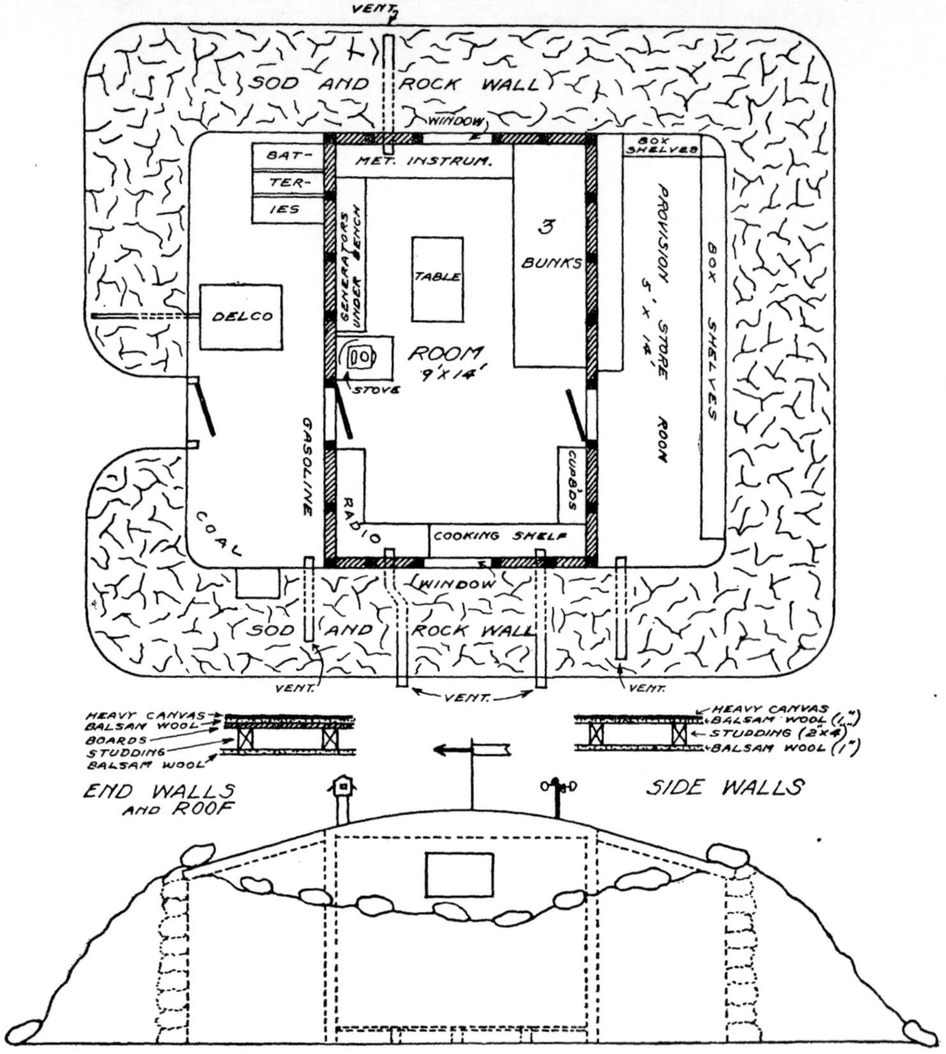

FIG. 8. Plan and elevation of the aërological observatory on Mount Evans

Pl. I B) and later a board shield strongly braced and within a rock wall was set up upon the windward side for the protection of the observers, especially during the winter months (see Pl. II A).

The observatory site was distant about three miles from the summer base of the expedition at Camp Lloyd, where was the land-

PLATE II

A. Wind shield of the theodolite stand (photograph by David M. Potter)

B. Summer balloon-inflating station (photograph by Fred Herz)

PLATE III

A. Sending up a pilot-balloon from the Otto Nordenskjöld glacier (photograph by Laurence M. Gould)

B. Sending up a pilot-balloon at Camp Cooley (photograph by Laurence M. Gould)

PLATE IV

A. Winter view of the summit of Mount Evans. On skyline, left to right, balloon-inflating shelter, theodolite wind shield and station. In center near summit the storehouse. In foreground Lake Herz with snow house (photograph by L. R. Schneider)

B. Site of the Norwegian aërological station at Mackenzie Bay (photograph supplied by Dr. Th. Hesselberg)

PLATE V

A. Kite work at Camp Lloyd (photograph by David M. Potter)

B. Nephoscope stand in use (photograph by David M. Potter)

PLATE VI

A. Following a pilot-balloon at Camp Michigan (photograph by W. H. Hobbs)

B. *Ballon sonde* ascensions with Rossby device at Camp Michigan (photograph by Laurence M. Gould)

ing place for supplies on the fjord (Fig. 6). A fairly good trail connected Camp Lloyd and Mount Evans. Both places were well supplied with good water (Fig. 7).

The observatory — The exposed position of the station in the sweep of winds off the inland-ice made it necessary to anchor the building strongly and to give the wind little opportunity to take a firm grip upon it. The building was designed by Mr. Fred Herz, the able mechanic and photographer of the Second Expedition. Its plan and the essentials of its construction are indicated in Figure 8. A view of the completed building is shown in Plate I A. It proved well adapted for its purpose and few changes were indicated as desirable by the Expedition during two years of occupation. In the summer of 1928 the outside door was widened by a few inches to permit the passage of pilot-balloons, which during the cold weather must be inflated inside the building. The generators were also moved from the central living and instrument room to the outer storehouse near the Delco generator.

The building withstood storms with wind velocities of 120 miles per hour, as measured by the single-register (three-cup) anemometer of Weather Bureau pattern, and remained firmly anchored in its place, though the radio mast outside was several times blown down (Pl. IV A).

Kite studies at Camp Lloyd. — On the expedition of 1928 kite equipment supplied by the United States Weather Bureau was brought in to the base and a kite house was erected near Camp Lloyd. On the terrace above Camp Lloyd the hand reel was put up and on favorable days kite ascensions were carried out, but to such moderate heights only — they did not reach above the level of the observatory — that little of value was learned from them (Pl. V A).

Summer expedition to the ice-cap in 1927. — In the summer of 1927 several trips across the tundra to the inland-ice were made, and one of these penetrated about eight miles over the Russell Glacier, where pilot-balloon ascensions were made. On this expedition Professor Church carried out three-hourly meteorological observations, as he had already done in the preceding summer. These studies are to be included in the comparative study of meteorological conditions upon the ice-cap margin, on the hinterland, and at coastal stations.

The ice-cap expedition of the winter of 1927–28. — In the months of January, February, and March, 1928, Mr. Helge Bangsted with

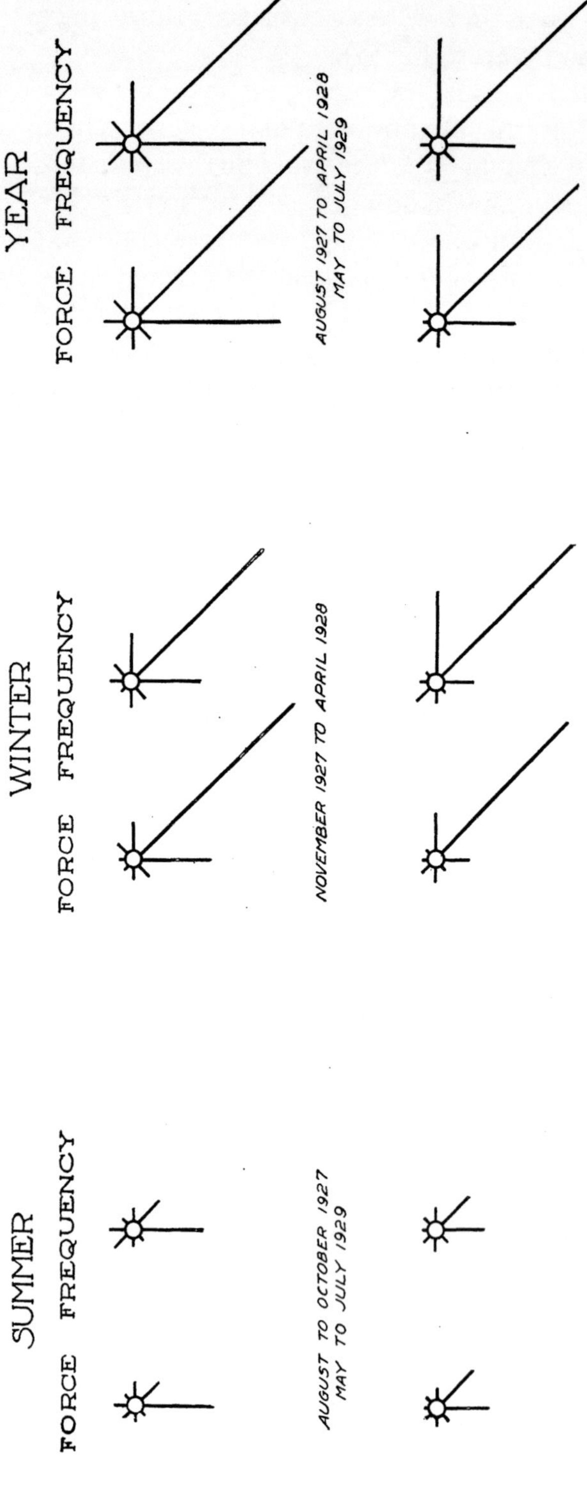

Fig. 9. Surface wind roses at Mount Evans for the strong winds (20 m.p.h. and more)

Professor James E. Church and the Eskimo Marius, went in from Camp Lloyd to the Russell Glacier outlet, and, despite exceptional difficulties due to the scant snowfall and to prevalent foehn storms, they carried out ground meteorological observations from a tent on the surface of the ice during a period of forty days. The observations by Mr. Bangsted and Professor Church will be included in the volume devoted to meteorology.

Domination by the glacial anticyclone. — Practically all the strong winds at Mount Evans come from the southeasterly quarters and blow down off the inland-ice lying to the eastward. The currents are deviated in the clockwise direction through earth rotation. This is clearly shown by the wind roses for all winds in excess of twenty miles per hour velocity (see Fig. 9). The importance of the southerly component is probably explained in part at least by winds off the long Knud Rasmussen ice-arm lying directly south of the station. The domination by the glacial anticyclone is most marked for the winter season and shows greatest variation in the month of May. Lighter winds often come from the northerly or even the westerly quadrants. The seasonal and yearly wind roses for all surface winds at Mount Evans are given in Figure 10.

To eliminate largely the effect of near topography, wind roses were prepared for the level of 750 meters above tide, which is about 400 meters (393 m.) above the station (Fig. 11). Already the winds from the westerly quadrants have largely disappeared. Here, however, the easterly and northeasterly winds, which must also come off the ice plateau, assume an importance almost equal to those of the southeasterly quadrant.

The lower outflowing and higher inflowing winds. — To determine the changes in the wind's force and direction at higher levels the diagrams of Figures 12–18 have been prepared, and these are of very special interest in showing at what levels the surface winds from the inland-ice give place to the inflowing currents above. For the midsummer months of July and August, 1927 (Fig. 12), the graphs betray a close relationship. The graph for July represents relatively few observations, since the runs began on the twenty-first; but except for the erratic changes observed between 1,000 and 2,500 meters, they follow the same course as do those of August. Above 4,000 meters the winds are all from the northwesterly quadrant, though from 2,000 to 4,000 meters southwesterly winds prevail.

In striking contrast to the midsummer currents are those of September and October, which much resemble each other. Here the

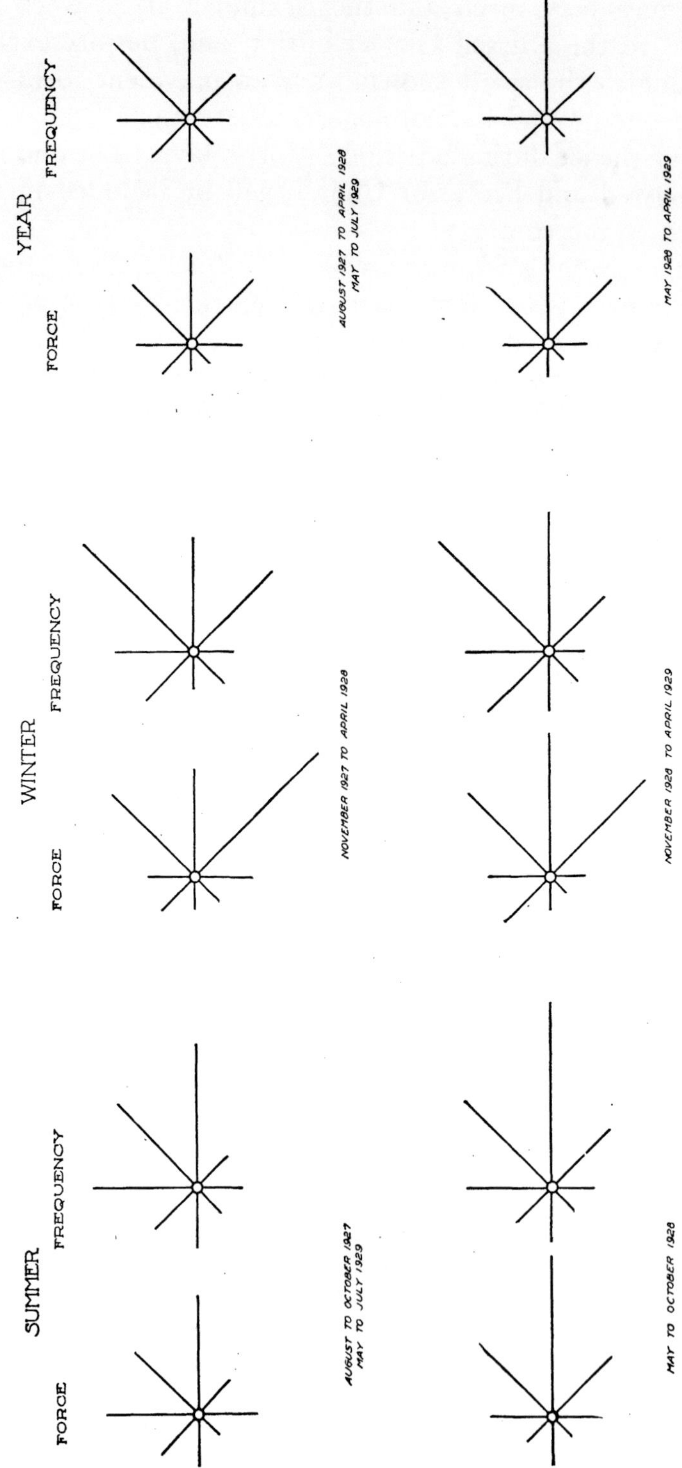

Fig. 10. Seasonal and annual wind roses for all surface winds at Mount Evans

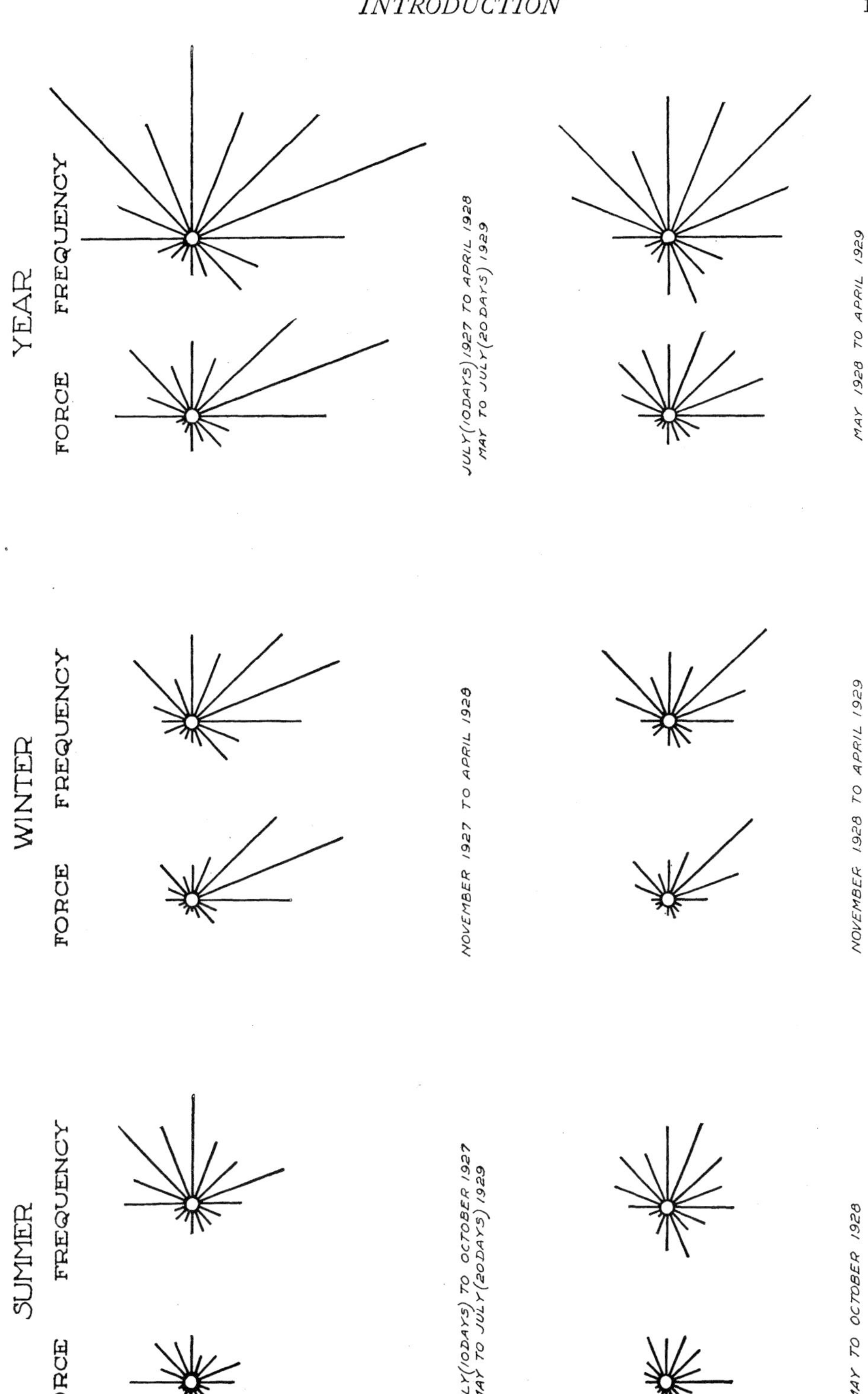

Fig. 11. Seasonal and annual winds at Mount Evans at 393 meters above the station

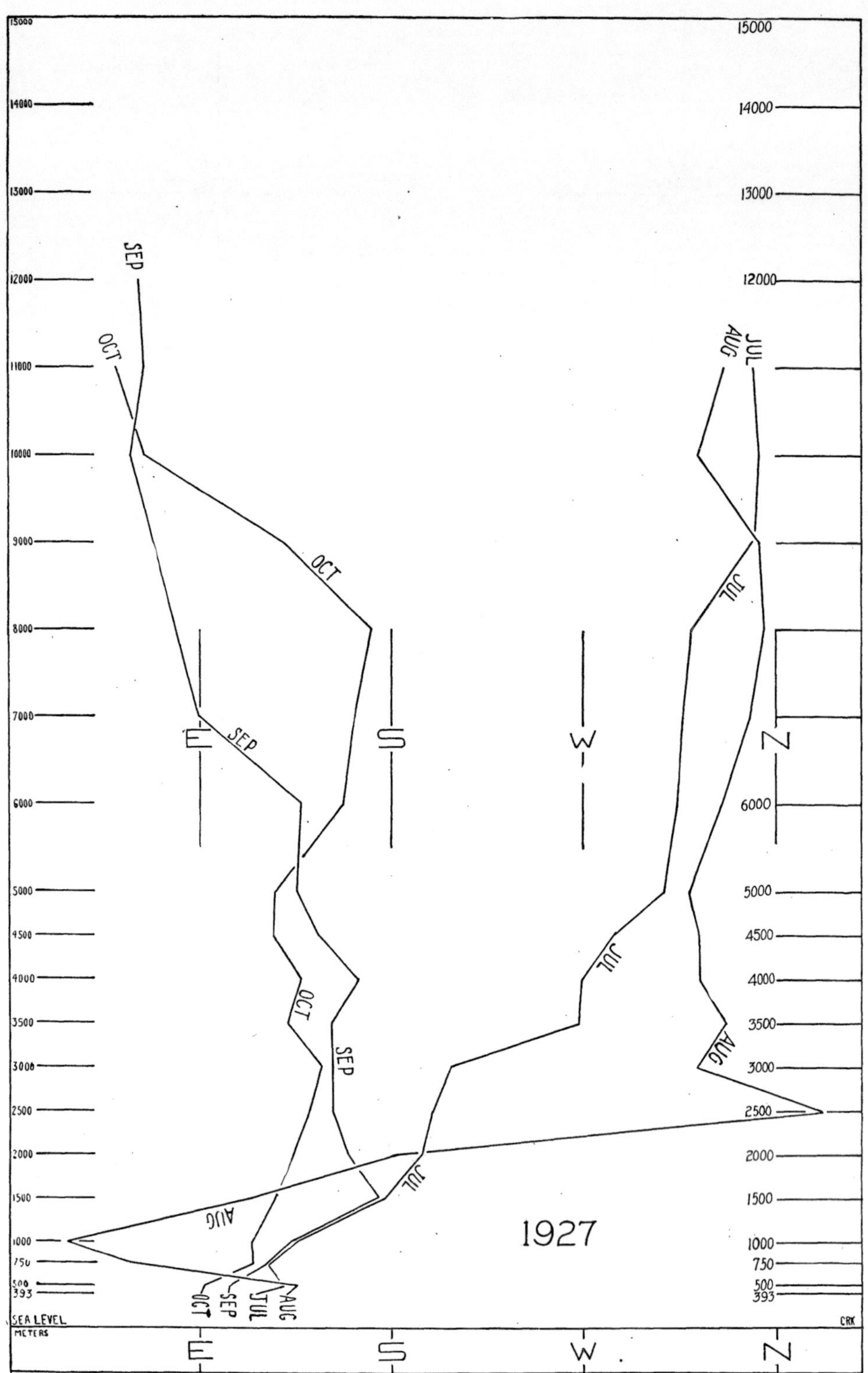

FIG. 12. Average upper air currents at Mount Evans for the months of July, August, September, and October, 1927

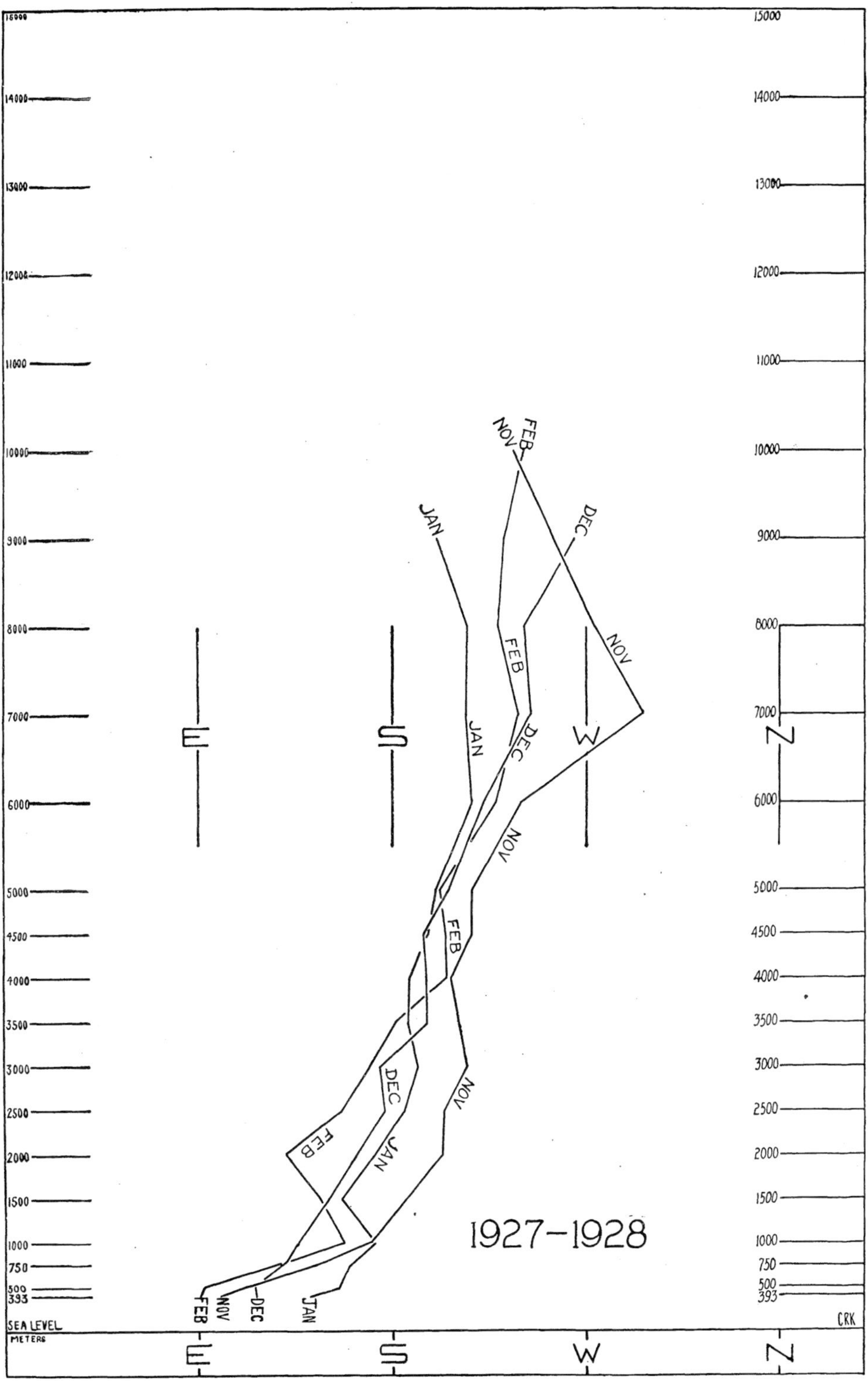

FIG. 13. Average upper air currents at Mount Evans for the months of November and December, 1927, and January and February, 1928

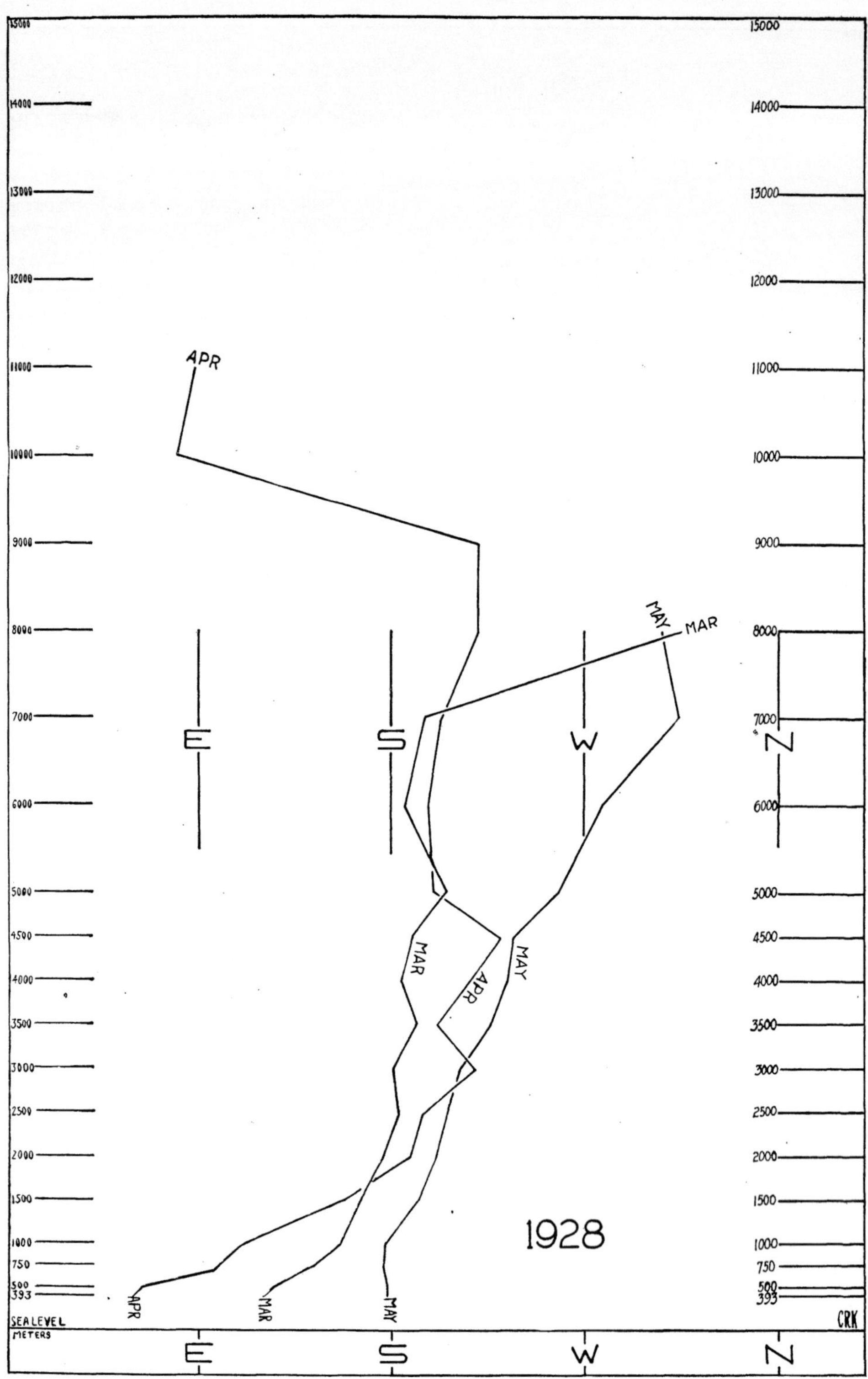

Fig. 14. Average upper air currents at Mount Evans for the months of March, April, and May, 1928

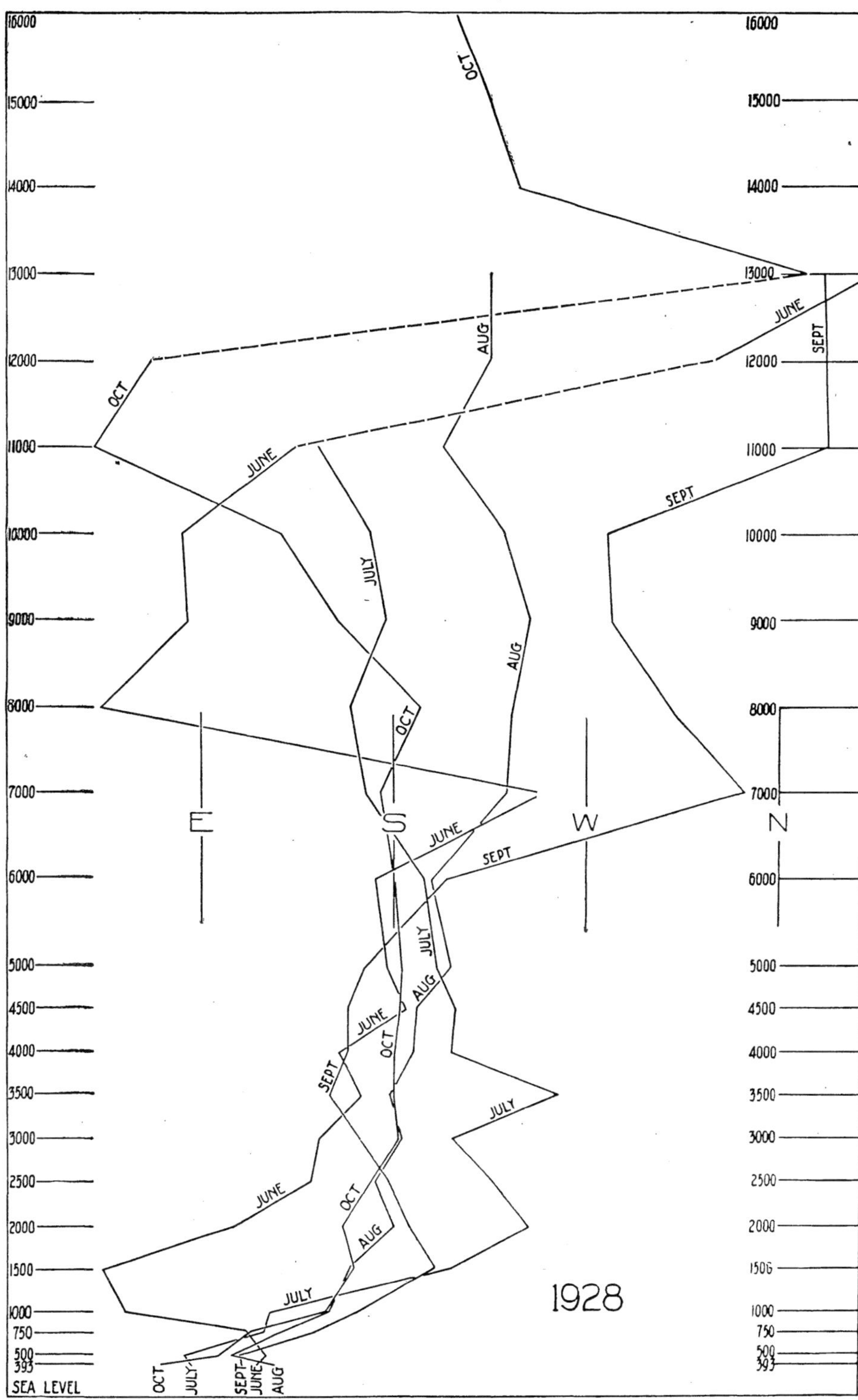

FIG. 15. Average upper air currents at Mount Evans for the months of June, July, August, September, and October, 1928

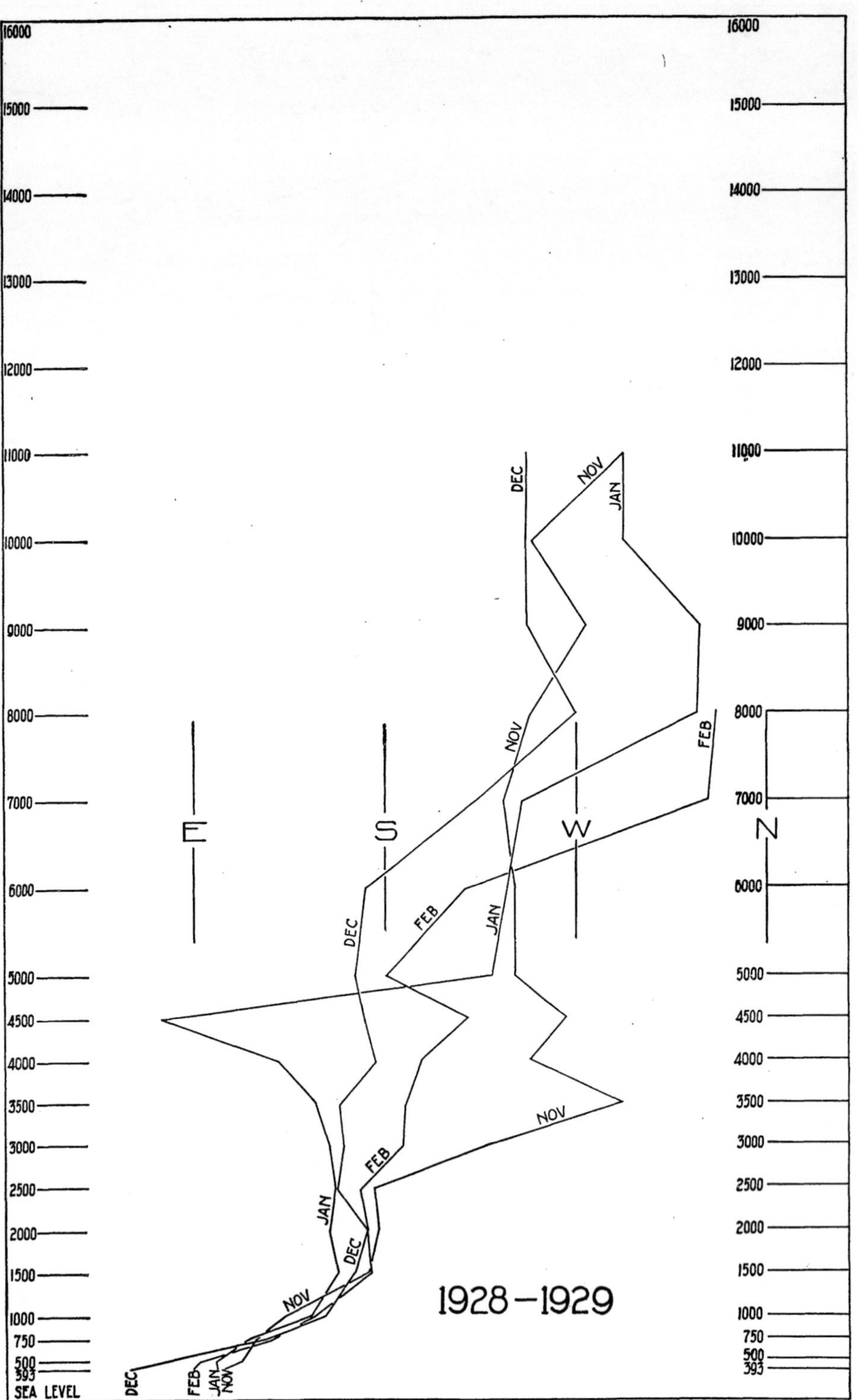

FIG. 16. Average upper air currents at Mount Evans for the months of November and December, 1928, and January and February, 1929

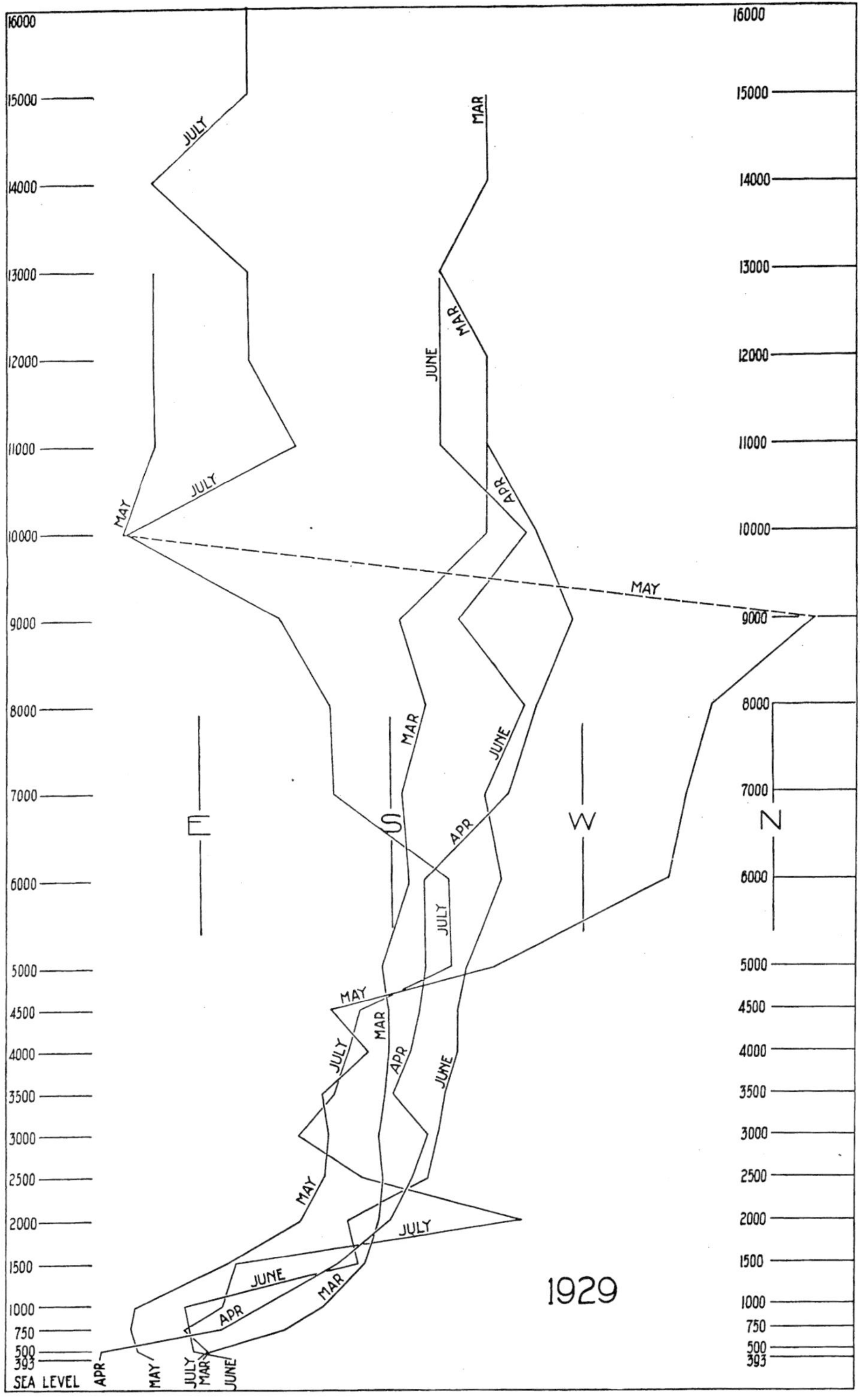

FIG. 17. Average upper air currents at Mount Evans for the months of March, April, May, June and July, 1929

reverse currents blowing inward toward the center of the glacial anticyclone first made their appearance at 7,000 and 9,500 meters, instead of at 2,500 and 4,000 meters, the values for July and August. As we shall see, these are very exceptional (Fig. 12).

Beginning with the month of November the winter conditions set in and remained notably constant for the four months of November, December, January, and February (Fig. 13). For these months in order the southerly surface currents swing into the southwesterly quadrant at altitudes of approximately 1,250, 3,200, 2,250, and 3,500 meters respectively; and the currents are then from the southwesterly quadrant up to near 10,000 meters, the highest points reached. Only in November and between 6,500 and 8,000 meters are the winds in the quadrant north of west, but it is still true that they blew throughout from the westerly quadrants above the first veering level.

For March the veering level from southeasterly to southwesterly winds appeared at 4,200 meters, for April 1,800 meters, and for May 1,200 meters (Fig. 14), whereas in March and May the winds swung from the southwesterly to the northwesterly quadrants above the altitudes of 7,500 and 5,500 meters respectively, those of April went back to the southwesterly quadrant above 9,500 meters.

The winds for June and July, 1928, were like those of late July and August, 1927, in contrast with those of the other months (Fig. 15). This is largely explained by their lower velocities at this season of the year when the domination of the glacial anticyclone is less pronounced. Something of this abnormal quality appears also in the graph for July winds in 1929 (Fig. 17). For the other months of 1928 and 1929 there is a fair correspondence (Figs. 15–17), which is well displayed in the yearly averages, August, 1927, to July, 1928, and August, 1928, to July, 1929 (Fig. 18). These yearly averages show greater differences within the levels above 8,000 meters, but this is to be accounted for largely by the small number of balloons whose course was followed above that altitude.

Below 3,000 meters the winds were generally from the southeast; from 5,000 meters to 8,000 meters the winds were from the southwest. Between 3,000 and 5,000 meters they were near south, though in the first year of observations they were within the southwesterly quadrant. Above 8,500 meters, with fewer observations, greater differences are noted, but with a strong tendency toward northwesterly winds with a swing in the reverse direction above 13,000 meters.

Strophs of the glacial anticyclone in their relation to storms on the

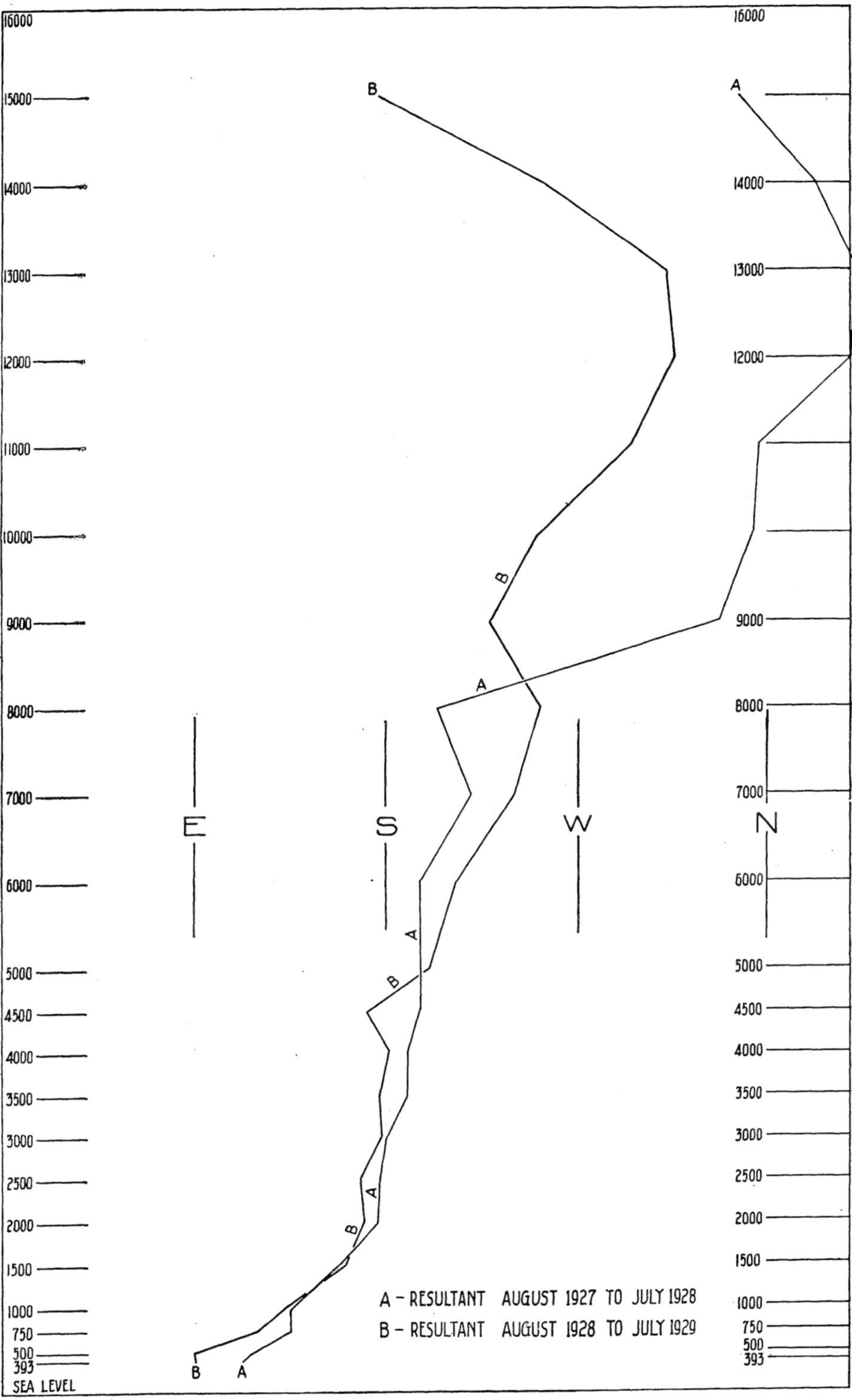

FIG. 18. Resultant upper air currents from August, 1927, to July, 1928, and from August, 1928, to July, 1929

north Atlantic Ocean. — The first recognition of a relationship between storms coming off the inland-ice and the weather at considerable distances away in lower latitudes was made by Sir Douglas Mawson as a consequence of his experiences in Adelie Land, Antarctic. His wireless connection with Australia through the substation upon Macquarie Island permitted him to note that the great strophs of the anticyclones in Adelie Land rather generally preceded by about 48 hours the arrival of storms on the south Australian coast.[3] Some of the weather observations made at Nanortalik and Angmagssalik were compared by the author with winter storms on the north Atlantic Ocean in the winter of 1926,[4] and these seem to confirm the relationship observed by Mawson for the other great area of inland-ice.

One of the chief difficulties in the way of bringing these phenomena into relationship with one another is at present the lack of sufficiently exact and complete data on the Atlantic storms. For the greater Atlantic disturbances the matter has been comparatively simple, for the columns of the New York *Times* and other metropolitan newspapers give tolerably full data sufficient for our purpose.

Now for some years the British Meteorological Office (Dr. George C. Simpson, director) has issued a daily "Chart of Weather in the Northern Hemisphere," of which the synoptic chart for the north Atlantic has been based on radiogram reports received from ships at sea. On these charts the grouping of ships for each day is largely an accidental one, and many of the charts are of necessity far more full and reliable than others. An indicated atmospheric depression may reveal a steep gradient, but have a central area 75 miles or more in diameter without any data whatever. In such cases, though the position of the depression is fairly well revealed, no numerical figure can be given for its extreme value. Though it appears later, the summary report on the weather of the north Atlantic by F. A. Young (more recently by F. G. Tingley) of the United States Weather Bureau [5] affords a perspective of north Atlantic weather for each month based upon a considerably larger number of reports from ships at sea.

Comparison of such data as we have indicates with some clearness that larger storms on the north Atlantic are generally preceded for an interval of four days or more by well-marked conditions ap-

[3] *The Home of the Blizzard*, Vol. 2.

[4] W. H. Hobbs, "The January Storms over the North Atlantic and the Strophs of the Greenland Anticyclone," *Monthly Weather Review*, July, 1926, pp. 286–288.

[5] "Weather of the Atlantic and Pacific Oceans," in the *Monthly Weather Review.*

pearing at Mount Evans, and with much probability that very similar conditions would be observed at other stations located in like positions with reference to the inland-ice margin of Greenland. In some instances where storms crowd together, the demonstration is naturally less complete and convincing. The strophs from the inland-ice are usually heralded by the appearance of lenticular foehn clouds lying to the eastward of the station above or near to the margin of the ice-cap, these remarkable forms being generally in flat mushroom shape and quite often piled above one another like coins. They are followed by high southeast winds having velocities from 30 miles per hour in the case of light strophs to 120 miles per hour for a five-minute interval in the case of the maximum disturbances. These winds are apt to be gusty with some evidence of periodicity at least in some instances.

High temperatures which accompany these winds are also remarkable, since they go to 50 degrees Fahrenheit and above in midwinter, usually though not always accompanied by a high humidity. Most noteworthy is it that there is little that is normal about the air pressure. It was early noted that these strophs arrive in some cases along with high pressure, but in other cases with low pressure areas arriving from the southward. The characteristic of the stroph is above all the foehn with its high winds coming down off the ice plateau, also its high temperature and (generally) high humidity. The harbingers of the storm are generally though not always present in the ominous purplish banners of the foehn in the eastern sky.

The relationship here pointed out between the strophs from the inland-ice of Greenland and the storms on the shipping lanes applies especially to the middle and eastern sections, and not to the western section in the neighborhood of the Newfoundland shore. Some storms of tropical origin, particularly the autumn hurricanes originating near the West India Islands, do not appear to be in a definite relationship with the strophs, though storms approaching the station from the southwest appear to correspond in time to disturbances over the ice-cap, as has already been indicated by earlier observers.[6]

The observers on the station at Mount Evans and Bangsted's party on the ice-cap during the winter of 1927–28 had many opportunities to observe how clouds approached the margin of the ice-cap from the west only to dissolve there and give place to clouds moving from easterly quadrants.

[6] *The Glacial Anticyclones*, pp. 154–155.

COMPARISON OF GREENLAND STROPHS AT MOUNT EVANS WITH STORMS ON THE NORTH ATLANTIC OCEAN WITHIN THE AREA BOUNDED ROUGHLY BY THE MERIDIANS OF FIFTEEN AND FORTY-FIVE WEST AND THE PARALLELS OF FORTY AND SIXTY NORTH

OBSERVATIONS AT MOUNT EVANS	STORMS ON THE NORTH ATLANTIC
1927	
Strong stroph	
July 18. Foehn clouds, wind 78 m. p. h. off ice-cap. Winds continued on 19th and 20th and on 21st reached gust velocities of 120 m. p. h.	*July 25–29.* Strongest gales of the month. Maximum winds with force 9 on 50th parallel between 20th and 24th meridians
Light stroph	
Sept. 7. Foehn clouds. Strong southeast winds, warm and sultry, with falling barometer. On 8th winds continued from east and southeast with shift to west as wind fell off	*Sept. 7–9.* Well-developed low over middle section with moderate southwest gales between 40th and 45th parallels and 35th and 40th meridians
Light stroph	
Sept. 25. Foehn clouds. Falling barometer. On 20th strong southeast winds	*Sept. 24–27.* Hurricane of tropical origin
Light stroph	
Oct. 4. Foehn clouds. Barometer fell to low value of 29.12 inches. On 5th southeast wind which shifted to south	*Oct. 6–8.* A low pressure area near the Azores with moderate to full gales in southwest quadrant
Light stroph	
Oct. 14. Strong southeast winds. Warm and muggy	*Oct. 15.* Storm with force 11, 10 degrees west of the Azores
Light stroph	
Oct. 23. Southeast winds and rising temperature	*Oct. 27–28.* Storm centered in lat. 45° N. and long. 30° W.
Moderate stroph	
Nov. 2. Alto-cumulus clouds. Winds 28 m. p. h. On 3d wind 50–60 m. p. h. Temperature high. Foehn conditions extended to the 8th	*Nov. 5–10.* Storm in eastern section. On 7th and 8th at a maximum centered north of the shipping lane

Strong stroph *Dec. 1.* Falling barometer with rising temperature and humidity. Wind rose and on morning of 2d was 56 m. p. h. and gusts much stronger. Before noon 78 m. p. h.; high wind continued through 3d	*Dec. 3–9.* Severe storm on middle and eastern sections. Most severe on the 5th
Moderate stroph *Dec. 10.* Foehn clouds. Rising temperature. Southeast winds 48 m. p. h. These conditions continued through 13th	*Dec. 13–24.* Severe storms. Most severe between 15th and 18th between the Azores and Bermuda
Light stroph (?) *Dec. 24.* Foehn clouds. Light east and southeast winds. Temperature above 22° F. On 25th winds veered to west and became strong	*Dec. 28.* Storm centered in lat. 28° N. and long. 57° W. S.S. *President Hayes* from Marseilles for New York (arrived 29th) encountered gales on 21st which continued 24 hours; then after a lull a second and severe storm which smashed windows and flooded cabins on the promenade deck

1928

Exceptionally strong stroph *Jan. 14.* Foehn clouds and temperature rose from subzero values to 2° F. Barometer sinking and southeast winds rising. Wind rose on 15th to gale force with temperature still rising and barometer "pumping." After midnight of 16th barometer sank from 27.26 inches to 26.86 inches or four tenths of an inch in two hours. During this drop the southeast winds increased to hurricane force reaching 120 m. p. h. On 17th dropped to 78 m. p. h. A radiogram sent to Ann Arbor with some delay and warning issued through New York *Times*	*Jan. 18.* Entire shipping lanes swept by fierce gales. S.S. *De Grasse* Havre to New York arriving 21st recorded barometer reading of 28.27 inches. Second lowest mark observed by her commander in 27 years' experience in service on north Atlantic. At 10.30 bridge of the S.S. *Aquitania* 65 feet above the water line was struck by wave which broke windows and flooded navigating room. Worst weather captain of *Aquitania* had ever experienced. All vessels overdue. S.S. *Providence* to New York from Mediterranean ports arrived three days overdue battered by 65-mile gale. S.S. *Caspar* to New York from Gothenburg, Sweden, arrived 12 days overdue. Hove to for six days spreading oil on seas with little effect. Life-boats smashed

Light to moderate stroph *Feb. 22.* Foehn clouds. On 24th steady southeast winds with high temperature. On 25th more foehn clouds with increasing wind and still rising temperature. Relative humidity 98 percent. Barometer dropped in stages	*Feb. 27–28.* Storm over middle ocean lane. S.S. *Berengaria* reached New York on 28th after stormy passage on 24th; had ports smashed by waves and three tables torn loose. S.S. *Berlin* from Bremen on 27th reached New York a day late because of storms
Strong stroph *Feb. 29.* Foehn effect. Barometer sinking and "pumping." With each drop a gust of wind from southeast. Maximum wind velocity 88 m. p. h. Foehn clouds again. Storm continued to March 5 when wind velocity was 54 m. p. h.	*March 6.* Northwest gales on central section
March 8. Foehn clouds	
Light to moderate stroph *March 11.* High southeast winds with high temperatures and low humidity (28 per cent)	*March 17–20.* Stormy weather on central and eastern lanes
Light stroph *April 28.* Foehn clouds. Steady southeast winds	*May 2–3.* Storm centered 500 miles north of the Azores
Moderate stroph *May 12.* South and southeast winds reached gale force and continued on the 13th	*May 16.* Storm centered in lat. 42° N., long. 50° W.
Light stroph (?) *May 22.* Foehn clouds lying very low	*May 28.* Fairly severe storm localized in lat. 46° N., long. 35° W.
Weak stroph *Aug. 19.* Foehn clouds over inland-ice moving to west. On 20th wind aloft 30 m. p. h.	*Aug. 22–24.* Severe disturbances between 45th and 55th parallels and 20th and 30th meridians
Weak stroph *Aug. 29.* Wind at 9,000 m. 70 m. p. h.	*Sept. 5–6.* Steamer lanes between 15th and 45th meridians swept by heavy gales
Weak stroph *Sept. 6.* Lenticular alto-stratus clouds. Wind from southeast 20 m. p. h. Wind continued on 7th	*Sept. 9–11.* Moderate depression centered near 27° N., 51° W.
	Sept. 12. Hurricane of tropical origin

Strong stroph

Sept. 28. Foehn clouds. On 29th wind rose to 45 m. p. h. In evening at 1,000 m. altitude wind was 82 m. p. h.

Oct. 2–3. Depression in mid-Atlantic central in lat. 50° N. (972 mbs.). Five ships New York to Plymouth (*Caronia*, *Ascarnia*, *Amerika*, *Belgenland*, and *American Banker*) reported "terrific storm" in mid-Atlantic. On *Caronia* seas reached top deck and ship lost 100 m. in one day. *Carmania* and *Lapland* arrived New York on the 8th. On former men were swept off bridge. On 2d barometer 29.86 inches and on 3d had dropped to 28.86 inches, a drop of one inch in 24 hours. Five other ships arrived New York on 5th battered by storm. Wind of 85 m. p. h. reported and seas 45 feet high. *Albert Ballin* had bridge crushed and captain reported wind of 100 m. p. h.

Light Stroph

Oct. 13. Strong southeast winds off ice-cap fill air with much dust. Barometer falling since morning of 12th. On 15th continued low pressure and strong southeast wind

Oct. 14–15. Hurricane of tropical origin

Oct. 17–24. Very severe storms which moved eastward from the Newfoundland region

Strong stroph

Nov. 4. Foehn and southeast winds with velocity 60 m. p. h.; these conditions continued through to the 7th, but with less severe winds

Nov. 6–11. Moderate to strong gales on greater part of steamship lanes. S.S. *Mauretania* eastward bound from New York battled one of the worst storms in the captain's experience. For three days ships encountered high seas and winds of 80 m. p. h.

Strong stroph

Dec. 19. Foehn. Southeast winds 54 m. p. h. Barometer reached 27.40 inches

Dec. 22–23. Westerly gales on steamship lanes between 20th and 60th meridians

1929

Jan. 1–7. Severe gales swept steamer lanes. Four vessels overdue in New York

Strong stroph

Jan. 8. Foehn followed a fall of pressure. Winds from southeast 60 m. p. h. with gusts 70–75 m. p. h.

Fairly strong stroph

Jan. 13. Foehn with temperature of 45.5° F. On 14th wind velocity 54 m. p. h.

Jan. 14–19. Moderate to strong gales in middle section. S.S. *Majestic* on 14th encountered full gale with wind 65 m. p. h. and squalls of hurricane force. Flat-topped waves. One 45 feet high broke through steel hatch cover and flooded the hold. On 18th a strong depression central in mid-Atlantic. From 20th to 25th exceptionally severe storms. On 23d crew of S.S. *Florida* rescued by S. S. *America,* Captain Fried. On latter ship barometer gave low reading 28.58 inches with wind force of 11

Very strong strophs

Jan. 21. Foehn developed. On 22d wind 82 m. p. h. blew down the radio mast. Extreme low pressure of 27.40 inches. On 23d wind 40 m. p. h. and temperature 50° F. and on 24th wind velocity 120 m. p. h. On 24th at 2,000 m. altitude south-southeast wind had velocity of 104 m. p. h. Above this was a wedge of south-southwest wind with velocity 6 m. p. h. and above this a wedge of south-southeast wind with velocity of 30 m. p. h.

Abnormal moderate stroph

Jan. 29. Foehn. Wind with velocity 60 m. p. h., but with pressure high for 48 hours

Feb. 3. Strong gales between 45th and 50th parallels and 30th and 45th meridians

Feb. 12. Wind of hurricane force on ocean lanes near mid-ocean

Feb. 15–18. Severe storms widely extended over north Atlantic

Light stroph

Feb. 22. Extreme low pressure (27.40 inches). Maximum wind velocity 35 m. p. h. Gusty. Alto-cumulus clouds. Stroph continued to 26th

Feb. 24–27. Westerly gales on middle and eastern sections

No distinct strophs recorded at Mount Evans	*March 1–6.* Twin depressions of marked gradient, the eastern one central near 50th parallel and 40th meridian *March 25–28.* Gales over middle and eastern sections
April 2. After sudden rise of pressure temperature fell and humidity increased; wind shifted from southeast to west and blew with force 40 m.p.h. Great masses of strato-cumulus came from WSW and SSW. Once a row of fracto-stratus clouds moved with great rapidity toward NNW.	*April 6.* Twin moderate depressions in mid-Atlantic in lat. 57° N.
Stroph (?) *April 7.* Aloft the strong wind which on preceding day had been from south-southwest veered to south with greatly increased force. At 5,700 m. altitude wind had force of 87 m. p. h.	*April 11–14.* Strong depression and storms over eastern section of lanes. S.S. *Majestic* docked at New York on 17th after encountering storms and heavy seas all the way across
Unusual stroph *May 16.* Easterly winds with force 24 m. p. h. On 20th easterly winds veering to south and later northeast. At altitude of 12,000 m. wind force from east-northeast was 331 m. p. h. and a half hour earlier 241 m. p. h.	*May 22.* Moderate gales over central section of lanes
Moderate stroph *June 12.* Foehn with lenticular alto-stratus clouds. Pressure fell to near 28.00 inches. Wind strong from south-southeast and south. On 13th pressure rose with marked fluctuations. Wind strong with gust velocities of 60 m. p. h. Foehn clouds continued, but disappeared with the wind on 14th	No storm conditions reported on north Atlantic in central or eastern sections. Airplane *Yellow Bird* crossed, 13th to 14th
July 15. Wind from E and SE attained force of gale with fluctuations of barometer	No reports of storms on north Atlantic; but few ships within area

Results of pilot-balloon ascents made at Mygbukta (Mackenzie Bay), east Greenland. — At the request of the director of the Greenland Expeditions of the University of Michigan, Director Th. Hesselberg of the Meteorological Institute of Oslo was able to have pilot-balloon ascents carried out during the months of June, July, and August, 1927, at Mygbukta (Mackenzie Bay), east Greenland. These were to furnish comparative observations from the Greenland east coast, and the observations Director Hesselberg has generously turned over with permission to publish. (See Pl. IV B.)

Unfortunately, the Michigan Expedition of 1927 was delayed somewhat beyond expectation in getting its station into operation, so that balloon ascents were first started on July 21. For this reason somewhat less than one half of the balloon runs carried out at

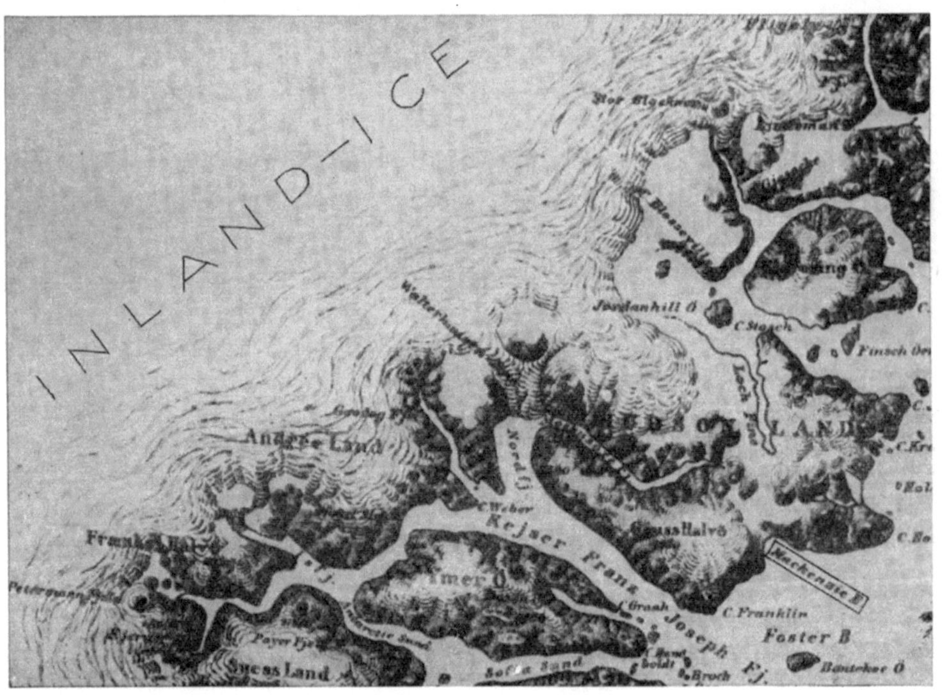

Fig. 19. Map of the Mackenzie Bay region of east Greenland

Mygbukta were within the period when both stations were in simultaneous operation.

The situation of the colony of Mackenzie Bay is indicated on the official Danish map reproduced in Figure 19, from which it will be seen that the inland-ice lies to the west, northwest, and north of the station. The thirty-five runs of pilot-balloons are reproduced on pages 33–34. Rather generally the surface currents

UPPER AIR CIRCULATION
AT
MYGBUKTA (MACKENZIE BAY), EAST GREENLAND
BETWEEN JUNE 18, AND AUGUST 10, 1927
OBSERVATIONS MADE BY THE METEOROLOGICAL INSTITUTE
AT
OSLO, NORWAY, DR. TH. HESSELBERG, DIRECTOR

WIND DIRECTIONS SHOWN BY ARROWS ORIENTED WITH REGARD TO CARDINAL DIRECTIONS; WIND VELOCITIES IN METERS PER SECOND. ALTITUDES IN KILOMETERS.

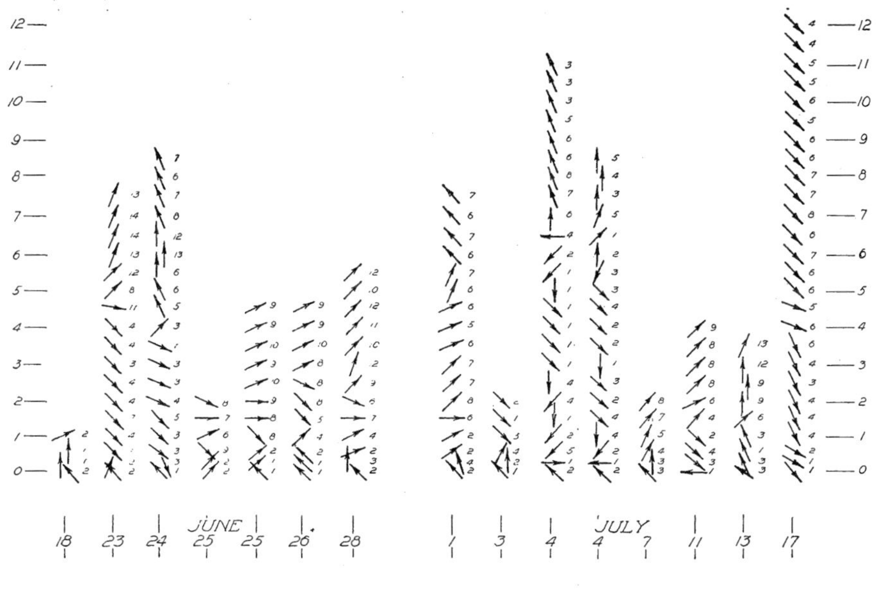

Fig. 20

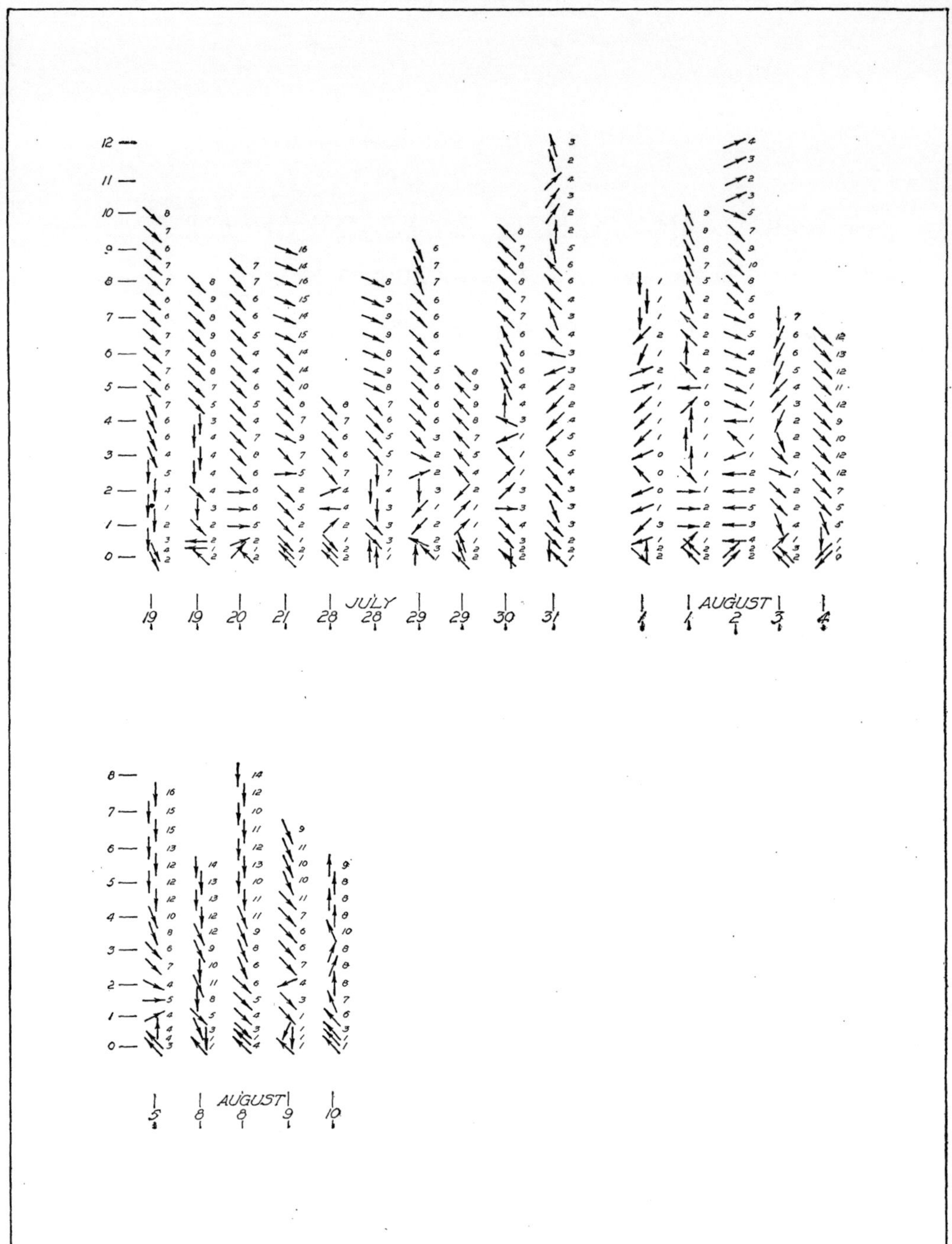

Fig. 21

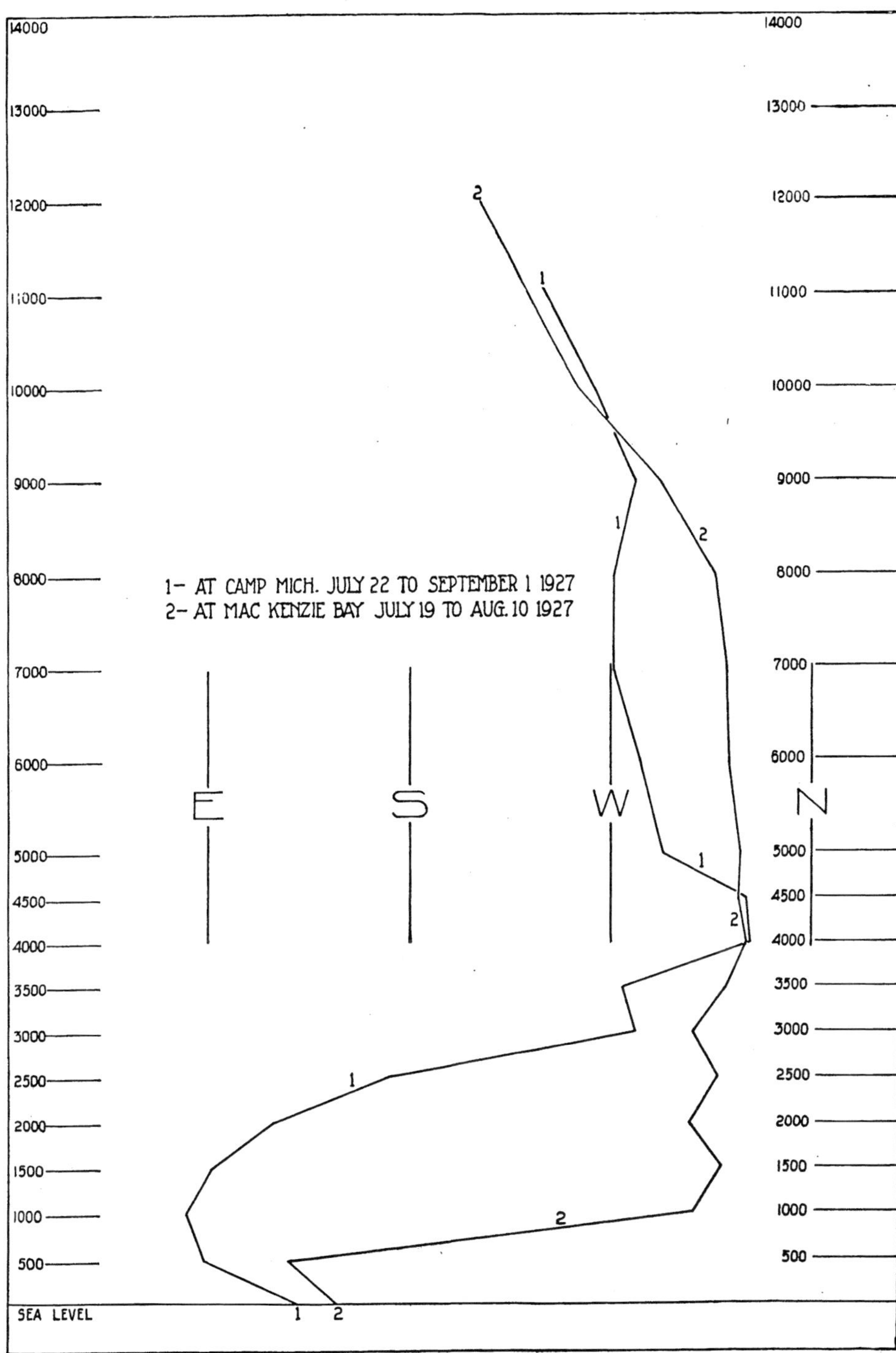

FIG. 22. Comparison of upper air currents at Mount Evans and at Mackenzie Bay in the months of July and August, 1927

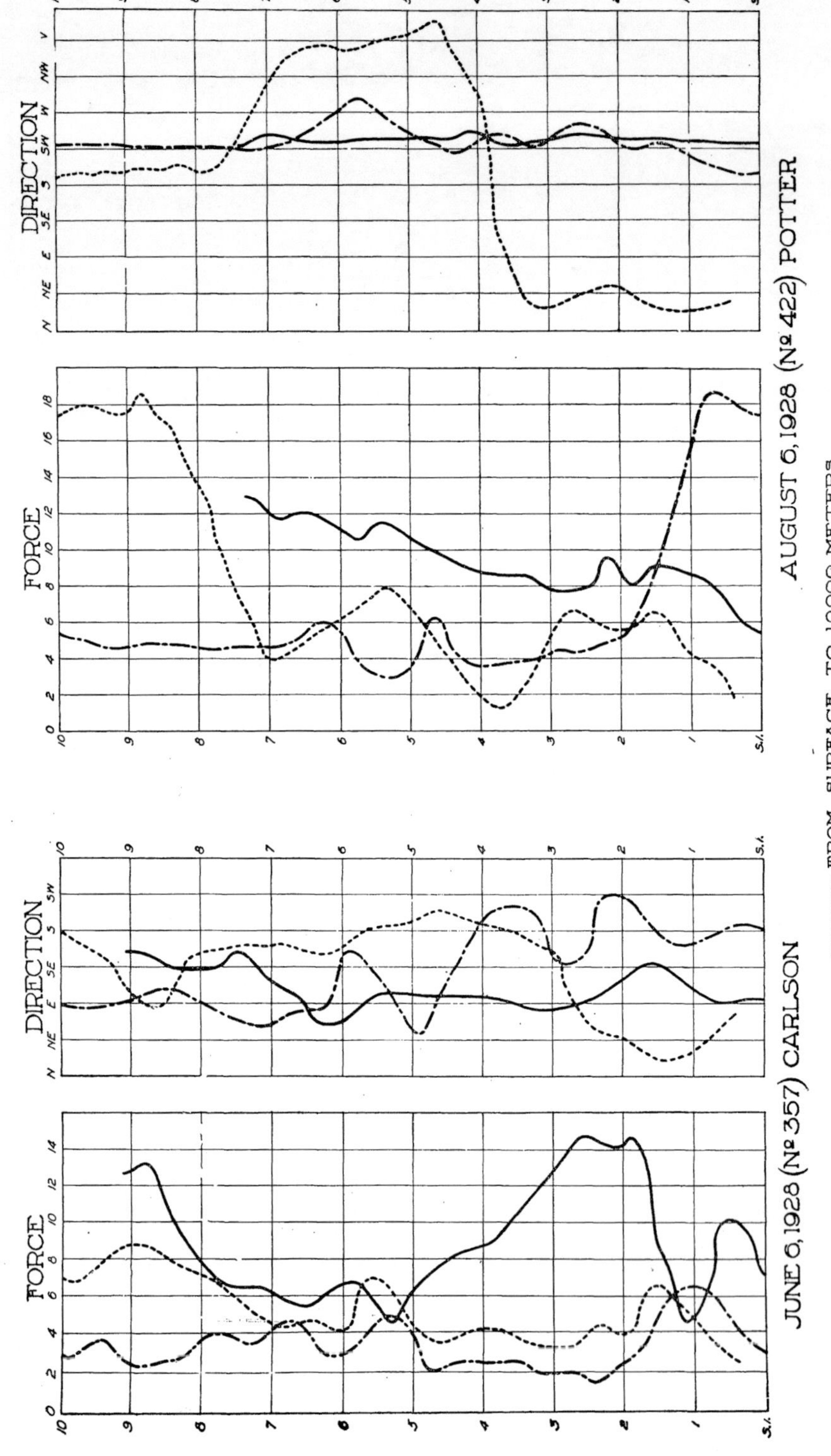

FIG. 23. Force and direction of upper air currents at Mount Evans at exceptional altitudes on June 6 and August 6, 1928

show faint zephyrs or light winds coming in from the Greenland sea lying to the south and east, and these have a force almost always less than 4; but at levels above 1,000 meters the air currents generally blow down strongly from the inland-ice, that is to say, from within the westerly or northerly quadrants. At still higher levels, often from 4,000 to 5,000 meters, sometimes at much higher levels, and in one or two instances at much lower levels, the upper inblowing circulation from the easterly quadrants makes itself manifest. Just as at the Mount Evans station on the west coast, the domination of the glacial anticyclone of Greenland is here manifest, but here on the east coast with directions almost exactly reversed. (See graph 2 in Fig. 22.)

Exceptionally high balloon runs at Mount Evans. — Two exceptionally long pilot-balloon runs were made at Mount Evans, one by William S. Carlson on June 6, 1928, and the other by David M. Potter, Jr., on August 6 of the same year (see Fig. 23). The conditions of visibility were on both occasions exceptionally favorable, and the Buff and Buff theodolite used was an exceptionally good one. Unless defects developed in the balloons the altitudes attained must have been 29,000 and 27,000 meters respectively. It is, however, a rather general rule to look with suspicion on such high runs when recorded from a single station, and they are presented in Figure 23 for what they are worth.

II. THE AËROLOGICAL SOUNDINGS

BY

S. P. FERGUSSON

DESCRIPTION OF APPARATUS AND METHODS; EXPLANATION OF TABLES AND GRAPHS

WITH the exception of materials and generators for producing hydrogen, the aërological apparatus employed by the Expeditions was supplied by the United States Weather Bureau from standard supplies used in regular work and the methods followed are in general use.

During the first, or reconnaissance, expedition, the needs of which could not be foreseen, the instruments, selected with a view of portability, comprised the following:

For current meteorological records

One mercurial barometer, in carrying case
Two Richard barographs
Two thermo-hygrographs of special light pattern
Three sling psychrometers
Two sets of maximum and minimum thermometers
Three special "hand" anemometers with post supports
One nephoscope

For the aërological soundings .

Two Bunge balloon theodolites
Three Fergusson balloon meteorographs
Three Rossby deflating valves
Two hydrogen generators designed by Sverdrup
One hundred and fifty 15-centimeter pilot-balloons
Twenty *ballons sondes* 40 centimeters in diameter
Thirty kilograms of calcium hydride
Accessories such as forms, tubing, fuse, materials for repairs, tools, etc.

Owing to an unforeseen advance of three weeks in the time of departure of the Expedition, a timing device and inflation balance were not ready and the staff was obliged to depend upon inflations and lifts measured by a spring-balance, which greatly increased the labor of computing. The Expeditions of 1927–29, however, with more time for preparation, and profiting by the experience gained during the summer of 1926, could be much better equipped, and carried the following:

For the current meteorological record

The same apparatus taken in 1926, with additional thermometers
One Robinson anemometer of the Weather Bureau standard type
One recording apparatus for velocity of the wind
One wind-vane with indicating dial for the ceiling of the observatory
Two water thermometers

For the aërological soundings

The balloon theodolites, meteorographs, deflating valves and accessories carried in 1926, with the following additions:
One Buff and Buff balloon theodolite
One timing clock
One plotting board for computing observations
One "definite-lift" balance for determining ascensional rates
One graphing board
Two hydrogen generators modified from Patterson's design and made in Washington
Sufficient 15-centimeter pilot-balloons for two ascensions daily during each year
One portable reel with 4,000 meters of wire for use in soundings with the "free-captive" balloon
One portable kite reel with 4,000 meters spare line
Five kites of standard Weather Bureau pattern
Two Marvin kite meteorographs
Forms, slide rules for computing, tools and other accessories
Calcium hydride sufficient for two years' use

The calcium-hydride process for the manufacture of hydrogen has been used extensively in Europe during the past twenty years, but apparently is little known in America, where the only instance of its use is in the work of Patterson in Canada in 1908–10, whose generator was modified for our work. The process is simple, consisting of dissolving lumps of calcium hydride in water inside a closed vessel from which the gas is conducted (preferably through a drying tube) to the balloon. The action is very rapid and inflation of a *ballon sonde* can be accomplished in almost the same time as by means of compressed gas; but the gas obtained from calcium hydride is of variable purity and the assumed ascensional rates of pilot-balloons are probably not so accurate as they would be if pure gas were available. However, with the definite-lift balance the same lift can always be assured although the quantity of gas may vary.

The Rossby deflating valve, so named for its inventor, Dr. Carl

G. Rossby, consists chiefly of a light valve kept closed by an elastic band until the latter is burned through by a fuse timed for a definite period during which the balloon and apparatus will attain a certain height limited so that the apparatus, after deflation of the balloon, will fall within sight or easy reach of the starting point. Three successful ascensions were made with this device and in two others the valve failed and the equipment was lost; by chance, however, the apparatus lost during the first ascension was recovered with perhaps the first record of a *ballon sonde* accomplished in Greenland (Pl. VI B). Details of the ascensions with balloons carrying meteorographs are given on page 48 and in Table III (p. 241).

The technique of the pilot-balloon has been given very careful study throughout the world, with the important result that the methods and apparatus are now of such dependable quality that data from observations at one station are sufficiently accurate for study and use in forecasting. With the technique developed by the Aërological Division of the Weather Bureau, a trained observer and computer can make, compute and plot for use all observations during a long ascension within an hour or less time after the ascension is completed. When it is suspected that unusual behavior of a balloon may be due to defects, the ascension is repeated.

Since the primary purpose of the meteorological and aërological work of the expeditions was the synoptic study of the atmosphere at all possible heights, the program included regular ascensions of pilot-balloons at definite hours, observations of clouds in detail, including measures of heights and velocities, etc., and, as time permitted, other soundings with captive balloons and kites. The data, in detail, are presented in Tables I and II (pp. 49–224) and in the accompanying diagrams (Figs. 24 to 29). The data from ascensions of *ballons sondes* are plotted in Figure 30.

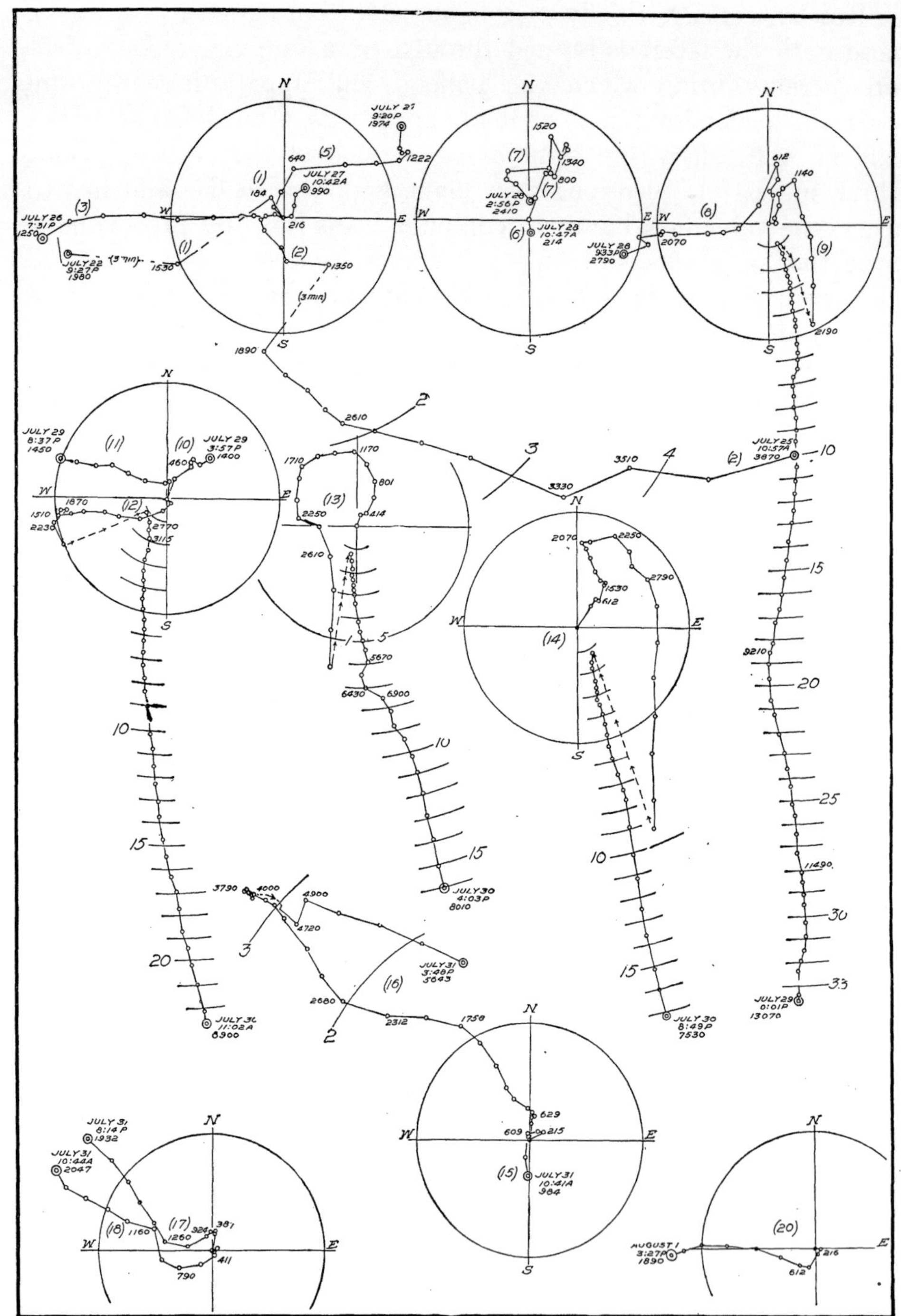

FIG. 24. Graphs of balloon runs at Camp Michigan (1926) in horizontal projection

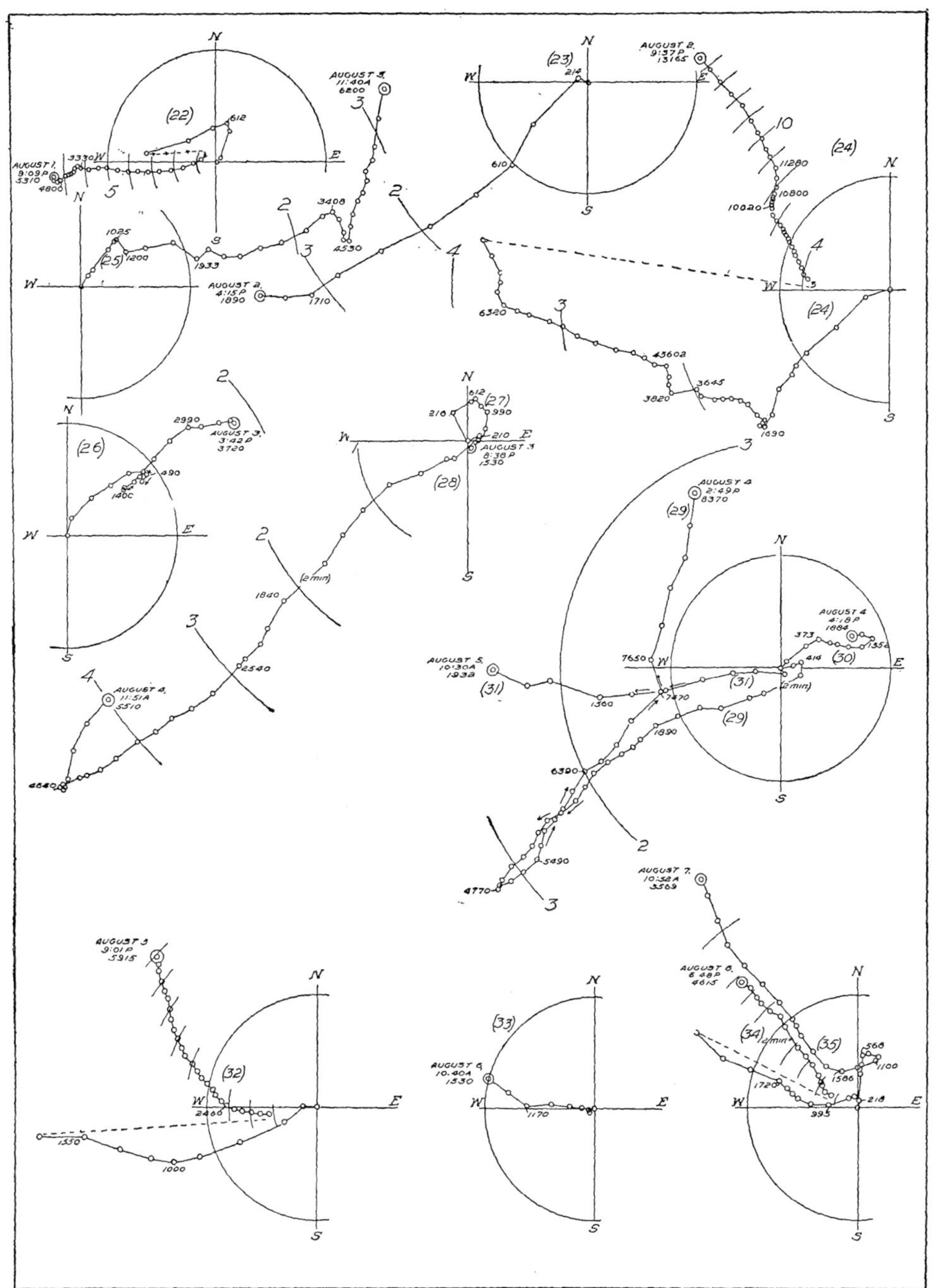

Fig. 25. Graphs of balloon runs at Camp Michigan (1926) in horizontal projection

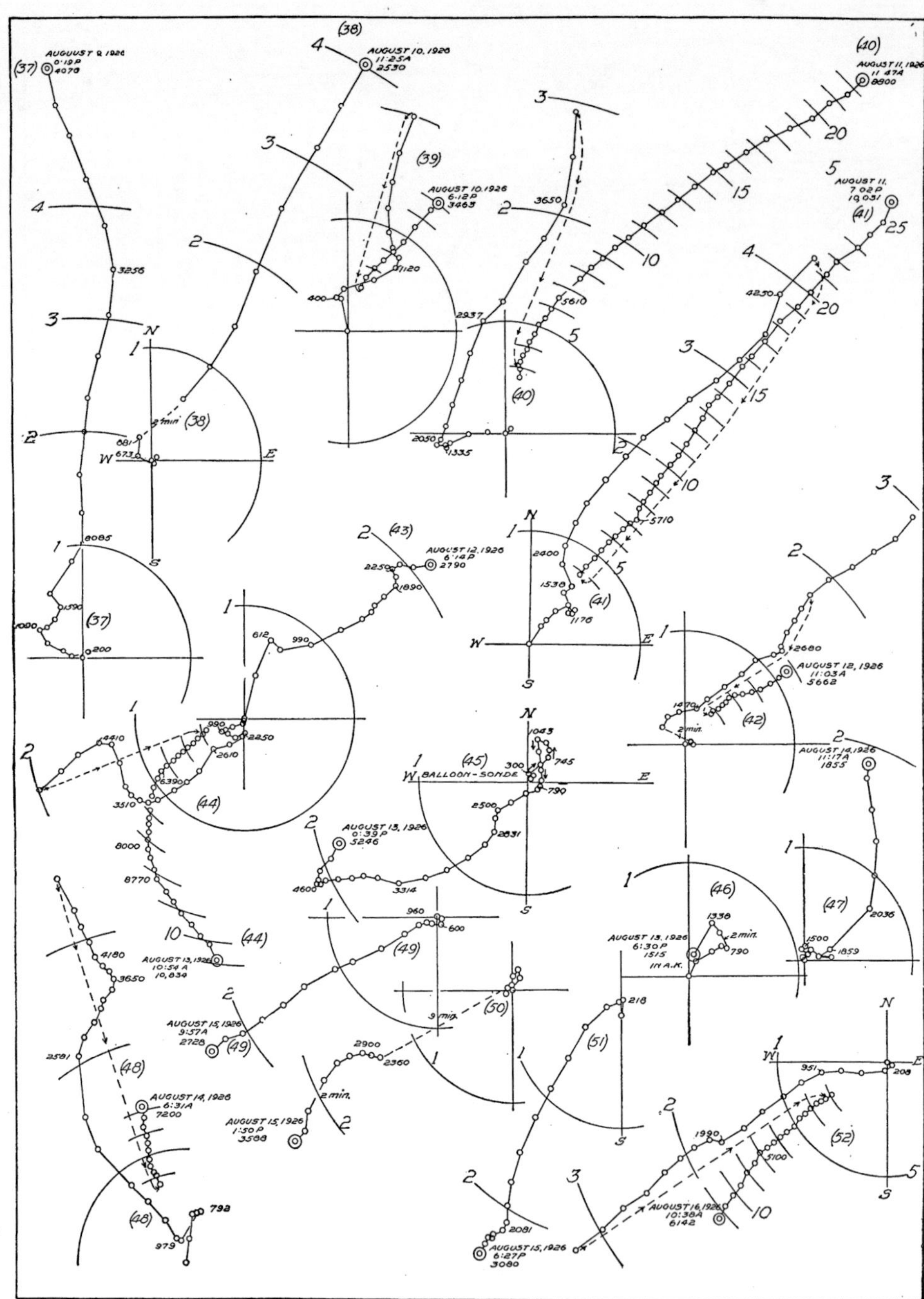

FIG. 26. Graphs of balloon runs at Camp Michigan (1926) in horizontal projection

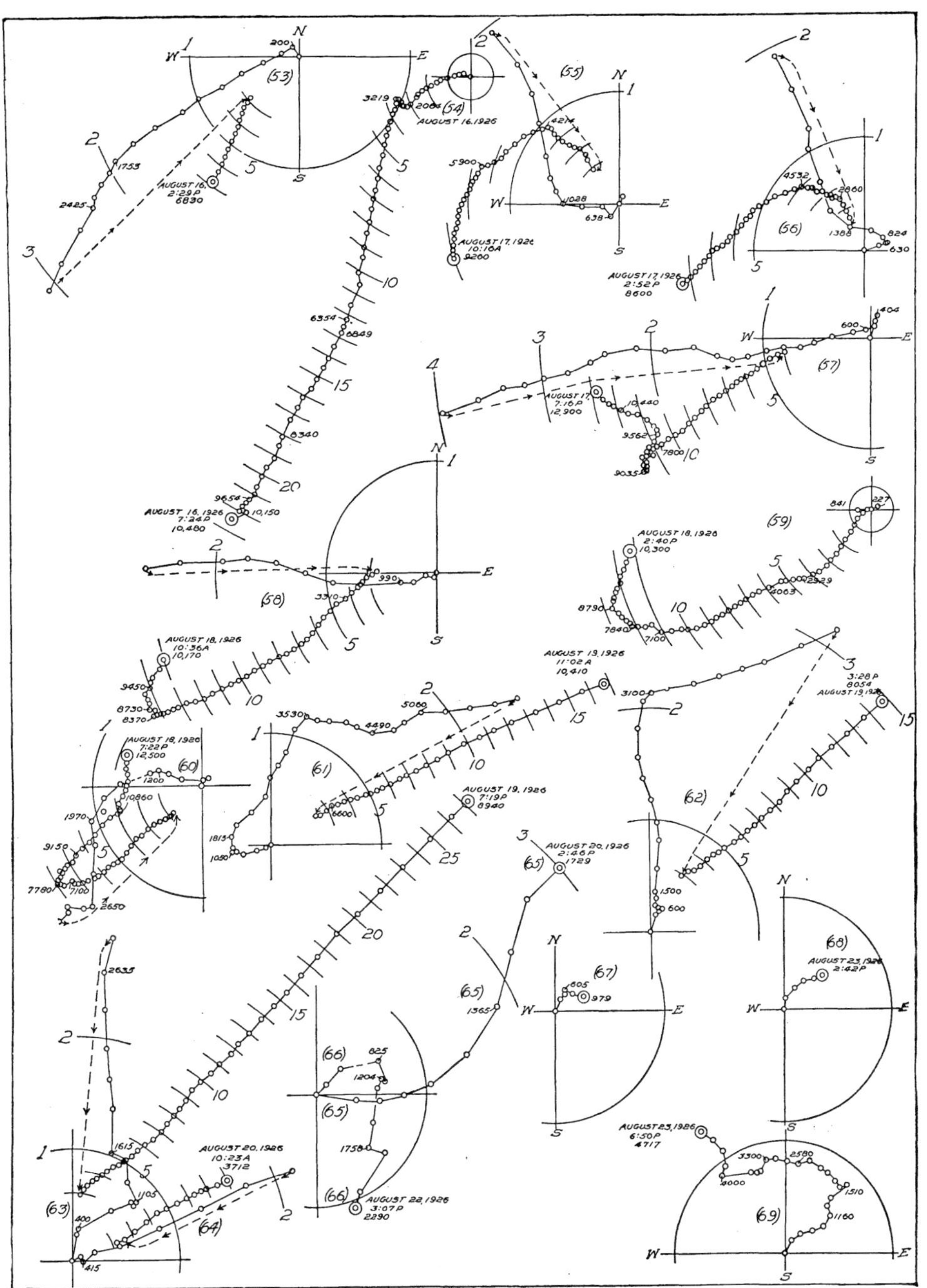

FIG. 27. Graphs of balloon runs at Camp Michigan (1926) in horizontal projection

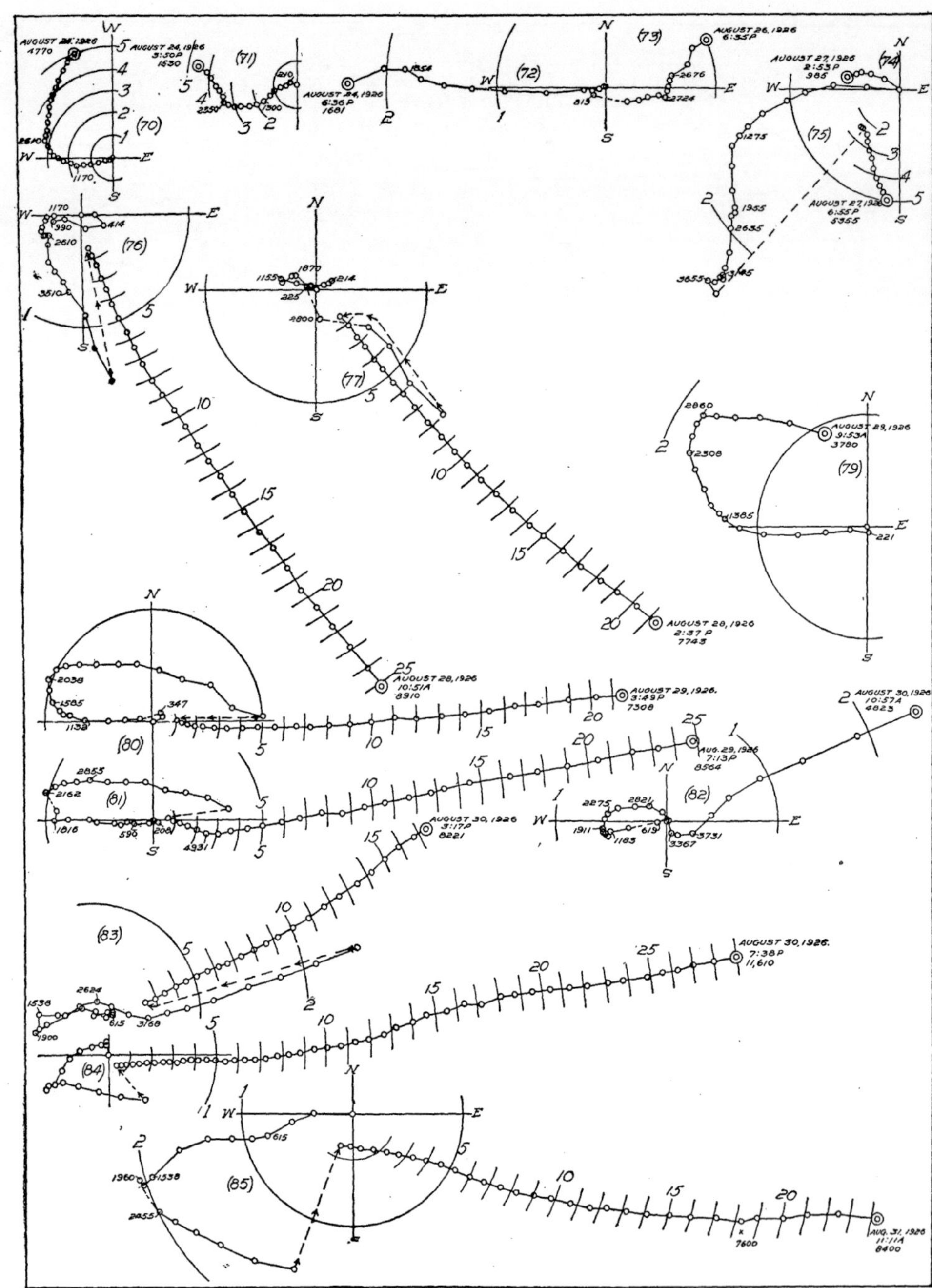

Fig. 28. Graphs of balloon runs at Camp Michigan (1926) in horizontal projection

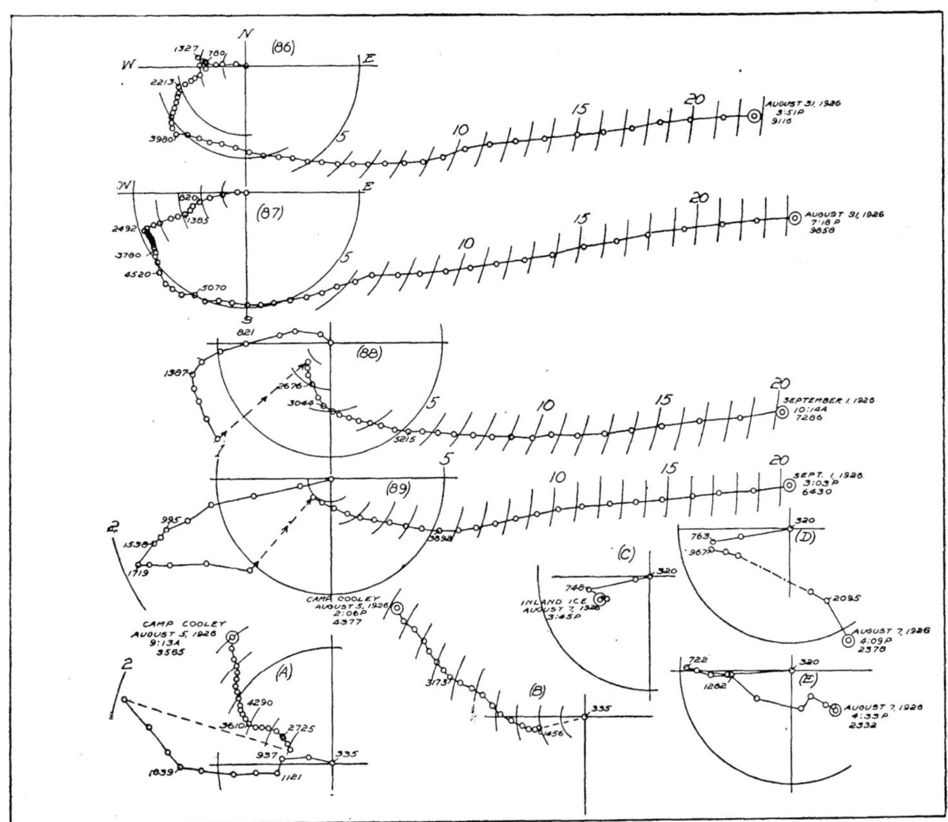

FIG. 29. Graphs of balloon runs at Camp Michigan (1926) in horizontal projection

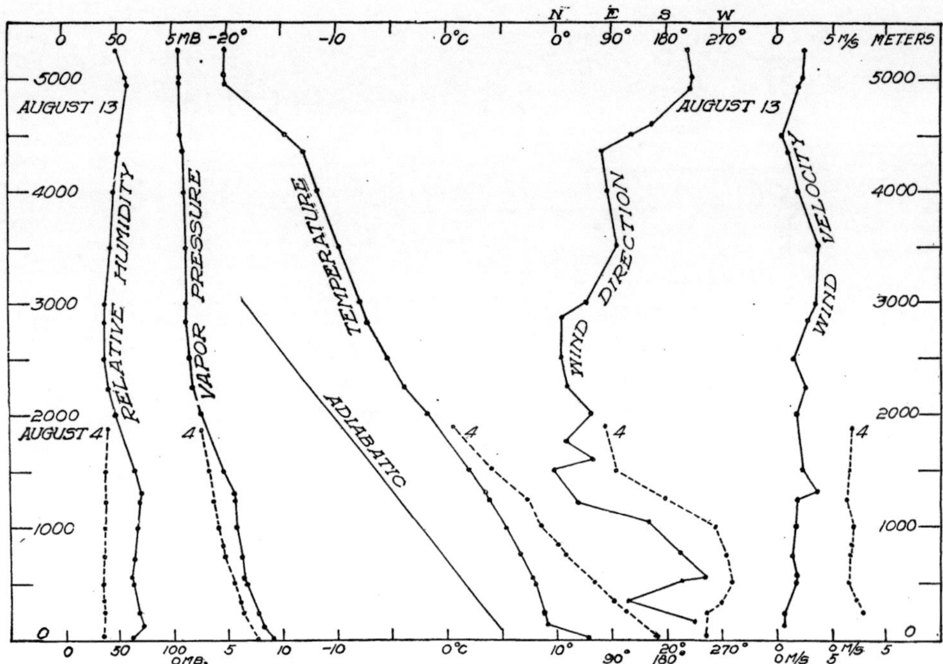

FIG. 30. Ascensions of *ballons sondes* on August 4 and 13, 1926. Relative humidity, vapor pressure, temperature, wind direction, and wind velocity

III. HEIGHTS, AZIMUTHS (DIRECTIONS), AND VELOCITIES OF PILOT-BALLOONS AT CAMP MICHIGAN FROM JULY 22 TO SEPTEMBER 3, 1926

BY

S. P. FERGUSSON

TABLE I (1926)

(THE FIGURES IN PARENTHESES REFER TO THE SERIATIM NUMBERS OF BALLOON RUNS)

Date and time	Height in meters	Azimuth from; degrees	Velocity m/s
July 22			
9:18P	414	327	1
:24	1530	235	3
27	2070	273	5
25			
11:37A	216	106	1
38	414	75	1
39	612	62	1
40	801	338	2
41	990	311	3
42	1170	333	1
43	1350	274	6
46	1890	38	5
47	2070	285	15
48	2250	279	12
49	2430	276	9
50	2610	285	6
51	2790	262	13
52	2970	260	15
53	3150	261	13
54	3330	268	20
55	3510	235	20
56	3690	248	17
57	3870	251	15
26			
7:25P	200	90	5
26	380	84	6
27	564	86	5
28	740	93	5
29	915	86	6
30	1080	50	6
31	1245	35	7

Date and time	Height in meters	Azimuth from; degrees	Velocity m/s
July 27			
10:38A	216	251	1
39	414	190	0
40	610	178	1
41	800	202	1
42	990	235	2
27			
9:11P	225	139	2
12	433	207	2
13	640	234	4
14	835	266	5
15	1030	260	8
16	1220	206	3
17	1410	225	3
18	1600	114	2
19	1785	151	3
20	1975	158	10
28			
10:47A	214	354	1
28			
2:44P	214	195	3
45	414	205	4
46	610	270	1
47	800	303	1
48	965	20	5
49	1160	345	1
50	1340	38	1
51	1520	152	2
52	1700	3	4
53	1875	264	6
54	2060	15	1
55	2235	295	1
56	2410	312	3

Date and time	Height in meters	Azimuth from; degrees	Velocity m/s
July 28			
9:19P	216	187	4
20	414	140	1
21	610	212	1
22	800	172	1
23	990	22	6
24	1170	42	6
25	1350	75	2
26	1530	85	2
27	1710	88	5
28	1890	100	2
29	2070	31	1
30	2250	79	3
31	2430	305	1
33	2790	68	2
29			
10:48A	210	265	1
49	404	154	1
50	596	198	2
51	780	178	2
52	965	221	1
53	1140	234	2
54	1315	349	2
55	1490	344	3
56	1655	335	4
57	1837	356	3
58	2015	358	4
59	2190	0	5
11:01	2543	338	4
02	2720	317	4
03	2896	230	2
04	3071	355	8
05	3247	359	5
06	3423	15	5

TABLE I (1926)

Date and time	Height in meters	Azimuth from; degrees	Velocity m/s	Date and time	Height in meters	Azimuth from; degrees	Velocity m/s	Date and time	Height in meters	Azimuth from; degrees	Velocity m/s
July **29**				July **29**				July **30**			
11:07A	3598	347	4	11:55A	12020	351	13	10:35A	4180	8	7
08	3773	348	5	56	12190	7	10	36	4360	355	7
09	3950	8	5	57	12370	6	10	37	4540	354	8
10	4125	353	6	58	12545	23	8	38	4717	359	8
11	4300	355	6	59	12720	9	8	39	4895	5	8
12	4475	6	5	0:01P	13070	358	9	40	5070	0	9
13	4650	355	6					41	5250	358	9
14	4825	8	6	**29**				42	5430	358	9
15	5000	4	6	3:51P	222	204	3	43	5607	355	9
16	5175	358	6	52	455	237	3	44	5785	350	11
17	5351	17	7	53	673	185	1	45	5960	352	10
18	5527	35	9	54	881	230	1	46	6140	357	10
19	5700	11	7	55	1090	302	1	47	6320	351	11
20	5880	356	8	56	1400	223	1	48	6500	356	13
21	6055	357	8					49	6675	353	13
22	6230	2	8	**29**				50	6853	351	12
23	6405	352	8	8:30P	212	186	2	51	7030	354	10
24	6580	359	8	31	410	72	1	52	7210	351	16
25	6755	8	9	32	605	93	3	53	7390	348	12
26	6930	4	10	33	791	112	2	54	7570	343	15
27	7105	6	9	34	969	116	3	55	7645	343	12
28	7281	5	10	35	1147	85	2	56	7825	350	12
29	7456	7	12	36	1325	98	3	57	8000	351	16
30	7632	9	9	37	1450	98	2	58	8180	341	13
31	7807	15	13					59	8360	345	12
32	7983	7	14	**30**				11:00A	8535	341	15
33	8158	9	13	10:13A	214	297	1	01	8710	345	14
34	8333	10	12	14	408	42	2	02	8890	355	15
35	8509	12	11	15	605	71	3				
36	8684	17	10	16	790	95	3	**30**			
37	8860	11	15	17	979	105	2	3:20P	216	195	2
38	9035	21	7	18	1160	90	3	21	414	258	1
39	9211	11	8	19	1335	78	2	22	612	209	3
40	9387	348	11	20	1510	83	2	23	801	182	2
41	9562	353	16	21	1690	201	1	24	990	155	3
42	9737	342	11	22	1870	260	1	25	1170	132	3
43	9913	345	14	23	2050	44	2	26	1350	88	3
44	10089	339	14	24	2225	335	3	27	1530	75	2
45	10264	351	12	25	2400	340	3	28	1710	59	3
46	10440	348	13	26	2580	348	4	29	1890	13	2
47	10615	347	13	27	2767	352	5	30	2070	7	3
48	10790	354	7	28	2937	359	5	31	2250	354	3
49	10965	351	15	29	3115	355	6	32	2430	293	3
50	11140	353	14	30	3293	11	8	33	2610	339	4
51	11315	349	12	31	3470	16	8	34	2790	355	5
52	11490	354	7	32	3650	8	8	35	2970	5	6
53	11670	345	11	33	3827	1	7	36	3150	1	5
54	11840	356	7	34	4000	358	8	37	3330	354	5

TABLE I (1926)

Date and time	Height in meters	Azimuth from; degrees	Velocity m/s	Date and time	Height in meters	Azimuth from; degrees	Velocity m/s	Date and time	Height in meters	Azimuth from; degrees	Velocity m/s
July				**July**				**July**			
30				**30**				**31**			
3:38P	3510	356	6	8:27P	3870	358	7	3:35P	3238	142	5
39	3690	3	4	28	4050	0	4	36	3422	141	3
40	3870	350	4	30	4410	355	5	37	3608	123	1
41	4050	355	3	31	4590	353	5	38	3792	108	2
42	4230	351	4	32	4770	348	2	39	3978	328	1
43	4410	357	3	33	4950	347	9	40	4162	314	1
44	4590	352	6	34	5130	346	7	41	4348	180	0
45	4770	346	7	35	5310	348	8	42	4532	291	1
46	4950	348	6	36	5490	348	8	43	4718	307	8
47	5130	353	8	37	5670	348	9	44	4902	204	4
48	5310	347	7	38	5850	344	8	45	5088	288	5
49	5490	343	8	39	6030	340	12	46	5272	287	6
50	5670	345	8					47	5458	292	7
51	5850	27	10	**30**				48	5643	293	7
52	6030	346	9	8:40P	6210	339	12				
53	6210	301	16	41	6390	341	7	**31**			
54	6390	325	11	42	6570	353	20	8:04P	202	186	3
55	6570	346	11	43	6750	354	20	05	387	35	0
56	6750	322	11	44	6930	350	18	06	571	103	1
57	6930	334	14	45	7110	353	23	07	747	45	1
58	7110	343	12	46	7290	351	19	08	924	62	3
59	7290	344	14	47	7370	345	25	09	1192	98	3
4:00P	7470	349	17	48	7450	347	14	10	1260	147	3
01	7650	345	18	49	7530	343	20	11	1428	145	4
02	7830	348	19					12	1596	149	4
03	8010	345	17	**31**				13	1764	141	4
				10:37A	215	238	2	14	1932	129	5
30				38	412	95	1				
8:07P	216	207	3	39	609	79?	1	**Aug.**			
08	414	218	1	40	796	5	3	**1**			
09	612	298	0	41	984	353	3	10:34A	213	228	1
10	801	204	2					35	411	25	1
11	990	205	0	**31**				36	605	59	3
12	1170	121	1	3:19P	222	182	2	37	790	84	3
13	1350	149	1	20	426	181	1	38	979	110	3
14	1530	158	2	21	629	224	0	39	1157	155	3
15	1710	157	2	22	824	170	1	40	1352	128	5
16	1890	143	1	23	1018	134	1	41	1505	119	3
17	2070	267	1	24	1204	111	2	42	1690	115	4
18	2250	256	4	25	1388	145	3	43	1870	118	3
19	2430	316	3	26	1572	154	3	44	2047	154	3
20	2610	349	2	27	1758	142	4				
21	2790	309	3	28	1942	129	4	**1**			
22	2970	339	4	29	2128	132	3	3:18P	216	288	1
23	3150	359	5	30	2312	92	6	19	414	31	1
24	3330	5	5	32	2682	115	4	20	612	35	2
25	3510	358	6	33	2868	130	4	21	801	99	1
26	3690	6	5	34	3052	150	4	22	990	115	3

TABLE I (1926)

Date and time	Height in meters	Azimuth from; degrees	Velocity m/s	Date and time	Height in meters	Azimuth from; degrees	Velocity m/s	Date and time	Height in meters	Azimuth from; degrees	Velocity m/s
Aug.				**Aug.**				**Aug.**			
1				**2**				**2**			
3:23P	1170	110	4	8:25P	214	71	4	9:13P	8792	152	2
24	1350	95	4	26	410	45	6	14	8970	138	2
25	1530	91	4	27	605	53	6	15	9147	172	2
26	1710	71	3	28	795	36	2	16	9324	144	4
27	1890	70	2	29	978	28	2	17	9500	132	4
				30	1155	41	3	18	9677	146	6
				31	1343	15	4	19	9855	170	5
1				32	1518	34	2	20	10020	183	3
8:41P	216	201	5	33	1690	237	0	21	10200	174	2
42	414	321	0	34	1875	89	1	22	10377	189	2
43	612	132	1	35	2042	219	1	23	10553	202	2
44	801	50	1	36	2220	231	0	24	10733	197	3
45	990	63	4	37	2400	135	1	25	10920	206	5
46	1170	67	7	38	2580	141	2	26	11108	178	7
47	1350	64	7	39	2755	118	1	27	11278	170	7
48	1530	67	8	40	2938	99	1	28	11460	155	10
49	1710	77	8	41	3115	92	1	29	11636	154	6
50	1890	85	9	42	3290	88	1	30	11810	156	8
51	2070	89	8	43	3460	104	2	31	11990	145	5
52	2250	93	7	44	3645	140	1	32	12170	153	10
53	2430	87	8	45	3820	83	4	33	12353	150	13
54	2610	96	8	46	4010	166	1	34	12530	136	13
55	2790	105	8	47	4186	186	1	35	12710	138	13
56	2970	86	7	48	4363	168	2	36	12890	140	12
57	3150	83	8	49	4539	96	2	37	13165	138	11
58	3330	102	6	50	4717	123	2				
59	3510	127	3	51	4893	117	2	**3**			
9:00P	3690	88	1	52	5070	97	3	11:08A	222	210	2
01	3870	49	2	53	5247	112	3	09	402	222	1
02	4050	23	3	54	5426	108	3	10	628	217	3
03	4230	58	2	55	5607	121	3	11	822	219	2
04	4410	64	3	56	5786	108	2	12	1025	224	0
05	4590	69	3	57	5964	107	3	13	1205	293	1
06	4770	45	4	58	6142	106	2	14	1390	325	2
07	4950	63	4	59	6320	110	2	15	1573	257	3
08	5130	118	2	9:00P	6497	155	2	16	1756	259	4
09	5310	178	2	01	6673	194	1	17	1933	304	4
				02	6850	180	1	18	2117	234	2
				03	7027	152	3	19	2301	290	2
2				04	7105	153	3	20	2485	268	3
4:06P	214	114	2	05	7383	143	5	21	2670	249	3
07	416	45	10	06	7550	155	5	22	2855	257	3
08	610	25	7	07	7728	164	6	23	3040	245	4
09	805	52	7	08	7905	150	5	24	3224	229	3
10	990	57	8	09	8082	164	4	25	3408	253	2
11	1175	64	8	10	8259	156	4	26	3592	318	1
12	1350	61	8	11	8437	155	4	27	3786	340	2
13	1530	53	5	12	8615	153	2	28	3970	358	1

TABLE I (1926)

Date and time	Height in meters	Azimuth from; degrees	Velocity m/s
Aug. 3			
9:29A	4164	278	1
30	4348	187	2
31	4532	193	2
32	4717	201	2
33	4900	218	1
34	5080	195	2
35	5265	196	3
36	5450	194	1
37	5634	257	1
38	5818	190	2
39	6002	186	4
40	6186	191	5
3			
3:23P	218	193	3
24	420	224	4
25	655	235	4
26	825	237	3
27	1015	250	1
28	1190	321	1
29	1372	332	1
31	1720	1	1
32	1900	48	4
33	2082	210	1
34	2264	235	2
35	2448	229	2
36	2630	216	3
37	2810	222	3
38	2990	234	3
39	3172	265	2
40	3354	260	4
41	3537	261	2
42	3720	299	1
3			
8:31P	216	60	5
32	414	236	3
33	612	235	1
34	801	325	1
35	990	315	1
36	1170	6	3
37	1350	325	2
38	1530	75	1
4			
11:21A	210	252	2
22	403	48	5

Date and time	Height in meters	Azimuth from; degrees	Velocity m/s
Aug. 4			
11:23A	595	85	1
24	778	64	4
25	963	71	5
26	1137	45	6
27	1312	40	5
28	1487	32	5
30	1842	48	4
31	2012	31	5
32	2187	25	3
33	2362	46	3
34	2537	44	2
35	2712	45	6
36	2887	55	4
37	3062	65	3
38	3237	49	3
39	3412	62	3
40	3587	54	4
41	3762	55	3
42	3937	66	2
43	4112	69	1
44	4287	81	2
45	4462	42	1
46	4637	344	1
47	4812	15	0
48	4987	192	2
49	5162	195	5
50	5340	207	5
51	5510	221	5
4			
2:04P	216	258	2
05	414	249	1
06	612	5	2
08	990	64	3
09	1170	70	2
10	1350	71	5
11	1536	89	3
12	1710	71	4
13	1890	67	4
14	2070	47	3
15	2250	44	2
16	2430	62	2
17	2610	61	3
19	2970	52	1
20	3150	31	2
21	3330	33	3
22	3510	51	3
23	3690	63	2

Date and time	Height in meters	Azimuth from; degrees	Velocity m/s
Aug. 4			
2:24P	3870	36	2
25	4050	27	2
26	4230	38	2
27	4410	51	2
28	4590	34	4
29	4770	28	2
30	4950	223	1
31	5130	240	2
32	5310	238	2
33	5490	225	3
34	5670	198	2
35	5850	195	2
36	6030	222	2
37	6210	218	2
38	6390	209	3
39	6570	213	4
40	6750	241	5
41	6930	221	3
42	7110	214	3
43	7290	225	6
44	7470	163	5
45	7650	195	5
46	7830	194	6
47	8010	207	5
48	8190	190	5
49	8370	187	5
4			
3:09P	75	223	4
10	224	233	3
11	373	269	2
12	522	289	2
13	671	288	1
14	820	277	2
15	1086	249	2
16	1352	175	1
17	1618	95	2
18	1884	86	2
5			
10:21A	221	348	1
22	423	103	3
23	626	89	5
24	819	79	5
25	1012	74	5
26	1196	80	5
27	1380	88	6
28	1564	109	8

TABLE I (1926)

Date and time	Height in meters	Azimuth from; degrees	Velocity m/s
Aug. 5			
10:29A	1748	82	3
30	1932	115	6
5			
8:30P	218	93	2
31	419	50	4
32	619	65	7
33	810	72	7
34	1001	115	4
35	1183	100	3
36	1365	108	5
5			
8:37P	1547	109	6
38	1729	90	7
39	1911	90	7
40	2093	97	7
41	2275	108	7
42	2457	110	7
43	2639	136	5
44	2821	132	7
45	3003	148	4
46	3185	138	6
47	3367	137	6
48	3549	145	6
49	3731	143	6
50	3913	142	9
51	4090	154	5
52	4277	156	8
53	4459	154	7
54	4641	169	8
55	4823	168	8
56	5005	170	8
57	5187	154	6
58	5369	161	8
59	5551	164	8
9:00	5733	181	8
01	5915	172	6
6			
10:33A	216	23	1
34	414	130	1
36	801	100	1
37	990	95	1
38	1170	74	3
39	1350	125	3
40	1530	125	4

Date and time	Height in meters	Azimuth from; degrees	Velocity m/s
Aug. 6			
6:24P	218	204	1
25	416	131	1
26	615	76	1
27	806	72	3
28	995	92	3
29	1176	112	2
30	1357	132	1
31	1538	136	1
32	1719	131	1
33	1900	109	5
34	2081	151	5
35	2262	169	9
36	2455	190	12
37	2624	184	13
38	2805	172	10
39	2990	150	10
40	3170	150	9
42	3530	154	9
43	3712	153	9
44	3892	113	7
45	4073	137	10
46	4253	137	6
47	4434	137	9
48	4615	137	7
7			
10:33A	200	187	5
34	372	179	2
35	568	203	1
36	742	70	0
37	915	251	1
38	1087	287	2
39	1252	33	0
40	1419	71	3
41	1586	77	3
42	1753	81	1
43	1917	130	2
44	2085	138	3
45	2254	141	3
46	2425	151	3
47	2590	143	3
48	2755	112	4
49	2922	124	5
50	3089	136	4
51	3256	186	5
52	3423	181	5
53	3569	194	5

Date and time	Height in meters	Azimuth from; degrees	Velocity m/s
Aug. 7			
6:03P	215	5	7
04	270		
9			
11:56A	200	227	1
57	372		
58	568	77	3
59	742	124	1
12:00M	915	117	3
0:01P	1087	150	2
02	1252	253	1
03	1419	225	2
04	1586	209	2
05	1753	140	3
06	1917	37	3
07	2085	29	3
08	2254	179	5
09	2425	177	6
10	2590	188	7
11	2755	186	5
12	2922	197	6
13	3089	196	6
14	3256	187	7
15	3423	169	7
16	3590	158	8
17	3757	159	7
18	3924	156	5
19	4070	167	6
10			
11:12A	222	242	1
13	455	21	1
14	673	116	3
15	881	187	1
17	1220	231	4
18	1405	222	3
19	1595	213	7
20	1782	202	9
21	1970	202	10
22	2155	212	11
23	2340	210	7
25	2808	211	4
10			
5:54P	208	170	5
55	398	108	1
56	589	218	2

TABLE I (1926)

Date and time	Height in meters	Azimuth from; degrees	Velocity m/s	Date and time	Height in meters	Azimuth from; degrees	Velocity m/s	Date and time	Height in meters	Azimuth from; degrees	Velocity m/s
Aug. 10				**Aug. 11**				**Aug. 11**			
5:57P	770	255	5	11:28A	5430	221	8	6:33P	4980	225	8
58	952	240	4	29	5607	214	10	34	5161	226	7
59	1125	198	2	31	5960	228	11	35	5345	227	8
6:00P	1298	145	2	32	6140	224	12	36	5526	224	7
01	1472	125	3	33	6320	230	11	37	5707	241	6
02	1645	178	4	34	6500	229	12	38	5880	195	8
03	1818	191	4	35	6675	234	12	39	6068	207	6
04	1991	193	4	36	6853	230	15	40	6249	216	8
05	2164	203	6	37	7030	235	16	41	6430	212	8
06	2338	209	9	38	7210	232	16	42	6511	208	7
07	2511	225	9	39	7390	232	16	43	6692	224	11
08	2684	227	9	40	7570	232	17	44	6873	221	9
09	2857	228	10	41	7645	242	11	45	7054	213	10
10	3030	227	11	42	7825	236	17	46	7235	209	12
11	3204	216	14	43	8000	236	18	47	7416	211	11
12	3377	223	18	44	8180	245	19	48	7597	206	11
12:30	3463	228	15	45	8360	246	18	49	7678	227	9
				46	8535	230	18	50	7859	219	14
11				47	8710	243	15	51	8040	217	16
10:59A	214	280	0	47:30	8800	224	30	52	8221	221	10
11:00A	408	203	1					53	8402	219	15
01	605	220	1	**11**				54	8583	219	18
02	790	104	2	6:07P	218	214	3	55	8764	229	17
03	979	83	3	08	416	225	1	56	8945	220	14
04	1160	62	3	09	615	225	2	57	9126	221	18
05	1335	58	1	10	806	245	2	58	9307	220	13
06	1510	339	0	11	995	309	1	59	9488	236	20
07	1690	110	1	12	1176	36	1	7:00P	9669	223	14
08	1870	102	1	13	1357	99	1	01	9850	225	13
09	2050	215	1	14	1538	188	1	02	10031	214	17
10	2225	206	2	15	1719	163	2				
11	2400	197	3	16	1900	258	2	**12**			
12	2580	201	4	17	2081	152	4	10:33A	208	259	1
13	2767	199	4	18	2262	178	3	34	398	calm	calm
14	2937	203	5	19	2455	200	4	35	588	313	1
15	3115	226	4	20	2624	208	3	37	951	122	2
16	3293	210	7	21	2805	219	4	38	1124	210	1
17	3470	219	4	22	2990	220	5	39	1297	247	2
18	3650	212	6	23	3168	222	4	40	1470	261	3
19	3827	190	7	24	3349	233	4	41	1643	228	2
20	4000	184	7	25	3530	228	5	42	1816	235	3
21	4180	185	6	26	3712	233	5	43	1989	234	3
22	4360	203	5	27	3892	233	5	44	2162	224	3
23	4540	208	5	28	4073	226	6	45	2335	254	3
24	4717	199	6	29	4253	199	6	46	2508	246	1
25	4895	214	7	30	4434	224	7	47	2682	167	1
26	5070	201	7	31	4615	224	9	48	2855	203	2
27	5250	220	7	32	4800	215	8	49	3028	212	2

TABLE I (1926)

Date and time	Height in meters	Azimuth from; degrees	Velocity m/s	Date and time	Height in meters	Azimuth from; degrees	Velocity m/s	Date and time	Height in meters	Azimuth from; degrees	Velocity m/s
Aug.				**Aug.**				**Aug.**			
12				**13**				**13**			
10:50A	3201	212	2	10:17A	3330	80	1	1:14P	1341	353	4
51	3374	213	2	18	3510	103	1	15	1490	357	2
52	3547	229	4	19	3690	119	1	16	1639	75	0
53	3720	245	4	20	3870	146	2	17	1788	16	1
54	3893	235	4	21	4050	167	3	18	1937	58	1
55	4066	241	4	22	4230	154	3	19	2086	64	2
56	4239	219	4	23	4410	98	2	20	2235	58	3
57	4412	245	4	24	4590	60	4	21	2384	45	2
58	4585	261	6	25	4770	48	4	22	2533	10	2
59	4758	259	7	26	4950	49	4	23	2682	23	2
11:00A	4931	254	7	27	5130	46	5	24	2831	44	3
01	5104	240	9	28	5310	48	5	25	2992	55	3
02	5277	235	8	29	5490	49	5	26	3153	65	4
03	5450	233	7	30	5670	52	5	27	3314	76	4
				31	5850	49	5	28	3475	91	4
12				32	6030	47	5	29	3636	101	3
6:00P	216	194	7	33	6210	51	6	30	3797	92	2
01	414	208	6	34	6390	41	6	31	3958	83	2
02	612	328	2	35	6570	29	6	32	4119	87	2
03	801	261	5	36	6750	23	7	33	4280	79	1
04	990	245	5	37	6930	9	6	34	4441	65	1
05	1170	243	4	38	7110	11	6	35	4602	153	0
06	1350	233	2	39	7290	4	5	36	4763	196	1
07	1530	205	1	40	7470	10	6	37	4924	209	2
08	1710	229	2	41	7650	2	4	38	5085	215	3
09	1890	229	2	42	7830	9	6	39	5246	213	3
10	2070	182	1	43	8010	9	6				
11	2250	156	1	44	8190	356	7	**13**			
12	2430	237	1	45	8370	346	7	6:23P	222	201	2
13	2610	283	2	46	8550	340	9	24	420	241	3
14	2790	263	2	47	8730	342	8	25	610	240	2
				48	8910	323	12	26	790	292	1
13				49	9090	322	9	28	1160	152	1
10:00A	216	267	0	50	9270	324	10	29	1338	143	2
01	414	38	1	51	9450	319	11	30	1515	31	5
02	612	74	2	52	9630	321	10				
03	801	64	1	53	9810	326	11	**14**			
04	990	70	1	54	9990	328	12	11:03A	212	140	0
05	1170	285	1					04	406	240	1
06	1350	294	1	**13**				05	602	214	0
07	1530	262	1	1:06P	149	225	0	06	788	131	1
08	1710	236	0	07	298	115	0	07	974	250	0
11	2250	59	1	08	447	213	1	08	1151	305	1
12	2430	74	2	09	596	230	2	09	1328	308	2
13	2610	35	4	10	745	211	1	10	1505	269	2
14	2790	49	3	11	894	176	1	11	1682	92	2
15	2970	57	2	12	1043	150	1	12	1859	236	2
16	3150	61	3	13	1192	38	1	13	2036	226	9

TABLE I (1926)

Date and time	Height in meters	Azimuth from; degrees	Velocity m/s	Date and time	Height in meters	Azimuth from; degrees	Velocity m/s	Date and time	Height in meters	Azimuth from; degrees	Velocity m/s
Aug. 14				**Aug. 15**				**Aug. 16**			
11:14A	2213	190	5	9:46A	783	95	1	10:03A	208	304	1
15	2390	183	5	47	968	65	2	04	398	56	1
16	2567	171	5	48	1144	58	3	05	589	84	4
17	2744	170	5	49	1320	59	4	06	769	96	3
				50	1496	65	5	07	951	85	3
14				51	1672	67	3	08	1125	65	3
5:54P	214	190	4	52	1848	62	5	09	1298	53	3
55	409	187	3	53	2024	48	4	10	1471	56	4
56	605	254	1	54	2200	55	4	11	1643	50	4
57	792	349	1	55	2376	56	4	12	1816	60	4
58	979	28	5	56	2552	72	3	13	1989	95	2
59	1157	111	1	57	2728	48	2	14	2162	64	3
6:00P	1335	147	3					15	2335	52	3
01	1513	136	4					16	2508	50	3
02	1691	134	4	**15**				17	2682	40	4
04	2047	135	4	1:31P	212	162	1	18	2855	59	4
05	2225	157	5	32	402	187	1	19	3028	44	4
06	2403	181	3	33	595	220	1	20	3201	58	5
07	2581	171	6	34	778	351	1	21	3374	61	4
08	2759	197	3	35	962	224	3	22	3547	66	4
09	2937	217	3	36	1138	197	1	23	3720	54	6
10	3115	206	2	44	2538	107	1	24	3893	47	6
11	3293	197	1	45	2713	89	3	25	4066	51	5
12	3471	248	2	46	2888	60	2	26	4239	33	7
13	3649	202	1	47	3063	39	3	27	4412	53	7
14	3827	160	1	48	3238	30	5	28	4585	52	6
15	4005	149	1	49	3413	9	3	29	4758	55	7
16	4183	125	1	49:35	3588	45	4	30	4931	50	6
17	4361	140	2					31	5104	58	7
18	4539	156	2					32	5277	25	8
19	4717	155	3	**15**				33	5450	38	10
20	4895	156	3	6:12P	218	183	3	34	5623	27	11
21	5073	151	4	13	416	66	1	35	5796	36	9
22	5251	158	7	14	615	57	2	36	5969	40	10
23	5429	140	3	15	806	52	4	37	6142	25	11
24	5607	145	5	16	995	30	5				
25	5785	173	5	17	1176	31	5	**16**			
26	5963	175	6	18	1357	30	5	2:05P	200	136	2
27	6141	175	6	19	1538	25	5	06	372	63	2
28	6319	175	5	20	1719	17	5	07	568	63	3
29	6497	156	8	21	1900	8	4	08	742	66	4
30	6675	189	7	22	2081	3	3	09	915	61	4
31	6853	170	8	23	2262	30	1	10	1087	63	4
				24	2455	65	2	11	1252	51	3
15				25	2624	57	1	12	1419	62	5
9:43A	211	284	1	26	2805	282	0	13	1586	55	4
44	409	8	1	27	2990	48	1	14	1753	49	4
45	598	101	1	27:30	3080	30	3	15	1917	25	2

TABLE I (1926)

Date and time	Height in meters	Azimuth from; degrees	Velocity m/s	Date and time	Height in meters	Azimuth from; degrees	Velocity m/s	Date and time	Height in meters	Azimuth from; degrees	Velocity m/s
Aug. 16				**Aug. 16**				**Aug. 17**			
2:16P	2085	38	2	6:51P	5034	12	8	9:40A	2531	148	8
17	2254	31	2	52	5199	11	8	41	2618	154	4
18	2425	10	2	53	5364	8	10	42	2905	139	7
19	2590	30	4	54	5529	10	8	43	3092	97	4
20	2755	30	6	55	5694	10	11	44	3279	113	5
21	2922	25	5	56	5859	6	8	45	3466	115	6
22	3089	34	5	57	6024	15	12	46	3653	132	4
23	3256	20	4	58	6189	354	9	47	3840	137	4
24	3423	17	5	59	6354	25	17	48	4027	148	5
25	3590	15	5	7:00P	6519	12	9	49	4214	120	2
26	3757	25	6	01	6684	21	6	50	4401	68	6
27	3924	25	6	02	6849	22	5	51	4588	72	7
28	4091	23	7	03	7014	28	2	52	4775	64	8
29	4259	24	6	04	7179	27	59	53	4962	54	9
30	4427	19	6	05	7344	28	9	54	5149	48	9
31	4595	29	7	06	7509	31	0	55	5336	37	6
32	4763	30	8	07	7674	31	8	56	5523	64	6
				08	7839	43	9	57	5710	63	6
16				09	8004	27	8	58	5897	86	5
6:22P	198	116	3	10	8169	32	11	59	6084	30	6
23	379	91	3	11	8334	25	14	10:00A	6271	23	4
24	561	70	5	12	8499	18	7	01	6458	42	3
25	735	75	6	13	8664	22	11	02	6645	33	3
26	907	67	5	14	8829	41	9	03	6832	11	3
27	1074	62	3	15	8994	29	8	04	7019	27	4
28	1239	62	5	16	9159	25	9	05	7206	34	4
29	1404	68	4	17	9324	16	6	06	7393	23	4
30	1569	50	4	18	9489	45	6	07	7580	27	5
31	1734	49	4	19	9654	52	4	08	7767	13	5
32	1899	4	2	20	9819	48	4	09	7954	3	4
33	2064	43	8	21	9984	31	4	10	8141	17	4
34	2229	59	4	22	10149	209	4	11	8328	20	4
35	2394	93	2	23	10314	147	6	12	8515	357	4
36	2559	130	2	24	10480	68	12	13	8702	5	5
37	2724	170	2					14	8889	6	3
38	2889	132	3	**17**				15	9076	1	3
39	3054	94	2	9:28A	224	205	1	16	9263	348	3
40	3219	176	3	29	430	31	3				
41	3384	16	5	30	638	135	2	**17**			
42	3549	21	4	31	839	87	3	2:07P	222	254	2
43	3714	24	6	32	1028	100	3	08	426	249	1
44	3879	15	4	33	1215	147	2	09	629	69	0
45	4044	14	6	34	1402	153	2	10	824	152	1
46	4209	12	6	35	1589	167	3	11	1018	119	2
47	4374	8	8	36	1777	168	5	12	1204	100	3
48	4539	13	10	37	1970	168	5	13	1388	127	4
49	4704	10	8	38	2157	153	5	14	1572	160	5
50	4869	11	8	39	2344	140	6	15	1758	160	5

TABLE I (1926)

Date and time	Height in meters	Azimuth from; degrees	Velocity m/s	Date and time	Height in meters	Azimuth from; degrees	Velocity m/s	Date and time	Height in meters	Azimuth from; degrees	Velocity m/s
Aug. 17				**Aug. 17**				**Aug. 17**			
2:16P	1942	179	3	6:13P	1837	78	3	7:01P	10264	163	4
17	2128	156	7	14	2015	70	3	02	10440	210	6
18	2312	151	6	15	2190	86	2	03	10615	179	5
19	2497	157	4	16	2365	99	2	04	10790	234	5
20	2682	146	5	17	2545	109	4	05	10965	193	6
21	2868	122	4	18	2720	83	5	06	11141	202	6
22	3052	57	2	19	2896	95	4	07	11316	168	3
23	3238	111	1	20	3071	80	5	08	11492	143	4
24	3422	123	2	21	3247	56	3	09	11667	131	7
25	3608	103	3	22	3423	63	4	10	11842	119	8
26	3792	122	5	23	3598	79	4	11	12018	95	8
27	3978	95	4	24	3773	70	3	12	12193	118	6
28	4162	125	5	25	3950	81	3	13	12370	118	7
29	4348	90	3	26	4125	65	4	14	12545	116	4
30	4532	95	5	27	4300	70	5	15	12720	122	6
31	4718	76	5	28	4475	69	5	16	12895	139	8
32	4902	69	8	29	4650	64	6				
33	5088	70	8	30	4825	55	7	**18**			
34	5273	58	8	31	5000	46	6	9:40A	216	16	1
35	5458	59	7	32	5175	51	6	41	414	99	1
36	5643	51	5	33	5351	58	4	42	612	64	2
37	5828	37	5	34	5527	65	5	43	801	91	2
38	6013	55	5	35	5700	56	4	44	990	86	6
39	6198	39	6	36	5880	48	5	45	1170	98	4
40	6383	39	5	37	6055	54	5	46	1350	113	4
41	6568	48	6	38	6230	53	6	47	1530	107	5
42	6753	24	4	39	6405	53	7	48	1710	93	4
43	6938	50	5	40	6580	49	7	49	1890	85	4
44	7123	63	5	41	6755	40	5	50	2070	88	6
45	7308	64	4	42	6930	46	7	51	2250	82	5
46	7493	26	4	43	7105	44	7	52	2430	58	5
47	7678	45	6	44	7281	46	7	53	2610	63	4
48	7863	40	5	45	7456	57	6	54	2790	44	5
49	8048	40	5	46	7632	61	6	55	2970	40	6
50	8233	41	5	47	7807	53	7	56	3150	48	7
51	8418	44	3	48	7983	65	3	57	3330	53	6
52	8603	67	3	49	8158	42	4	58	3510	58	8
				50	8333	22	4	59	3690	40	7
17				51	8509	31	6	10:00A	3870	50	7
6:04P	210	196	3	52	8684	0	0	01	4050	35	7
05	404	205	0	53	8860	59	0	02	4230	40	8
06	596	25	1	54	9035	50	1	03	4410	36	6
07	780	46	2	55	9211	230	1	04	4590	49	5
08	965	79	2	56	9387	241	0	05	4770	57	7
09	1140	75	3	57	9562	336	3	06	4950	55	7
10	1315	74	2	58	9737	21	2	07	5130	65	6
11	1490	71	2	59	9913	8	0	08	5310	64	8
12	1655	88	3	7:00P	10089	57	2	09	5490	61	6

TABLE I (1926)

Date and time	Height in meters	Azimuth from; degrees	Velocity m/s	Date and time	Height in meters	Azimuth from; degrees	Velocity m/s	Date and time	Height in meters	Azimuth from; degrees	Velocity m/s
Aug. 18				**Aug. 18**				**Aug. 18**			
10:10A	5670	60	7	2:09P	4441	63	5	6:26P	2822	119	1
11	5850	55	7	10	4630	58	6	27	2993	94	2
12	6030	64	7	11	4819	60	5	28	3164	93	1
13	6210	63	7	12	5008	56	5	29	3335	45	2
14	6390	62	8	13	5197	50	5	30	3506	35	2
15	6570	61	9	14	5386	61	6	31	3677	47	4
16	6750	59	7	15	5575	65	5	32	3848	66	3
17	6930	71	8	16	5764	54	4	33	4019	70	5
18	7110	85	7	17	5953	66	6	34	4190	62	5
19	7290	65	7	18	6142	63	7	35	4361	54	6
20	7470	71	4	19	6331	84	7	36	4532	49	6
21	7650	63	4	20	6420	91	6	37	4703	53	6
22	7830	75	4	21	6709	77	8	38	4874	40	6
23	8010	83	2	22	6898	87	8	39	5045	34	6
24	8190	68	3	23	7087	126	9	40	5216	36	5
25	8370	79	3	24	7276	70	5	41	5387	55	4
26	8550	193	4	25	7465	92	6	42	5558	66	3
27	8730	94	4	26	7654	93	4	43	5729	55	5
28	8910	180	4	27	7843	150	3	44	5900	51	5
29	9090	159	3	28	8032	126	4	45	6071	54	4
30	9270	151	4	29	8221	134	4	46	6242	64	4
31	9450	221	6	30	8410	141	4	47	6413	62	4
32	9630	192	4	31	8599	135	6	48	6584	72	5
33	9810	202	4	32	8788	184	5	49	6755	96	3
34	9990	225	7	33	8977	195	6	50	6925	102	3
35	10170	201	8	34	9166	195	4	51	7097	116	1
				35	9355	196	3	52	7268	64	7
18				36	9544	211	5	53	7439	95	3
1:47P	227	249	1	37	9733	175	5	54	7610	117	2
49	640	80	1	38	9922	208	6	55	7781	140	2
51	1039	64	4	39	10111	200	6	56	7952	182	1
52	1228	42	6	40	10300	196	8	57	8123	238	1
53	1417	35	6					58	8294	316	0
54	1606	27	5	**18**				59	8465	316	0
55	1795	30	6	6:11P	205	205	1	7:00P	8636	203	3
56	1984	33	6	12	393	280	1	01	8807	192	3
57	2173	34	6	13	582	84	1	02	8978	112	1
58	2362	39	6	14	761	93	3	03	9149	235	8
59	2551	34	5	15	941	112	4	04	9320	224	4
2:00P	2740	51	7	16	1112	108	2	05	9491	218	5
01	2929	65	5	17	1283	46	1	06	9662	220	8
02	3118	65	7	19	1625	76	3	07	9833	228	8
03	3307	84	3	20	1796	50	3	08	10004	223	7
04	3496	83	5	21	1967	28	4	09	10175	216	9
05	3685	84	6	22	2138	352	4	10	10346	232	8
06	3874	81	4	23	2309	3	4	11	10517	242	5
07	4063	62	10	24	2480	2	5	12	10688	214	3
08	4252	68	8	25	2651	32	1	13	10859	169	1

TABLE I (1926)

Date and time	Height in meters	Azimuth from; degrees	Velocity m/s	Date and time	Height in meters	Azimuth from; degrees	Velocity m/s	Date and time	Height in meters	Azimuth from; degrees	Velocity m/s
Aug.				**Aug.**				**Aug.**			
18				**19**				**19**			
7:14P	11030	174	4	10:46A	7354	245	5	3:13P	5399	234	6
15	11201	207	6	47	7545	247	5	14	5576	239	9
16	11372	208	5	48	7736	246	5	15	5753	235	6
17	11543	184	3	49	7927	243	7	16	5930	230	7
18	11714	204	3	50	8118	242	6	17	6107	233	7
19	11885	181	5	51	8309	244	9	18	6284	227	8
20	12056	159	4	52	8500	243	9	19	6461	229	9
21	12227	195	4	53	8691	242	9	20	6638	230	9
22	12498	179	4	54	8882	241	10	21	6815	217	9
				55	9073	243	13	22	6992	225	11
19				56	9264	246	13	23	7169	224	13
10:09A	229	54	0	57	9455	245	12	24	7346	224	11
10	439	63	1	58	9646	247	15	25	7523	229	14
11	650	68	1	59	9837	249	20	26	7700	230	15
12	850	79	2	11:00A	10028	249	20	27	7877	229	17
13	1050	107	1	01	10219	244	18	28	8054	230	19
14	1242	220	0	02	10410	252	18				
15	1433	90	3					**19**			
16	1624	103	1	**19**				6:29P	204	192	4
17	1815	172	2	2:44P	212	195	3	30	391	198	3
18	2006	205	2	45	406	225	1	31	578	241	6
19	2197	231	3	46	602	135	1	32	756	247	3
20	2388	221	3	47	788	309	0	33	935	328	0
21	2579	194	3	48	974	316	1	34	1105	234	1
22	2770	213	2	49	1151	316	0	35	1275	335	4
23	2961	210	2	50	1328	162	2	36	1445	77	4
24	3152	202	4	51	1505	165	1	37	1615	113	2
25	3343	221	3	52	1682	186	3	38	1785	183	7
26	3534	288	2	53	1859	184	4	39	1955	174	5
27	3725	273	2	54	2036	175	4	40	2125	176	5
28	3916	278	2	55	2213	164	4	41	2295	177	6
29	4107	291	2	56	2390	158	4	42	2465	178	6
30	4298	290	3	57	2567	168	3	43	2635	196	5
31	4489	263	3	58	2744	179	4	44	2805	214	4
32	4680	235	2	59	2921	191	5	45	2975	221	4
33	4871	240	3	3:00P	3098	228	2	46	3145	232	5
34	5062	269	4	01	3275	256	3	47	3315	243	4
35	5253	264	4	02	3452	261	4	48	3485	237	5
36	5444	261	4	03	3629	256	4	49	3655	231	5
37	5635	255	4	04	3806	253	4	50	3825	231	6
38	5826	234	6	05	3903	254	4	51	3995	248	4
39	6017	233	5	06	4160	247	4	52	4165	236	4
40	6208	244	5	07	4337	244	6	53	4335	238	7
41	6399	244	5	08	4514	261	5	54	4505	227	7
42	6590	258	5	09	4691	239	6	55	4675	233	7
43	6781	253	5	10	4868	228	7	56	4845	240	7
44	6972	251	7	11	5045	234	6	57	5015	229	9
45	7163	255	3	12	5222	246	4	58	5185	216	7

TABLE I (1926)

Date and time	Height in meters	Azimuth from; degrees	Velocity m/s	Date and time	Height in meters	Azimuth from; degrees	Velocity m/s	Date and time	Height in meters	Azimuth from; degrees	Velocity m/s
Aug.				**Aug.**				**Aug.**			
19				**20**				**23**			
6:59P	5355	218	8	2:39P	619	259	4	6:34P	1870	139	1
7:00P	5525	226	7	40	810	247	5	35	2050	136	2
01	5695	215	8	41	1001	232	7	36	2225	125	2
02	5865	211	9	42	1183	213	9	37	2400	119	2
03	6035	228	7	43	1365	196	8	38	2580	73	2
04	6205	230	11	44	1547	198	9	39	2767	104	2
05	6375	224	10	45:35	1729	227	12	40	2937	98	2
06	6545	229	11					41	3115	85	2
07	6715	217	10	**22**				42	3293	26	2
08	6885	227	11	2:57P	222	222	2	43	3470	69	1
09	7055	222	14	58	426	224	3	45	3827	89	0
10	7325	228	13	3:00P	824	259	6	46	4000	85	4
11	7495	223	14	01	1018	343	3	47	4180	19	1
12	7665	220	16	02	1204	132	0	48	4360	170	2
13	7835	223	18	03	1388	32	2	49	4540	139	2
14	8005	220	22	04	1573	8	5	50	4717	123	2
15	8175	227	20	05	1758	9	4				
16	8345	224	24	06	1942	325	3	**24**			
17	8515	225	24	07	2128	32	5	10:15A	216	35	2
18	8685	225	29	07:53	2290	15	2	16	414	80	4
19	8855	224	23					17	612	72	6
19:30	8940	228	34	**23**				18	801	75	7
				11:34A	214	204	4	19	990	82	5
20				35	408	218	2	20	1170	90	5
10:03A	218	239	1	36	605	188	0	21	1350	116	5
04	416	333	1	37	790	281	1	22	1530	112	5
05	615	232	2	38	979	286	2	23	1710	117	5
06	806	258	4					24	1890	122	4
07	995	247	7	**23**				25	2070	140	4
08	1176	245	4	2:36P	224	188	2	26	2250	155	2
09	1357	244	8	37	430	223	2	27	2430	148	3
10	1538	252	7	38	638	227	2	28	2610	153	4
11	1719	245	6	39	839	260	2	29	2790	198	4
12	1900	233	10	40	1028	244	0	30	2970	183	4
13	2081	233	11	41	1215		0	31	3150	191	6
14	2262	234	10	42	1402		0	32	3330	184	6
15	2455	245	8	42:38	1526		0	33	3510	185	6
16	2624	242	9					34	3690	194	5
17	2805	246	8	**23**				35	3870	198	5
18	2990	245	6	6:25P	214	206	2	36	4050	201	6
19	3168	244	6	26	408	238	2	37	4230	208	6
20	3349	249	4	27	605	257	2	38	4410	203	8
21	3530	250	9	28	790	249	2	39	4590	213	7
22:55	3712	246	8	29	979	215	2	40	4770	215	8
				30	1160	163	2				
20				31	1335	202	1	**24**			
2:37P	218	278	5	32	1510	232	3	2:32P	209	101	4
38	419	273	4	33	1690	74	1	33	402	57	3

TABLE I (1926)

Date and time	Height in meters	Azimuth from; degrees	Velocity m/s	Date and time	Height in meters	Azimuth from; degrees	Velocity m/s	Date and time	Height in meters	Azimuth from; degrees	Velocity m/s
Aug. 24				**Aug. 27**				**Aug. 28**			
2:34P	592	78	5	2:50P	412	18	2	10:12A	1890	20	1
35	774	58	6	51	609	11	2	13	2070	327	1
36	957	38	5	52	797	68	1	14	2250	235	0
37	1131	47	7	53	985	88	2	15	2430	265	0
38	1305	61	6					16	2610	283	0
39	1480	83	6	**27**				17	2790	2	1
40	1654	87	6	6:25P	204	93	5	18	2970	358	3
41	1828	93	5	26	391	91	5	19	3150	325	2
42	2002	99	5	27	578	74	4	20	3330	317	2
43	2176	126	4	28	756	65	3	21	3510	330	2
44	2350	178	5	29	935	63	4	22	3690	325	5
45	2524	141	5	30	1105	50	3	23	3870	351	5
46	2698	147	4	31	1275	42	2	24	4050	333	5
47	2872	149	6	32	1445	35	2	25	4230	333	6
48	3046	141	6	33	1615	0	3	26	4410	333	8
49	3220	126	5	34	1785	1	4	27	4590	342	11
50	3394	111	4	35	1955	349	2	28	4770	335	9
				36	2125	359	1	29	4950	340	10
24				37	2295	50	1	30	5130	337	12
6:28P	212	80	3	38	2465	1	2	31	5310	330	12
29	407	88	6	39	2635	4	3	32	5490	333	13
30	604	91	6	40	2805	16	2	33	5670	331	14
31	787	96	5	41	2975	18	1	34	5850	330	14
32	973	97	4	42	3145	69	0	35	6030	327	13
33	1150	106	4	43	3315	277	0	36	6210	322	15
34	1327	121	3	44	3485	43	1	37	6390	324	14
35	1504	93	4	45	3655	104	1	38	6570	329	16
36	1681	68	6	46	3825	326	2	39	6750	322	15
				47	3995	328	5	40	6930	322	15
26				48	4165	348	5	41	7110	324	16
6:20P	221	93	0	49	4335	347	6	42	7290	322	16
21	423	60	1	50	4505	349	7	43	7470	327	19
22	625	75	1	51	4675	353	8	44	7650	320	15
23	819	341	1	52	4845	342	8	45	7830	325	19
25	1197	280	2	53	5015	332	6	46	8010	324	20
26	1384	262	2	54	5185	337	5	47	8190	318	18
27	1569	253	1	55	5355	332	7	48	8370	320	18
28	1754	257	1					49	8550	321	20
29	1940	288	1	**28**				50	8730	324	21
30	2124	200	1	10:03A	216	273	2	51	8910	318	16
31	2308	196	1	04	414	319	2				
32	2492	194	1	05	612	81	3	**28**			
33	2676	205	1	06	801	115	3	1:55P	214	242	2
34	2860	237	3	07	990	82	2	56	410	234	1
35	3044	216	5	08	1170	185	1	57	605	65	1
				09	1350	81	1	58	795	86	2
27				10	1530	35	1	59	978	96	2
2:49P	215	128	2	11	1710	20	1	2:00P	1155	103	2

TABLE I (1926)

Date and time	Height in meters	Azimuth from; degrees	Velocity m/s	Date and time	Height in meters	Azimuth from; degrees	Velocity m/s	Date and time	Height in meters	Azimuth from; degrees	Velocity m/s
Aug. 28				**Aug. 29**				**Aug. 29**			
2:01P	1333	234	—	9:46A	2492	190	3	3:41P	6115	271	17
02	1511	268	0	47	2676	198	2	42	6266	265	19
03	1690	239	1	48	2860	220	2	43	6417	265	18
04	1870	270	1	49	3044	274	2	44	6568	263	19
05	2047	301	3	50	3228	275	3	45	6719	260	19
06	2224	107	2	51	3412	270	4	46	6870	265	18
08	2589	341	2	52	3596	280	5	47	7021	263	22
10	3115	278	4	53	3780	107	5	48	7172	265	22
12	3293	311	4					48:55	7308	264	22
13	3471	330	6	**29**							
14	3650	313	7	3:02P	181	239	2	**29**			
15	3828	318	9	03	347	159	1	6:25P	208	286	1
16	4006	321	10	05	673	75	2	26	398	90	1
17	4184	320	9	06	830	89	2	27	589	85	2
18	4362	322	11	07	981	88	3	28	769	105	1
19	4540	332	11	08	1132	92	4	29	951	62	1
20	4718	319	13	09	1283	108	3	30	1125	96	4
21	4896	319	12	10	1434	99	1	31	1298	110	1
22	5074	319	13	11	1585	47	1	32	1471	90	3
23	5252	319	15	12	1736	141	2	33	1643	84	2
24	5430	316	15	13	1887	163	2	34	1816	195	2
25	5608	317	15	14	2038	178	2	36	2162	149	2
26	5786	316	14	15	2189	218	2	37	2335	239	2
27	5964	315	16	16	2340	251	2	38	2508	252	1
28	6142	313	16	17	2491	265	2	39	2682	263	1
29	6320	313	17	18	2642	269	3	40	2855	263	2
30	6498	310	17	19	2793	272	3	41	3028	276	3
31	6676	305	14	20	2944	269	3	42	3201	269	3
32	6854	306	19	21	3095	285	4	43	3374	271	3
33	7032	309	18	22	3246	291	4	44	3547	294	3
34	7209	305	18	23	3397	301	4	45	3720	276	3
35	7387	303	15	24	3548	299	5	46	3893	281	4
36	7565	305	18	25	3699	284	5	47	4066	298	4
37	7743	309	19	26	3850	287	5	48	4239	298	4
				27	4001	287	5	49	4412	288	5
29				28	4152	282	5	50	4585	287	7
9:34A	221	354	1	29	4303	273	6	51	4758	291	7
35	423	97	3	30	4454	272	10	52	4931	286	8
36	625	83	4	31	4605	272	10	53	5104	274	9
37	819	84	4	32	4756	271	12	54	5277	256	10
38	1013	91	5	33	4907	269	11	55	5450	262	13
39	1197	104	4	34	5058	269	13	56	5623	257	11
40	1385	122	3	35	5209	269	13	57	5796	260	15
41	1570	137	1	36	5360	271	14	58	5969	258	15
42	1754	132	2	37	5511	266	14	59	6142	255	17
43	1940	158	3	38	5662	266	15	7:00P	6315	271	15
44	2124	155	3	39	5813	263	16	01	6488	256	17
45	2308	160	3	40	5964	261	19	02	6661	257	16

TABLE I (1926)

Date and time	Height in meters	Azimuth from; degrees	Velocity m/s	Date and time	Height in meters	Azimuth from; degrees	Velocity m/s	Date and time	Height in meters	Azimuth from; degrees	Velocity m/s
Aug.				**Aug.**				**Aug.**			
29				**30**				**30**			
7:03P	6834	257	19	2:40P	1719	1	2	6:43P	1710	31	2
04	7007	258	17	41	1900	50	1	44	1890	54	1
05	7180	250	10	42	2081	56	2	45	2070	202	1
06	7353	257	20	43	2262	67	3	46	2250	258	1
07	7526	260	31	44	2455	64	3	47	2430	251	2
08	7699	260	21	45	2634	73	3	48	2610	291	3
09	7872	260	23	46	2805	290	2	49	2790	287	3
10	8045	258	23	47	2990	292	3	50	2970	279	4
11	8218	257	27	48	3168	278	3	51	3150	278	4
12	8391	261	20	49	3349	258	3	52	3350	269	4
13	8564	261	27	50	3530	258	3	53	3510	264	5
				51	3712	254	4	54	3690	229	9
30				52	3892	250	6	55	3870	232	10
10:32A	218	160	0	53	4073	255	5	56	4050	253	9
33	418	25	0	54	4253	250	6	57	4230	261	8
34	619	76	1	55	4434	249	7	58	4410	266	6
36	1001	82	2	56	4615	247	8	59	4590	282	5
37	1183	77	3	57	4800	243	9	7:00P	4770	241	5
38	1365	36	1	58	4980	241	10	01	4950	247	6
39	1547	95	0	59	5161	245	9	02	5130	267	7
40	1729	164	1	3:00P	5345	243	9	03	5310	274	7
41	1911	182	0	01	5526	248	10	04	5490	275	7
42	2093	179	1	02	5707	245	9	05	5670	274	8
43	2275	214	1	03	5880	256	7	06	5850	265	8
44	2457	240	2	04	6068	246	12	07	6030	268	8
45	2639	264	3	05	6249	243	11	08	6210	257	8
46	2821	269	2	06	6430	245	10	09	6390	258	8
47	3003	292	3	07	6511	240	10	10	6570	260	8
48	3185	332	1	08	6692	241	14	11	6750	253	9
49	3367	240	2	09	6873	234	14	12	6930	253	9
50	3549	196	1	10	7054	235	14	13	7110	254	10
51	3731	173	2	11	7233	234	14	14	7290	260	10
52	3913	136	4	12	7416	237	16	15	7470	258	10
53	4095	135	4	13	7597	231	14	16	7650	249	12
54	4277	150	5	14	7678	228	14	17	7830	258	11
55	4459	157	7	15	7859	231	14	18	8010	248	13
56	4641	155	9	16	8040	238	14	19	8190	241	12
57	4823	157	10	17	8221	239	11	20	8370	243	13
								21	8550	247	14
30				**30**				22	8730	253	13
2:32P	218	253	2	6:35P	216	173	2	23	8910	247	16
33	416	269	1	36	414	357	1	24	9090	271	14
34	615	2	0	37	612	300	0	25	9270	241	16
35	806	67	1	38	801	95	1	26	9450	265	12
36	995	110	2	39	990	88	2	27	9630	263	12
37	1176	110	1	40	1170	75	2	28	9810	252	14
38	1357	75	4	41	1350	53	2	29	9990	274	16
39	1538	89	3	42	1530	35	2	30	10170	262	13

TABLE I (1926)

Date and time	Height in meters	Azimuth from; degrees	Velocity m/s	Date and time	Height in meters	Azimuth from; degrees	Velocity m/s	Date and time	Height in meters	Azimuth from; degrees	Velocity m/s
Aug.				**Aug.**				**Aug.**			
30				**31**				**31**			
7:31P	10350	268	15	11:05A	7416	271	19	3:40P	7169	258	15
32	10530	262	16	06	7597	279	19	41	7346	250	16
33	10710	262	15	07	7678	262	12	42	7523	260	18
34	10890	266	11	08	7859	271	20	43	7700	262	19
35	11070	261	11	09	8040	264	18	44	7877	264	22
36	11250	260	12	10	8221	268	24	45	8054	264	24
37	11430	261	17	11	8402	278	31	46	8231	264	20
38	11610	258	17					47	8408	263	22
				31				48	8585	258	21
31				3:01P	212	99	7	49	8762	265	24
10:25A	217	89	6	02	407	87	10	50	8939	265	25
26	416	70	4	03	604	87	5	51	9116	267	23
27	615	63	4	04	787	100	7				
28	806	79	3	05	973	105	1	**31**			
29	995	90	3	06	1150	0	2	6:26P	221	87	7
30	1176	89	3	07	1327	336	3	27	423	80	11
31	1357	70	5	08	1505	42	6	28	625	77	10
32	1538	47	6	09	1682	58	4	29	819	71	8
33	1719	44	2	10	1859	54	4	30	1013	59	6
34	1900	141	1	11	2036	59	4	31	1197	28	4
37	2455	328	2	12	2213	69	5	32	1386	52	5
38	2624	303	3	13	2390	12	3	33	1570	74	6
39	2805	296	4	14	2537	359	2	34	1754	76	5
40	2990	296	4	15	2744	12	3	35	1940	66	9
41	3168	294	6	16	2921	13	4	36	2124	67	6
42	3349	283	6	17	3098	15	4	37	2308	68	5
43	3530	279	7	18	3275	20	3	38	2492	51	3
44	3712	286	8	19	3452	14	4	39	2676	297	2
45	3892	274	10	20	3629	355	5	40	2860	328	2
46	4073	278	8	21	3806	353	4	41	3044	333	3
47	4253	283	10	22	3983	330	5	42	3228	326	3
48	4434	280	10	23	4160	272	9	43	3412	341	3
49	4615	283	11	24	4337	288	8	44	3596	345	4
50	4800	285	11	25	4514	289	9	45	3780	345	3
51	4980	291	12	26	4691	277	9	46	3964	357	3
52	5161	293	13	27	4868	282	11	47	4148	338	5
53	5345	291	12	28	5045	291	11	48	4334	348	7
54	5526	287	13	29	5222	280	11	49	4518	328	9
55	5707	288	13	30	5399	275	12	50	4702	307	6
56	5880	281	14	31	5576	274	11	51	4886	301	8
57	6068	280	14	32	5753	281	11	52	5070	270	10
58	6249	283	15	33	5930	274	11	53	5254	301	9
59	6430	282	15	34	6107	276	12	54	5438	271	10
11:00A	6511	272	12	35	6284	270	12	55	5622	275	10
01	6692	278	15	36	6461	269	12	56	5804	277	11
02	6873	278	16	37	6638	267	13	57	5988	275	10
03	7054	274	17	38	6815	268	14	58	6172	264	11
04	7285	276	16	39	6991	264	14	59	6356	261	11

TABLE I (1926)

Date and time	Height in meters	Azimuth from; degrees	Velocity m/s	Date and time	Height in meters	Azimuth from; degrees	Velocity m/s	Date and time	Height in meters	Azimuth from; degrees	Velocity m/s
Aug.				**Sept.**				**Sept.**			
31				**1**				**1**			
7:00P	6540	259	13	9:47A	2308	352	4	2:32P	806	57	4
01	6724	254	11	48	2492	346	6	33	995	68	3
02	6908	248	12	49	2676	343	7	34	1176	35	2
03	7102	254	14	50	2860	336	11	35	1357	46	1
04	7286	247	13	51	3044	324	8	36	1538	38	4
05	7470	271	20	52	3228	300	8	37	1719	269	2
06	7654	267	17	53	3412	295	6	38	1900	271	3
07	7834	265	18	54	3596	288	7	39	2081	267	5
08	8018	262	19	55	3780	287	8	40	2262	279	7
09	8202	262	20	56	3964	282	9	41	2455	297	9
10	8386	262	22	57	4148	284	8	42	2654	299	10
11	8570	260	21	58	4334	287	11	43	2805	291	11
12	8754	255	24	59	4518	275	11	44	2990	286	10
13	8938	260	24	10:00A	4702	272	11	45	3168	280	11
14	9122	259	24	01	4846	274	11	46	3349	278	12
15	9306	260	27	02	5070	274	12	47	3530	283	13
16	9490	262	29	03	5254	275	13	48	3712	277	12
17	9674	263	25	04	5438	273	14	49	3892	277	13
18	9858	266	29	05	5622	272	14	50	4073	268	14
				06	5804	271	16	51	4253	260	13
Sept.				07	5988	260	14	52	4434	261	14
1				08	6172	272	19	53	4615	257	16
9:36A	221	123	2	09	6356	266	20	54	4800	259	16
37	423	97	4	10	6540	261	21	55	4980	258	17
38	626	76	2	11	6724	261	22	56	5161	263	17
39	821	77	5	12	6908	262	28	57	5345	264	21
40	1017	71	4	13	7102	265	30	58	5526	268	19
41	1205	63	3	14	7286	262	33	59	5707	265	19
42	1387	37	2					3:00P	5889	266	22
43	1572	355	2	**1**				01	6069	263	19
44	1756	339	2	2:29P	218	74	5	02	6250	267	25
45	1940	336	3	30	416	77	7	03	6430	263	27
46	2124	331	4	31	615	79	6				

IV. DIRECTION AND VELOCITY OF THE WIND FROM OBSERVATIONS OF PILOT-BALLOONS AND FROM THE ANEMOGRAPH AT THE CAMP ON THE MALIGIAKFJORD (CAMP MICHIGAN) FROM JULY TO SEPTEMBER, 1926

BY

S. P. FERGUSSON

(Wind directions shown by arrows oriented with regard to cardinal directions; wind velocities in meters per second. Beneath each run there is given in order from top to bottom the time on 45th meridian, the air pressure at the surface in millibars with indication of falling (F), stationary (S) or rising (R) barometer; and the air temperature in degrees centigrade, with the same abbreviations for falling stationary or rising temperature. Elevations in thousands of meters.)

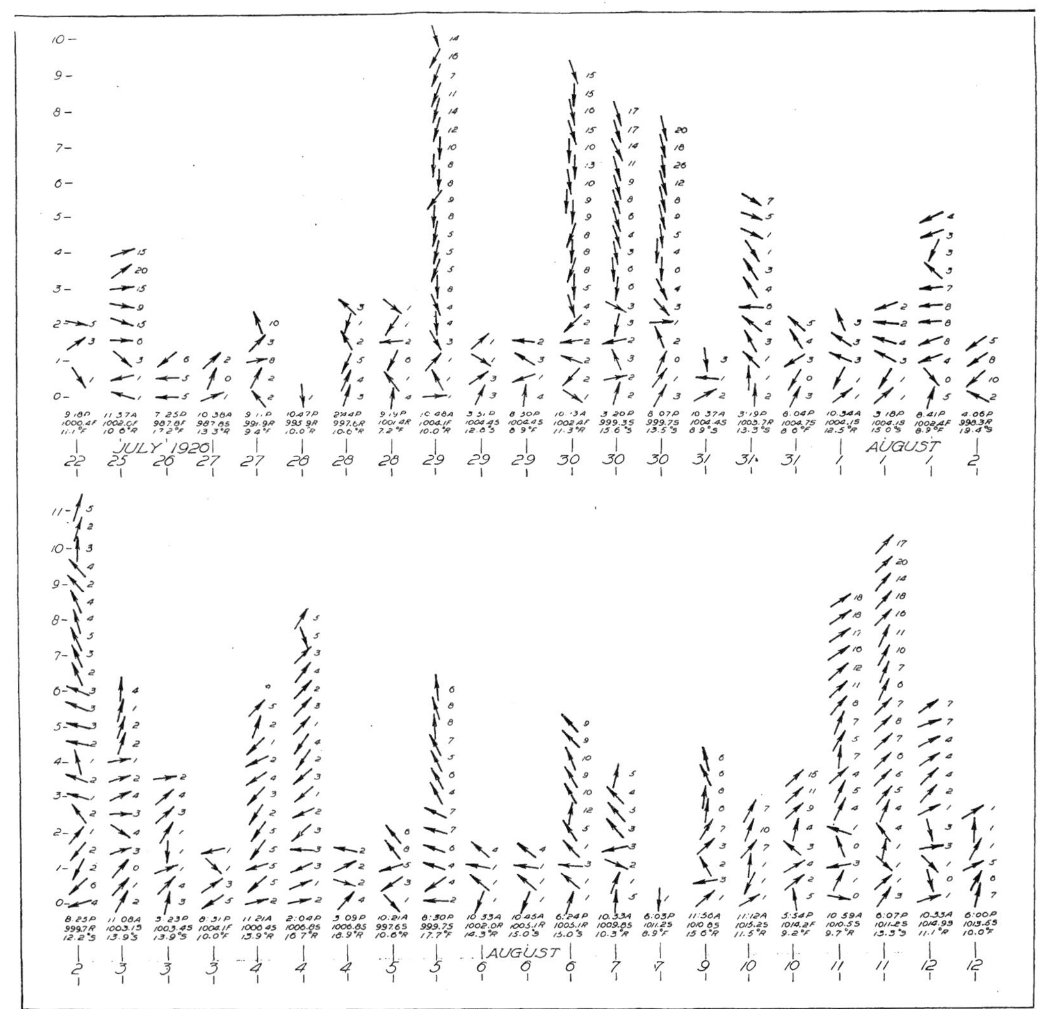

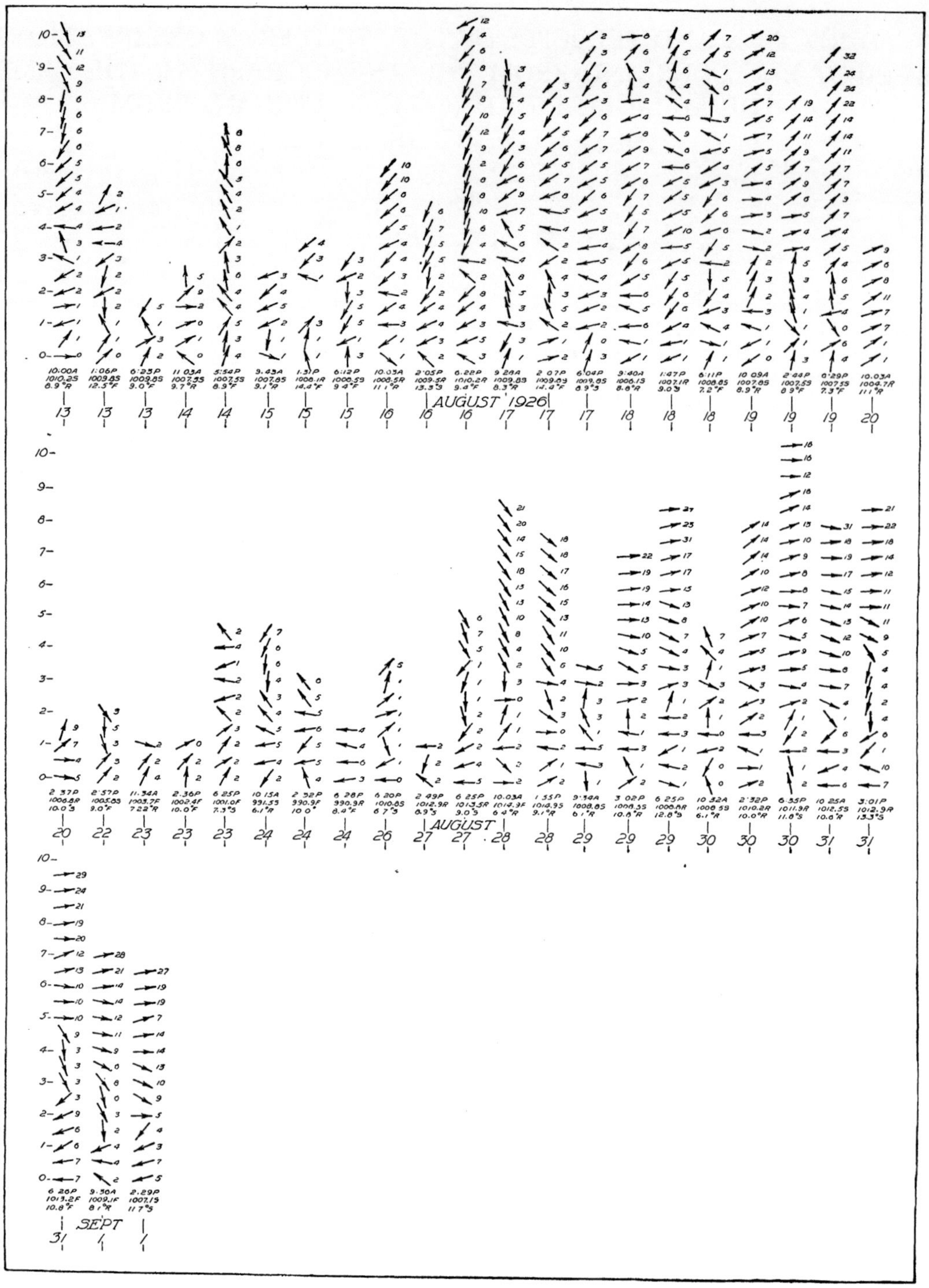

AUGUST 1926

10:00A 1010.2S 8.9°R — 13
1:06P 1009.8S 12.5°F — 13
6:25P 1009.8S 9.0°F — 13
11 03A 1007.5S 9.7°R — 14
5:54P 1007.5S 8.9°F — 14
9.43A 1007.8S 9.1°R — 15
1:31P 1008.1R 14.4°F — 15
6:12P 1008.5S 9.4°F — 15
10:03A 1008.5R 11.1°R — 16
2:05P 1009.5R 13.3°S — 16
6:22P 1010.2R 9.4°F — 16
9 28A 1009.8S 8.3°R — 17
2 07P 1009.8S 14.4°F — 17
6:04P 1009.8S 8.9°S — 17
9:40A 1008.1S 8.8°R — 18
1:47P 1007.1R 9.0°S — 18
6:11P 1008.8S 7.2°F — 18
10 09A 1007.8S 8.9°R — 19
2 44P 1007.5S 8.9°F — 19
6:29P 1007.5S 7.3°F — 19
10.03A 1004.7R 11.1°R — 20

AUGUST

2 37P 1006.8R 10.0°S — 20
2:57P 1005.8S 9.0°F — 22
11.34A 1003.7F 7.22°R — 23
2.36P 1002.4F 10.0°F — 23
6 25P 1001.0F 7.3°S — 23
10 15A 991.5S 6.1°R — 24
2 32P 990.9F 10 0° — 24
6 28P 990.9R 8.4°F — 24
6 20P 1010.8S 6 7°S — 26
2 49P 1012.9R 6.9°S — 27
6 25P 1013.5R 9.0°S — 27
10.03A 1014.9F 6 4°R — 28
1 35P 1014.9S 9.1°R — 28
9:34A 1008.8S 6 1°R — 29
3 02P 1008.5S 10.8°R — 29
6 25P 1008.8R 12.8°S — 29
10 32A 1008 5S 6.1°R — 30
2:32P 1010.2R 10.0°R — 30
6:35P 1011.9R 11.8°S — 30
10 25A 1012.5S 10.6°R — 31
3:01P 1012.9R 13.3°S — 31

SEPT

6 20P 1013.2F 10.8°F — 31
9:30A 1009.1F 8.1°R — 1
2.29P 1007.1S 11.7°S — 1

V. HEIGHTS, AZIMUTHS AND VELOCITIES OF PILOT-BALLOONS AT MOUNT EVANS FROM JULY 21, 1927, TO JULY 19, 1929

BY

CLARENCE R. KALLQUIST, July 21, 1927, to May 27, 1928
WILLIAM S. CARLSON, May 28, to July 10, 1928
LEONARD R. SCHNEIDER, July 11, 1928, to July 19, 1929

TABLE II (1927)

(THE FIGURES IN PARENTHESES REFER TO THE SERIATIM NUMBERS OF BALLOON RUNS)

Date and time	Height in meters	Azimuth from; degrees	Velocity m/s	Date and time	Height in meters	Azimuth from; degrees	Velocity m/s	Date and time	Height in meters	Azimuth from; degrees	Velocity m/s
July				**July**				**July**			
21	(1)			**23**				**23**			
2:22P	395	182	8	2:04P	1196	90	2	2:35P	6785	314	7
23	587	206	15	05	1385	172	2	36	6965	307	8
24	763	195	10	06	1565	177	3	37	7145	308	8
25	939	188	10	07	1745	171	3	38	7325	307	9
26	1107	187	16	08	1925	183	5	39	7505	307	10
27	1275	183	18	09	2105	184	5	40	7685	305	11
28	1435	185	17	10	2285	184	5	41	7865	304	10
29	1595	188	17	11	2465	198	4	42	8045	302	9
30	1755	195	18	12	2645	220	2	43	8225	299	9
31	1915	194	19	13	2825	238	2	44	8405	297	9
32	2075	193	21	14	3005	238	2	45	8585	298	11
33	2255	197	22	15	3185	291	2	46	8765	298	13
				16	3365	306	3				
22	(2)			17	3545	280	3	**27**	(4)		
1:15P	395	312	5	18	3725	272	3	1:56P	395	136	5
16	611	277	1	19	3905	277	3	57	611	121	1
17	809	256	3	20	4085	265	4	58	809	74	1
18	1007	281	3	21	4265	266	5				
19	1196	298	3	22	4445	279	5	**29**	(5)		
20	1385	303	3	23	4625	284	5	11:45A	395	135	7
21	1565	265	1	24	4805	293	6	46	611	114	4
22	1745	231	2	25	4985	308	6	47	809	118	3
23	1925	229	2	26	5165	312	6	48	1007	134	3
24	2105	237	2	27	5345	310	6	49	1196	152	4
25	2285	242	2	28	5525	313	6	50	1385	165	3
				29	5705	315	6	51	1565	168	2
23	(3)			30	5885	326	7	52	1745	169	2
2:00P	395	82	7	31	6065	320	7	53	1925	167	2
01	611	78	6	32	6245	308	7	54	2105	155	3
02	809	74	5	33	6425	306	7	55	2285	170	2
03	1007	72	5	34	6605	311	7	56	2465	221	2

TABLE II (1927)

Date and time	Height in meters	Azimuth from; degrees	Velocity m/s	Date and time	Height in meters	Azimuth from; degrees	Velocity m/s	Date and time	Height in meters	Azimuth from; degrees	Velocity m/s
July 29				**July 29**				**July 30**			
11:57A	2645	220	3	0:45P	11285	335	16	2:56P	7865	1	14
58	2825	222	4	46	11465	337	16	57	8045	354	16
59	3005	221	7	47	11645	338	15	58	8225	352	17
0:00P	3185	217	7	48	11825	338	14	59	8405	355	16
01	3365	222	6					3:00P	8585	353	17
02	3545	237	5					01	8765	351	21
03	3725	251	3	**30**	(6)			02	8945	348	22
04	3905	278	2	2:15P	395	70	7	03	9125	347	21
05	4085	326	1	16	611	66	6	04	9305	348	21
06	4265	337	4	17	809	62	5	05	9485	349	22
07	4445	343	5	18	1007	73	4	06	9665	353	21
08	4625	334	5	19	1196	125	5	07	9845	355	21
09	4805	344	6	20	1385	148	3	08	10025	355	24
10	4985	339	8	21	1565	151	3	09	10205	355	25
11	5165	337	10	22	1745	159	2	10	10385	355	22
12	5345	340	11	23	1925	209	1	11	10565	355	22
13	5525	344	10	24	2105	211	1	12	10745	356	23
14	5705	346	11	25	2285	94	1	13	10925	0	22
15	5885	341	12	26	2465	69	2	14	11105	0	24
16	6065	338	13	27	2645	67	3	15	11285	0	22
17	6245	340	12	28	2825	65	3	16	11465	0	18
18	6425	344	12	29	3005	65	4				
19	6605	344	12	30	3185	71	4				
20	6785	344	11	31	3365	74	3	**31**	(7)		
21	6965	342	13	32	3545	67	3	2:00P	395	81	3
22	7145	341	14	33	3725	50	3	01	611	77	3
23	7325	343	13	34	3905	40	4	02	809	72	3
24	7505	341	15	35	4085	20	3	03	1007	86	2
25	7685	341	17	36	4265	10	3	04	1196	119	2
26	7865	342	17	37	4445	5	5	05	1385	156	4
27	8045	341	17	38	4625	358	5	06	1565	180	2
28	8225	336	18	39	4805	5	5	07	1745	188	2
29	8405	337	18	40	4985	13	6	08	1925	173	2
30	8585	337	18	41	5165	11	6	09	2105	203	5
31	8765	337	18	42	5345	6	7	10	2285	209	8
32	8945	336	18	43	5525	3	6	11	2465	202	8
33	9125	335	19	44	5705	3	6	12	2645	195	9
34	9305	334	18	45	5885	4	7	13	2825	197	9
35	9485	332	22	46	6065	0	7	14	3005	208	8
36	9665	332	24	47	6245	0	8	15	3185	228	6
37	9845	333	22	48	6425	0	11	16	3365	260	5
38	10025	333	21	49	6605	356	13	17	3545	268	8
39	10205	337	22	50	6785	357	13	18	3725	252	11
40	10385	336	19	51	6965	357	13	19	3905	244	12
41	10565	333	20	52	7145	357	13	20	4085	239	12
42	10745	332	21	53	7325	358	13	21	4265	239	14
43	10925	335	25	54	7505	354	13	22	4445	239	16
44	11105	335	20	55	7685	359	14	23	4625	232	15

TABLE II (1927)

Date and time	Height in meters	Azimuth from; degrees	Velocity m/s	Date and time	Height in meters	Azimuth from; degrees	Velocity m/s	Date and time	Height in meters	Azimuth from; degrees	Velocity m/s
July 31				**Aug. 3**				**Aug. 4**	(10)		
2:24P	4805	236	14	1:41P	1925	10	3	2:20P	395	89	6
25	4985	239	12	42	2105	356	5	21	611	71	7
26	5165	237	10	43	2285	336	6	22	809	56	5
27	5345	245	9	44	2465	320	4	23	1007	62	4
28	5525	240	9	45	2645	260	4	24	1196	68	3
29	5705	231	11	46	2825	249	5	25	1385	90	1
30	5885	232	11	47	3005	268	7	26	1565	143	2
31	6065	237	10	48	3185	301	5	27	1745	162	3
32	6245	242	11	49	3365	313	4	28	1925	156	3
33	6425	243	11	50	3545	327	3	29	2105	145	3
34	6605	245	13	51	3725	301	4	30	2285	124	3
35	6785	249	13	52	3905	276	5	31	2465	96	3
36	6965	247	14	53	4085	288	6	32	2645	93	4
37	7145	249	15	54	4265	291	6	33	2825	109	4
38	7325	249	15	55	4445	284	5	34	3005	117	3
39	7505	251	17	56	4625	286	7	35	3185	123	3
40	7685	254	14	57	4805	291	8	36	3365	117	3
41	7865	254	14	58	4985	292	9	37	3545	110	3
42	8045	261	13	59	5165	290	9	38	3725	105	4
43	8225	265	16	2:00P	5345	286	9	39	3905	113	3
44	8405	269	16	01	5525	287	10	40	4085	119	4
				02	5705	285	10	41	4265	113	5
				03	5885	282	9	42	4445	110	7
				04	6065	281	9	43	4625	108	8
Aug. 2	(8)			05	6245	272	9	44	4805	103	9
1:55P	395	273	5	06	6425	272	9	45	4985	102	9
56	611	261	6	07	6605	274	7	46	5165	110	8
57	809	257	4	08	6785	299	6	47	5345	111	7
58	1007	266	3	09	6965	318	7	48	5525	105	7
59	1196	257	2	10	7145	304	9	49	5705	99	6
2:00P	1385	246	3	11	7325	294	10	50	5885	97	7
01	1565	229	3	12	7505	295	9	51	6065	101	10
02	1745	210	6	13	7685	295	9	52	6245	102	11
03	1925	194	8	14	7865	298	9	53	6425	99	11
04	2105	179	9	15	8045	304	9	54	6605	94	13
05	2285	178	10	16	8225	313	9	55	6785	90	11
06	2465	184	7	17	8405	319	11	56	6965	82	10
07	2645	195	5	18	8585	315	11	57	7145	84	10
				19	8765	313	11	58	7325	88	12
3	(9)			20	8945	312	11	59	7505	91	14
1:33P	395	90	6	21	9125	312	12	3:00P	7685	93	14
34	611	88	8	22	9305	309	11	01	7865	90	14
35	809	76	7	23	9485	304	11	02	8045	90	13
36	1007	58	5	24	9665	308	12	03	8225	87	14
37	1196	69	3	25	9845	311	13	04	8405	83	14
38	1385	118	1	26	10025	308	14	05	8585	82	15
39	1565	89	1	27	10205	306	14	06	8765	82	15
40	1745	35	2					07	8945	85	15

TABLE II (1927)

Date and time	Height in meters	Azimuth from; degrees	Velocity m/s	Date and time	Height in meters	Azimuth from; degrees	Velocity m/s	Date and time	Height in meters	Azimuth from; degrees	Velocity m/s
Aug. 4				**Aug. 4**				**Aug. 16**			
3:08P	9125	89	17	4:07P	4085	142	5	2:30P	2285	352	2
09	9305	91	16	08	4265	133	4	31	2465	348	2
10	9485	93	17	09	4445	121	4	32	2645	1	3
11	9665	91	18	10	4625	120	4	33	2825	10	3
12	9845	80	18	11	4805	124	5	34	3005	10	5
13	10025	70	16	12	4985	132	7	35	3185	7	7
14	10205	69	16	13	5165	125	9	36	3365	6	7
15	10385	74	15	14	5345	114	11	37	3545	4	7
16	10565	80	14	15	5525	108	11	38	3725	3	8
17	10745	79	13	16	5705	110	10	39	3905	1	8
18	10925	77	13	17	5885	108	11	40	4085	1	8
19	11105	74	13	18	6065	103	11	41	4265	9	8
20	11285	71	14	19	6245	106	9	42	4445	8	7
21	11465	69	16	20	6425	105	8	43	4625	8	7
22	11645	73	11	21	6605	101	9	44	4805	9	8
23	11825	84	6	22	6785	96	11	45	4985	10	9
24	12005	54	2	23	6965	98	11	46	5165	11	11
25	12185	55	4	24	7145	103	11	47	5345	12	14
26	12365	53	6	25	7325	109	11	48	5525	12	16
27	12545	52	4	27	7505	111	12	49	5705	10	17
28	12725	47	4	28	7685	105	13	50	5885	9	18
29	12905	54	1	29	7865	100	15	51	6065	8	19
30	13085	159	1	30	8045	96	17	52	6245	8	18
31	13265	64	2	31	8225	91	17	53	6425	16	17
32	13445	64	1	32	8405	92	17	54	6605	17	16
33	13625	261	1	33	8585	98	18	55	6785	18	15
				34	8765	102	20	56	6965	16	14
4	(11)			35	8945	99	20	57	7145	14	15
3:47P	395	74	5	36	9125	97	20	58	7325	15	16
48	611	76	6	37	9305	95	20	59	7505	11	16
49	809	97	4	38	9485	99	21	3:00P	7685	6	15
50	1007	123	3	39	9665	99	22	01	7865	7	16
51	1196	151	2	40	9845	95	22	02	8045	10	15
52	1385	131	2	41	10025	95	23	03	8225	10	17
53	1565	131	2	42	10205	94	24	04	8405	11	17
54	1745	185	1	43	10385	93	23	05	8585	14	17
55	1925	180	2					06	8765	14	17
56	2105	191	5	**16**	(12)			07	8945	12	16
57	2285	192	6	2:20P	395	135	3	08	9125	11	16
58	2465	183	6	21	611	92	1	09	9305	11	17
59	2645	170	5	22	809	9	1	10	9485	11	20
4:00P	2825	152	5	23	1007	338	2				
01	3005	143	4	24	1196	1	2	**17**	(13)		
02	3185	162	4	25	1385	12	2	1:27P	395	66	3
03	3365	170	4	26	1565	22	1	28	611	63	2
04	3545	154	3	27	1745	10	1	29	809	20	3
05	3725	143	4	28	1925	351	1	30	1007	8	4
06	3905	147	5	29	2105	352	2	31	1196	10	3

TABLE II (1927)

Date and time	Height in meters	Azimuth from; degrees	Veloc-ity m/s	Date and time	Height in meters	Azimuth from; degrees	Veloc-ity m/s	Date and time	Height in meters	Azimuth from; degrees	Veloc-ity m/s
Aug. 17				**Aug. 17**				**Aug. 19**			
1:32P	1385	14	3	2:20P	10025	247	20	6:32P	4445	32	6
33	1565	9	2	21	10205	247	19	33	4625	35	6
34	1745	339	3	22	10385	264	17	34	4805	46	6
35	1925	327	2	23	10565	280	12	35	4985	47	6
36	2105	346	1	24	10745	270	7	36	5165	40	5
37	2285	332	1	25	10925	258	7	37	5345	43	5
38	2465	322	1					38	5525	48	5
39	2645	321	2	**17**	(14)			39	5705	48	6
40	2825	310	2	2:10P	395	150	5	40	5885	45	7
41	3005	302	2	11	611	202	1	41	6065	43	7
42	3185	295	2	12	809	12	1	42	6245	41	9
43	3365	291	3	13	1007	25	1	43	6425	45	11
44	3545	280	2	14	1196	92	1	44	6605	44	11
45	3725	256	2	15	1385	140	1	45	6785	44	11
46	3905	245	3	16	1565	153	2	46	6965	46	11
47	4085	244	3	17	1745	157	2	47	7145	52	10
48	4265	239	3	18	1925	171	2	48	7325	60	10
49	4445	248	4	19	2105	190	1	49	7505	63	10
50	4625	253	5	20	2285	200	3	50	7685	60	10
51	4805	248	6	21	2465	210	2	51	7865	58	11
52	4985	247	6	22	2645	253	1	52	8045	56	10
53	5165	253	6	23	2825	311	2	53	8225	51	8
54	5345	255	6	24	3005	318	2	54	8405	48	8
55	5525	259	9	25	3185	315	2	55	8585	50	7
56	5705	261	12					56	8765	54	8
57	5885	257	13	**19**	(15)			57	8945	56	8
58	6065	257	14	6:10P	395	312	5	58	9125	52	8
59	6245	256	15	11	611	322	8	59	9305	49	7
2:00P	6425	254	14	12	809	336	6	7:00P	9485	48	7
01	6605	255	14	13	1007	12	3	01	9665	36	8
02	6785	255	14	14	1196	65	2	02	9845	31	7
03	6965	255	15	15	1385	60	3	03	10025	36	7
04	7145	260	15	16	1565	24	3	04	10205	43	7
05	7325	263	15	17	1745	353	4	05	10385	36	5
06	7505	262	15	18	1925	357	5	06	10565	2	5
07	7685	263	15	19	2105	12	7	07	10745	358	8
08	7865	263	17	20	2285	17	8	08	10925	358	10
09	8045	262	17	21	2465	12	9	09	11105	345	10
10	8225	260	17	22	2645	14	9	10	11285	337	10
11	8405	261	18	23	2825	16	7	11	11465	342	10
12	8585	260	19	24	3005	15	5	12	11645	354	10
13	8765	259	19	25	3185	27	3	**21**	(16)		*
14	8945	261	19	26	3365	27	3	3:57P	395	208	7
15	9125	262	18	27	3545	25	3	58	611	199	7
16	9305	259	19	28	3725	27	4	59	809	152	3
17	9485	255	18	29	3905	30	4	4:00P	1007	93	5
18	9665	250	15	30	4085	33	5	01	1196	101	5
19	9845	251	17	31	4265	32	6	01:40	1322	102	4

TABLE II (1927)

Date and time	Height in meters	Azimuth from; degrees	Velocity m/s	Date and time	Height in meters	Azimuth from; degrees	Velocity m/s	Date and time	Height in meters	Azimuth from; degrees	Velocity m/s
Aug.				**Sept.**				**Sept.**			
27	(17)			**2**				**2**	(19)		
2:00P	395	125	3	9:33A	3005	7	1	10:46A	395	73	4
01	611	136	4	34	3185	335	1	47	611	82	2
02	809	136	4	35	3365	311	2	48	809	76	2
03	1007	138	5	36	3545	321	3	49	1007	58	3
04	1196	141	4	37	3725	317	4	50	1196	63	3
05	1385	141	5	38	3905	311	5	51	1385	66	3
06	1565	137	5	39	4085	310	5	52	1565	76	1
07	1745	142	3	40	4265	318	5	53	1745	103	1
08	1925	189	2	41	4445	328	4	54	1925	101	2
09	2105	194	4	42	4625	327	4	55	2105	94	2
10	2285	196	4	43	4805	330	4	56	2285	79	1
11	2465	189	4	44	4985	354	5	57	2465	51	1
12	2645	179	5	45	5165	2	5	58	2645	56	1
13	2825	186	4	46	5345	350	7	59	2825	45	1
14	3005	205	3	47	5525	342	7	11:00A	3005	331	1
15	3185	224	4	48	5705	338	8	01	3185	338	1
16	3365	244	4	49	5885	338	8	02	3365	350	2
17	3545	262	4	50	6065	331	8	03	3545	337	3
18	3725	266	5	51	6245	327	9	04	3725	323	3
19	3905	272	6	52	6425	323	10	05	3905	313	3
20	4085	280	6	53	6605	343	11	06	4085	323	4
21	4265	279	7	54	6785	343	14	07	4265	332	4
22	4445	276	8	55	6965	353	17	08	4445	334	4
23	4625	272	8	56	7145	356	19	09	4625	336	5
24	4805	270	9	57	7325	349	19	10	4805	338	5
25	4985	271	8	58	7505	352	20	11	4985	335	5
26	5165	276	7	59	7685	354	20	12	5165	333	7
27	5345	278	7	10:00A	7865	355	20	13	5345	329	7
28	5525	280	7	01	8045	356	20	14	5525	328	8
29	5705	281	7	02	8225	353	22	15	5705	329	11
				03	8405	351	22	16	5885	331	12
				04	8585	354	22	17	6065	335	12
Sept.				05	8765	354	23	18	6245	345	14
2	(18)			06	8945	355	22	19	6425	349	16
9:19A	395	76	4	07	9125	355	23	20	6605	348	16
20	611	81	4	08	9305	354	25	21	6785	348	16
21	809	77	4	09	9485	355	23	22	6965	349	17
22	1007	72	4	10	9665	354	24	23	7145	348	17
23	1196	49	4	11	9845	354	24				
24	1385	49	4	12	10025	353	23	**3**	(20)		
25	1565	57	4	13	10205	351	23	2:40P	395	167	2
26	1745	66	3	14	10385	351	21	41	611	225	1
27	1925	84	2	15	10565	355	20	42	809	251	3
28	2105	99	3	16	10745	2	19	43	1007	226	3
29	2285	98	4	17	10925	356	18	44	1196	214	4
30	2465	94	3	18	11105	336	17	45	1385	222	5
31	2645	88	2	19	11285	321	16	46	1565	234	7
32	2825	58	1					47	1745	251	7

TABLE II (1927)

Date and time	Height in meters	Azimuth from; degrees	Velocity m/s
Sept. 3			
2:48P	1925	270	5
49	2105	313	3
50	2285	245	4
51	2465	275	8
52	2645	280	7
53	2825	282	6
54	3005	282	6
55	3185	280	7
56	3365	280	8
57	3545	295	9
58	3725	302	10
59	3905	297	10
3:00P	4085	298	12
01	4265	298	12
02	4445	293	11
03	4625	293	12
04	4805	292	12
05	4985	299	11
06	5165	299	11
07	5345	293	11
08	5525	291	11
09	5705	292	12
4	(21)		
1:59P	395	87	5
2:00P	611	120	4
01	809	163	5
02	1007	181	5
03	1196	190	5
04	1385	196	5
05	1565	211	4
06	1745	239	4
07	1925	244	4
08	2105	233	4
09	2285	218	4
10	2465	199	4
11	2645	175	3
12	2825	180	3
5	(22)		
2:42P	395	80	4
43	611	86	5
44	809	123	3
45	1007	171	3
46	1196	196	2
47	1385	191	3
48	1565	188	4
49	1745	198	4
50	1925	211	3

Date and time	Height in meters	Azimuth from; degrees	Velocity m/s
Sept. 6	(23)		
1:15P	395	80	1
16	611	88	1
17	809	115	2
18	1007	127	3
19	1196	131	4
20	1385	143	4
21	1565	168	5
22	1745	183	8
23	1925	182	10
24	2105	181	11
25	2285	176	10
26	2465	173	11
27	2645	174	11
28	2825	173	10
29	3005	169	10
30	3185	158	11
31	3365	157	13
32	3545	162	15
33	3725	164	16
6	(24)		
1:45P	395	71	5
46	611	79	3
47	809	124	2
48	1007	139	4
49	1196	148	5
50	1385	170	6
51	1565	177	8
52	1745	179	8
53	1925	172	10
54	2105	166	11
55	2285	164	9
56	2465	167	9
57	2645	171	11
58	2825	171	12
59	3005	171	11
7	(25)		
2:03P	395	75	4
04	611	79	6
05	809	80	7
06	1007	82	6
07	1196	73	4
08	1385	68	2
09	1565	111	2
10	1745	159	3
11	1925	152	5
12	2105	139	7

Date and time	Height in meters	Azimuth from; degrees	Velocity m/s
Sept. 7			
2:13P	2285	135	6
14	2465	129	6
15	2645	128	6
16	2825	135	5
17	3005	139	6
18	3185	141	7
19	3365	144	7
20	3545	143	8
7	(26)		
3:45P	395	90	1
46	611	49	1
47	809	80	3
48	1007	99	3
49	1196	110	4
50	1385	109	5
51	1565	116	6
52	1745	111	7
53	1925	107	6
54	2105	108	6
55	2285	109	7
56	2465	117	7
57	2645	119	6
58	2825	107	6
59	3005	97	7
4:00P	3185	100	7
01	3365	101	6
02	3545	101	5
03	3725	103	5
04	3905	100	4
05	4085	97	6
06	4265	100	8
07	4445	104	10
08	4625	109	12
09	4805	113	13
10	4985	116	14
11	5165	118	13
12	5345	120	12
13	5525	124	12
14	5705	126	13
15	5885	129	13
16	6065	130	13
8	(27)		
8:07A	395	175	13
08	611	160	10
09	809	158	14
10	1007	161	18

TABLE II (1927)

Date and time	Height in meters	Azimuth from; degrees	Velocity m/s	Date and time	Height in meters	Azimuth from; degrees	Velocity m/s	Date and time	Height in meters	Azimuth from; degrees	Velocity m/s
Sept.				**Sept.**				**Sept.**			
8				**9**				**10**			
8:11A	1196	163	20	9:16A	3365	172	17	4:35P	4085	183	14
12	1385	170	24	17	3545	171	15	36	4265	180	16
13	1565	173	27	18	3725	176	13	37	4445	180	16
14	1745	176	24	19	3905	179	13	38	4625	181	17
15	1925	176	19	20	4085	176	14	39	4805	181	18
16	2105	174	15	21	4265	172	14	40	4985	183	18
17	2285	168	14	22	4445	172	15	41	5165	188	17
18	2465	162	16	23	4625	172	14	42	5345	189	16
19	2645	162	16	24	4805	177	14	43	5525	192	15
20	2825	154	14	25	4985	179	14	44	5705	195	16
21	3005	146	14	26	5165	180	14	45	5885	194	18
				27	5345	180	13				
9	(28)			28	5525	178	13	**11**	(31)		
2:45P	395	153	8	29	5705	183	11	2:00P	395	Calm	Calm
46	611	154	11	30	5885	182	13	01	611	224	1
47	809	150	10	31	6065	180	14	02	809	138	1
48	1007	147	8	32	6245	180	13	03	1007	135	3
49	1196	143	8	33	6425	180	14	04	1196	140	5
50	1385	141	8	34	6605	176	13	05	1385	141	6
51	1565	140	8	35	6785	169	15	06	1565	146	7
52	1745	133	7	36	6965	171	18	07	1745	152	8
53	1925	141	6	37	7145	172	18	08	1925	156	7
54	2105	161	11	38	7325	171	20	09	2105	163	5
55	2285	161	16	39	7505	172	22	10	2285	168	5
56	2465	155	19	40	7685	173	21	11	2465	165	6
57	2645	154	19	41	7865	173	22	12	2645	157	8
58	2825	154	18					13	2825	153	10
59	3005	149	18	**10**	(30)			14	3005	154	11
3:00P	3185	143	18	4:15P	395	61	7	15	3185	160	10
01	3365	142	19	16	611	71	6	16	3365	170	8
				17	809	73	4	17	3545	169	8
9	(29)			18	1007	72	4	18	3725	165	9
9:00A	395	Calm	Calm	19	1196	100	2	19	3905	169	13
01	611	166	1	20	1385	162	4	20	4085	171	16
02	809	156	2	21	1565	170	6				
03	1007	134	2	22	1745	168	7	**13**	(32)		
04	1196	156	3	23	1925	169	8	2:49P	395	97	2
05	1385	172	5	24	2105	175	8	50	611	31	2
06	1565	173	7	25	2285	182	8	51	809	336	2
07	1745	173	7	26	2465	188	7	52	1007	305	2
08	1925	177	7	27	2645	181	9				
09	2105	181	7	28	2825	172	11	**13**	(33)		
10	2285	179	6	29	3005	168	13	3:07P	395	82	5
11	2465	171	7	30	3185	168	15	08	611	32	2
12	2645	167	9	31	3365	169	15	09	809	334	2
13	2825	168	13	32	3545	173	13	10	1007	245	2
14	3005	170	17	33	3725	185	11	11	1196	217	4
15	3185	172	18	34	3905	189	13	12	1385	214	4

TABLE II (1927)

Date and time	Height in meters	Azimuth from; degrees	Velocity m/s	Date and time	Height in meters	Azimuth from; degrees	Velocity m/s	Date and time	Height in meters	Azimuth from; degrees	Velocity m/s
Sept. 13				**Sept. 14**				**Sept. 15**			
3:13P	1565	204	3	1:57P	3545	31	12	1:59P	2285	239	5
14	1745	181	3	58	3725	23	13	2:00P	2465	254	4
15	1925	157	3	59	3905	12	12	01	2645	267	4
16	2105	124	2	2:00P	4085	5	12				
17	2285	94	2	01	4265	3	12	**16**	(36)		
18	2465	98	1	02	4445	358	11	1:30P	395	95	5
19	2645	213	2	03	4625	353	9	31	611	91	2
20	2825	215	4	04	4805	350	9	32	809	41	1
21	3005	227	6	05	4985	349	10	33	1007	24	2
22	3185	241	8	06	5165	353	12	34	1196	55	3
23	3365	243	10	07	5345	353	13	35	1385	113	3
24	3545	243	9	08	5525	354	14	36	1565	152	5
25	3725	240	9	09	5705	351	14	37	1745	164	6
26	3905	238	9	10	5885	345	12	38	1925	147	6
27	4085	234	9	11	6065	345	12	39	2105	129	6
28	4265	228	10	12	6245	346	12	40	2285	127	4
29	4445	228	11	13	6425	346	11	41	2465	132	3
30	4625	231	11	14	6605	325	14	42	2645	134	2
31	4805	234	11	15	6785	317	17	43	2825	131	2
32	4985	234	12	16	6965	325	16	44	3005	118	5
33	5165	237	12	17	7145	322	17	45	3185	114	8
34	5345	238	14	18	7325	321	19	46	3365	114	10
35	5525	235	15	19	7505	319	19	47	3545	111	11
36	5705	233	15	20	7685	316	18	48	3725	109	12
37	5885	233	17	21	7865	311	20	49	3905	111	11
38	6065	238	19	22	8045	310	19	50	4085	110	11
39	6245	244	19	23	8225	311	21	51	4265	108	13
40	6425	243	20	24	8405	307	19	52	4445	110	13
41	6605	240	21	25	8585	310	19	53	4625	108	13
				26	8765	317	18	54	4805	107	14
14	(34)			27	8945	326	19	55	4895	104	14
1:40P	395	84	4	28	9125	324	19	56	5165	95	12
41	611	54	4	29	9305	320	15	57	5345	93	11
42	809	37	5	30	9485	320	14	58	5525	97	12
43	1007	32	7	31	9665	317	14	59	5705	91	12
44	1196	37	8	32	9845	312	13	2:00P	5885	90	13
45	1385	39	10					01	6065	91	13
46	1565	42	9	**15**	(35)			02	6245	89	16
47	1745	45	9	1:49P	395	248	5	03	6425	90	16
48	1925	38	11	50	611	268	4	04	6605	91	16
49	2105	35	12	51	809	280	6	05	6785	95	16
50	2285	32	11	52	1007	263	7	06	6965	96	14
51	2465	29	11	53	1196	249	6	07	7145	91	16
52	2645	25	11	54	1385	244	6	08	7325	93	15
53	2825	23	10	55	1565	250	7	09	7505	95	13
54	3005	21	10	56	1745	261	8	10	7685	92	14
55	3185	24	9	57	1925	259	8	11	7865	88	13
56	3365	34	9	58	2105	243	6	12	8045	87	12

TABLE II (1927)

Date and time	Height in meters	Azimuth from; degrees	Velocity m/s	Date and time	Height in meters	Azimuth from; degrees	Velocity m/s	Date and time	Height in meters	Azimuth from; degrees	Velocity m/s
Sept. 16				**Sept. 16**				**Sept. 17**			
2:13P	8225	84	12	5:00P	8585	97	22	2:39P	6605	79	8
				01	8765	95	21	40	6785	71	8
16	(37)			02	8945	95	20	41	6965	66	8
4:15P	395	92	5	03	9125	95	21	42	7145	71	7
16	611	66	6	04	9305	95	21	43	7325	78	7
17	809	47	6	05	9485	93	20	44	7505	76	7
18	1007	28	5	06	9665	89	21	45	7685	77	8
19	1196	16	3	07	9845	88	20	46	7865	77	9
20	1385	124	1	08	10025	90	18	47	8045	76	9
21	1565	169	1	09	10205	90	15	48	8225	73	9
22	1745	129	4	10	10385	95	14	49	8405	72	10
23	1925	120	6	11	10565	96	14	50	8585	75	10
24	2105	116	6					51	8765	74	9
25	2285	106	6	**17**	(38)			52	8945	71	8
26	2465	93	6	2:05P	395	122	2	53	9125	68	7
27	2645	81	4	06	611	179	2	54	9305	62	8
28	2825	76	4	07	809	190	2	55	9485	61	9
29	3005	91	5	08	1007	195	2	56	9665	74	11
30	3185	107	7	09	1196	197	3	57	9845	69	10
31	3365	111	9	10	1385	184	4	58	10025	77	13
32	3545	115	10	11	1565	173	6	59	10205	74	13
33	3725	116	11	12	1745	161	9	3:00P	10385	68	13
34	3905	109	11	13	1925	164	9	01	10565	67	13
35	4085	105	12	14	2105	168	10	02	10745	65	13
36	4265	104	12	15	2285	163	10	03	10925	66	11
37	4445	105	13	16	2465	158	9	04	11105	68	9
38	4625	105	14	17	2645	151	8	05	11285	57	8
39	4805	103	14	18	2825	146	8	06	11465	49	8
40	4985	100	14	19	3005	141	7	07	11645	50	8
41	5165	97	15	20	3185	123	5	08	11825	38	5
42	5345	96	16	21	3365	103	5	09	12005	14	3
43	5525	101	17	22	3545	109	7	10	12185	18	2
44	5705	103	17	23	3725	108	7	11	12365	40	2
45	5885	105	18	24	3905	96	8				
46	6065	106	19	25	4085	100	7	**17**	(39)		
47	6245	104	19	26	4265	111	6	5:21P	395	81	2
48	6425	106	19	27	4445	100	8	22	611	75	2
49	6605	104	18	28	4625	101	9	23	809	59	3
50	6785	104	18	29	4805	100	9	24	1007	48	3
51	6965	102	18	30	4985	101	9	25	1196	53	2
52	7145	101	19	31	5165	101	9	26	1385	127	2
53	7325	101	20	32	5345	104	10	27	1565	172	3
54	7505	98	18	33	5525	100	10	28	1745	173	4
55	7685	94	20	34	5705	90	11	29	1925	160	5
56	7865	92	21	35	5885	84	12	30	2105	143	5
57	8045	89	21	36	6065	83	11	31	2285	149	5
58	8225	92	24	37	6245	87	10	32	2465	167	4
59	8405	95	22	38	6425	85	9	33	2645	161	4

TABLE II (1927)

Date and time	Height in meters	Azimuth from; degrees	Veloc- ity m/s	Date and time	Height in meters	Azimuth from; degrees	Veloc- ity m/s	Date and time	Height in meters	Azimuth from; degrees	Veloc- ity m/s
Sept. 17				**Sept. 17**				**Sept. 18**			
5:34P	2825	141	5	6:22P	11465	20	8	2:13P	4625	7	9
35	3005	121	5	23	11645	11	8	14	4805	5	9
36	3185	98	5	24	11825	22	10	15	4985	6	10
37	3365	106	8	25	12005	40	8	16	5165	5	11
38	3545	116	8	26	12185	39	8	17	5345	3	12
39	3725	129	5	27	12365	41	9	18	5525	3	13
40	3905	140	5	28	12545	35	11	19	5705	8	13
41	4085	141	4	29	12725	40	15	20	5885	12	13
42	4265	128	4	30	12905	48	18	21	6065	14	13
43	4445	110	5	31	13085	50	19	22	6245	17	14
44	4625	101	5	32	13265	52	24	23	6425	19	14
45	4805	91	5	33	13445	47	21	24	6605	26	14
46	4985	82	4	34	13625	34	16	25	6785	31	15
47	5165	76	5	35	13805	30	19	26	6965	30	17
48	5345	73	6	36	13985	30	20	27	7145	33	17
49	5525	80	6	37	14165	32	18	28	7325	36	16
50	5705	80	6	38	14345	35	20	29	7505	44	17
51	5885	64	7	39	14525	39	22	30	7685	49	21
52	6065	54	7	40	14705	42	25	31	7865	54	24
53	6245	46	9	41	14885	45	20	32	8045	53	26
54	6425	48	9	42	15065	47	17	33	8225	50	26
55	6605	48	8	43	15245	49	22	34	8405	50	24
56	6785	49	7	44	15424	52	24	35	8585	53	24
57	6965	34	5					36	8765	55	27
58	7145	28	5	**18**	(40)			37	8945	56	27
59	7325	50	6	1:50P	395	148	2	38	9125	58	27
6:00P	7505	44	4	51	611	197	1	39	9305	59	28
01	7685	17	3	52	809	192	1	40	9485	58	28
02	7865	24	4	53	1007	269	1	41	9665	54	29
03	8045	21	4	54	1196	305	1	42	9845	52	29
04	8225	19	5	55	1385	329	2	43	10025	53	30
05	8405	24	6	56	1565	331	2	44	10205	59	27
06	8585	39	5	57	1745	353	2	45	10385	61	30
07	8765	52	4	58	1925	355	3	46	10565	61	33
08	8945	47	5	59	2105	351	3	47	10745	63	32
09	9125	48	5	2:00P	2285	17	3	48	10925	63	28
10	9305	49	5	01	2465	38	3	49	11105	65	27
11	9485	47	5	02	2645	28	5	50	11285	67	26
12	9665	45	6	03	2825	26	6	51	11465	62	28
13	9845	45	5	04	3005	21	6	52	11645	61	32
14	10025	47	5	05	3185	10	5	53	11825	75	25
15	10205	52	5	06	3365	14	6	54	12005	95	16?
16	10385	53	4	07	3545	16	5				
17	10565	48	4	08	3725	15	5	**19**	(41)		
18	10745	36	5	09	3905	11	6	2:02P	395	319	4
19	10925	24	6	10	4085	6	7	03	611	11	1
20	11105	22	6	11	4265	7	8	04	809	120	3
21	11285	33	7	12	4445	9	8	05	1007	121	2

TABLE II (1927)

Date and time	Height in meters	Azimuth from; degrees	Velocity m/s	Date and time	Height in meters	Azimuth from; degrees	Velocity m/s	Date and time	Height in meters	Azimuth from; degrees	Velocity m/s
Sept. 19				**Sept. 19**				**Sept. 20**			
2:06P	1196	139	2	2:54P	9845	27	30	3:06P	6965	98	12
07	1385	175	3	55	10025	27	30	07	7145	96	13
08	1565	203	3	56	10205	27	29	08	7325	100	14
09	1745	263	2	57	10385	28	29	09	7505	102	16
10	1925	280	2	58	10565	27	27	10	7685	104	18
11	2105	266	3	59	10745	24	25	11	7865	100	19
12	2285	258	2	3:00P	10925	23	26	12	8045	95	17
13	2465	198	1	01	11105	23	27	13	8225	98	18
14	2645	103	1	02	11285	24	24	14	8405	100	18
15	2825	45	2	03	11465	24	22	15	8585	98	19
16	3005	17	3					16	8765	100	20
17	3185	6	4	**20**	(42)						
18	3365	8	5	2:30P	395	102	1	**21**	(43)		
19	3545	7	7	31	611	36	1	1:59P	395	100	2
20	3725	7	6	32	809	29	2	2:00P	611	136	2
21	3905	15	6	33	1007	56	1	01	809	170	1
22	4085	28	7	34	1196	91	1	02	1007	174	1
23	4265	38	8	35	1385	162	1	03	1196	164	2
24	4445	42	9	36	1565	160	2	04	1385	143	2
25	4625	43	10	37	1745	137	2	05	1565	123	2
26	4805	44	11	38	1925	122	2	06	1745	138	4
27	4985	43	12	39	2105	136	3	07	1925	137	5
28	5165	42	14	40	2285	139	5	08	2105	139	5
29	5345	41	14	41	2465	137	6	09	2285	144	5
30	5525	39	17	42	2645	139	6	10	2465	138	7
31	5705	38	20	43	2825	136	6	11	2645	139	8
32	5885	37	18	44	3005	134	7	12	2825	139	7
33	6065	36	17	45	3185	135	8	13	3005	133	8
34	6245	32	17	46	3365	135	9	14	3185	131	9
35	6425	31	17	47	3545	132	11	15	3365	130	10
36	6605	30	17	48	3725	130	12	16	3545	128	11
37	6785	29	17	49	3905	128	12	17	3725	129	11
38	6965	28	17	50	4085	126	12	18	3905	127	10
39	7145	28	18	51	4265	123	13	19	4085	125	10
40	7325	28	19	52	4445	119	11	20	4265	122	12
41	7505	28	19	53	4625	120	9	21	4445	120	11
42	7685	26	21	54	4805	120	12	22	4625	118	12
43	7865	27	22	55	4985	120	10	23	4805	118	13
44	8045	23	22	56	5165	113	8	24	4985	118	13
45	8225	20	21	57	5345	118	9	25	5165	118	13
46	8405	21	21	58	5525	118	10	26	5345	116	12
47	8585	21	21	59	5705	115	11	27	5525	112	11
48	8765	21	22	3:00P	5885	114	12	28	5705	109	11
49	8945	23	22	01	6065	110	12	29	5885	103	10
50	9125	26	24	02	6245	108	13	30	6065	98	10
51	9305	26	29	03	6425	106	11	31	6245	97	11
52	9485	26	27	04	6605	103	11	32	6425	97	14
53	9665	26	26	05	6785	103	12	33	6605	99	15

TABLE II (1927)

Date and time	Height in meters	Azimuth from; degrees	Velocity m/s	Date and time	Height in meters	Azimuth from; degrees	Velocity m/s	Date and time	Height in meters	Azimuth from; degrees	Velocity m/s
Sept. **21**				Sept. **22**				Sept. **23**			
2:34P	6785	98	14	4:03P	5885	141	11	2:23P	4625	160	10
35	6965	97	15	04	6065	141	12	24	4805	161	9
36	7145	97	16	05	6245	142	11	25	4985	165	11
37	7325	96	18	06	6425	142	11	26	5165	165	12
38	7505	95	18	07	6605	139	10	27	5345	160	11
39	7685	94	20	08	6785	139	11	28	5525	158	11
40	7865	93	21	09	6965	144	13	29	5705	158	10
41	8045	93	21	10	7145	141	14	30	5885	160	10
42	8225	95	26	11	7325	138	14	31	6065	166	10
43	8405	94	26	12	7505	136	14	32	6245	168	11
44	8585	93	24	13	7685	133	14	33	6425	168	11
45	8765	93	25	14	7865	127	14	34	6605	166	12
46	8945	93	27	15	8045	121	14	35	6785	166	13
47	9125	95	28	16	8225	122	14	36	6965	168	12
48	9305	96	30	17	8405	124	16	37	7145	169	12
49	9485	96	30	18	8585	120	17	38	7325	166	13
				19	8765	123	18	39	7505	164	13
22	(44)			20	8945	115	19	40	7685	158	13
3:33P	395	142	5	21	9125	108	19	41	7865	151	14
34	611	207	3	22	9305	113	19	42	8045	149	13
35	809	241	4	23	9485	111	20	43	8225	152	14
36	1007	256	5	24	9665	111	20	44	8405	154	11
37	1196	260	5	25	9845	112	21	45	8585	154	11
38	1385	246	4					46	8765	156	14
39	1565	232	4	**23**	(45)			47	8945	157	15
40	1745	216	4	2:00P	395	68	2	48	9125	154	14
41	1925	199	5	01	611	85	1	49	9305	153	12
42	2105	195	5	02	809	171	2	50	9485	156	14
43	2285	198	6	03	1007	193	2	51	9665	150	14
44	2465	187	8	04	1196	215	2	52	9845	145	13
45	2645	184	10	05	1385	217	3	53	10025	145	15
46	2825	177	9	06	1565	220	3	54	10205	145	16
47	3005	162	9	07	1745	222	3	55	10385	149	16
48	3185	150	8	08	1925	216	2	56	10565	150	17
49	3365	144	8	09	2105	212	2	57	10745	151	16
50	3545	147	8	10	2285	204	3	58	10925	152	14
51	3725	149	8	11	2465	195	4				
52	3905	153	8	12	2645	183	4	**24**	(46)		
53	4085	155	9	13	2825	182	4	1:52P	395	87	3
54	4265	153	9	14	3005	169	4	53	611	88	2
55	4445	155	9	15	3185	163	5	54	809	108	1
56	4625	156	10	16	3365	170	5	55	1007	88	1
57	4805	153	10	17	3545	165	5	56	1196	13	1
58	4985	147	9	18	3725	164	6	57	1385	18	3
59	5165	144	9	19	3905	169	8	58	1565	17	4
4:00P	5345	145	10	20	4085	173	8	59	1745	14	5
01	5525	144	9	21	4265	170	10	2:00P	1925	16	5
02	5705	141	9	22	4445	165	10	01	2105	23	5

TABLE II (1927)

Date and time	Height in meters	Azimuth from; degrees	Velocity m/s	Date and time	Height in meters	Azimuth from; degrees	Velocity m/s	Date and time	Height in meters	Azimuth from; degrees	Velocity m/s
Sept. 24				**Sept. 24**				**Sept. 25**			
2:02P	2285	27	5	2:50P	10925	200	11	5:29P	1565	202	6
03	2465	23	5	51	11105	205	10	30	1745	192	5
04	2645	23	3	52	11285	205	10	31	1925	178	4
05	2825	48	1	53	11465	200	11	32	2105	148	4
06	3005	167	2	54	11645	196	11	33	2285	144	5
07	3185	180	4	55	11825	199	11	34	2465	153	6
08	3365	187	4	56	12005	205	10	35	2645	156	7
09	3545	188	7					36	2825	156	7
10	3725	193	8	**25**	(47)			37	3005	155	8
11	3905	195	7	1:48P	395	85	3	38	3185	152	7
12	4085	193	7	49	611	87	4	39	3365	148	5
13	4265	192	7	50	809	129	2	40	3545	140	5
14	4445	194	7	51	1007	182	3	41	3725	130	7
15	4625	193	7	52	1196	168	3	42	3905	137	7
16	4805	189	7	53	1385	109	3	43	4085	150	8
17	4985	182	7	54	1565	93	4	44	4265	144	9
18	5165	182	7	55	1745	109	4	45	4445	138	9
19	5345	186	7	56	1925	124	5	46	4625	139	7
20	5525	188	7	57	2105	142	6	47	4805	146	7
21	5705	186	8	58	2285	140	6	48	4985	143	7
22	5885	186	9	59	2465	121	6	49	5165	147	8
23	6065	187	8	2:00P	2645	111	7	50	5345	157	8
24	6245	187	8	01	2825	116	7	51	5525	162	7
25	6425	187	8	02	3005	125	6	52	5705	160	7
26	6605	189	7	03	3185	128	6	53	5885	156	8
27	6785	177	6	04	3365	122	6	54	6065	157	8
28	6965	166	7	05	3545	128	5	55	6245	163	8
29	7145	175	7	06	3725	148	6	56	6425	166	8
30	7325	172	6	07	3905	163	8	57	6605	167	7
31	7505	170	5	08	4085	173	9	58	6785	169	7
32	7685	173	5	09	4265	175	9	59	6965	167	7
33	7865	168	5	10	4445	174	9	6:00P	7145	160	7
34	8045	168	6	11	4625	176	9	01	7325	160	8
35	8225	176	6	12	4805	176	10	02	7505	167	9
36	8405	183	6	13	4985	170	9	03	7685	167	10
37	8585	183	6	14	5165	164	9	04	7865	169	11
38	8765	209	7	15	5345	167	10	05	8045	173	11
39	8945	216	7	16	5525	173	9	06	8225	173	11
40	9125	210	8	17	5705	168	9	07	8405	174	11
41	9305	216	9	18	5885	162	9	08	8585	174	11
42	9485	217	11					09	8765	183	11
43	9665	220	11	**25**	(48)			10	8945	191	12
44	9845	222	10	5:23P	395	60	2	11	9125	192	13
45	10025	217	9	24	611	65	3	12	9305	195	13
46	10205	209	8	25	809	133	2	13	9485	195	12
47	10385	201	9	26	1007	169	4	14	9665	195	14
48	10565	200	11	27	1196	179	6	15	9845	193	15
49	10745	200	11	28	1385	196	6	16	10025	190	14

TABLE II (1927)

Date and time	Height in meters	Azimuth from; degrees	Velocity m/s	Date and time	Height in meters	Azimuth from; degrees	Velocity m/s	Date and time	Height in meters	Azimuth from; degrees	Velocity m/s
Sept.				**Sept.**				**Sept.**			
25				**26**				**28**	(51)		
6:17P	10205	185	14	2:37P	7145	203	1	4:11P	395	69	5
18	10385	180	14	38	7325	207	4	12	611	79	6
19	10565	176	13	39	7505	223	5	13	809	105	5
20	10745	171	15	40	7685	241	5	14	1007	115	5
21	10925	172	17	41	7865	251	6	15	1196	129	5
22	11105	165	14	42	8045	247	7	16	1385	147	3
23	11285	165	12	43	8225	238	7	17	1565	149	2
24	11465	174	13	44	8405	236	7	18	1745	123	2
25	11645	179	13	45	8585	236	7	19	1925	137	3
				46	8765	231	8	20	2105	145	5
26	(49)			47	8945	232	7	21	2285	142	6
2:00P	395	112	7	48	9125	228	7	22	2465	135	8
01	611	121	8	49	9305	225	7	23	2645	136	9
02	809	126	9	50	9485	221	7	24	2825	135	9
03	1007	127	9	51	9665	201	9	25	3005	125	8
04	1196	129	7	52	9845	195	7	26	3185	116	7
05	1385	138	5	53	10025	133	3	27	3365	135	4
06	1565	140	5	54	10205	124	6	28	3545	195	4
07	1745	132	5	55	10385	133	7	29	3725	213	5
08	1925	132	6	56	10565	142	8	30	3905	206	6
09	2105	140	8	57	10745	142	7	31	4085	197	6
10	2285	144	10	58	10925	139	9	32	4265	193	6
11	2465	149	12	59	11105	142	7	33	4445	191	6
12	2645	153	14	3:00P	11285	143	6	34	4625	184	6
13	2825	157	16	01	11465	143	9	35	4805	189	6
14	3005	161	15	02	11645	147	9	36	4985	181	5
15	3185	167	12	03	11825	156	7	37	5165	151	5
16	3365	170	9	04	12005	172	6	38	5345	142	6
17	3545	158	8	05	12185	177	10	39	5525	184	7
18	3725	146	7	06	12365	175	13	40	5705	162	7
19	3905	128	5					41	5885	156	6
20	4085	113	5					42	6065	152	6
21	4265	106	5	**27**	(50)			43	6245	149	6
22	4445	118	5	2:00P	395	190	5	44	6425	148	6
23	4625	142	4	01	611	283	4	45	6605	148	7
24	4805	153	5	02	809	295	6	46	6785	148	7
25	4985	147	5	03	1007	288	7				
26	5165	153	7	04	1196	281	9	**29**	(52)		
27	5345	159	9	05	1385	275	9	1:37P	395	135	3
28	5525	162	9	06	1565	270	8	38	611	81	2
29	5705	160	10	07	1745	263	6	39	809	38	3
30	5885	165	9	08	1925	251	4	40	1007	36	3
31	6065	168	7	09	2105	249	4	41	1196	35	3
32	6245	169	5	10	2285	254	4	42	1385	35	4
33	6425	174	3	11	2465	255	4	43	1565	36	5
34	6605	178	3	12	2645	245	4	44	1745	39	6
35	6785	182	2	13	2825	232	5	45	1925	40	6
36	6965	186	1	14	3005	227	5	46	2105	45	6

TABLE II (1927)

Date and time	Height in meters	Azimuth from; degrees	Velocity m/s	Date and time	Height in meters	Azimuth from; degrees	Velocity m/s	Date and time	Height in meters	Azimuth from; degrees	Velocity m/s
Sept.				**Sept.**				**Oct.**			
29				**30**	(53)			**1**			
1:47P	2285	51	7	2:22P	395	93	4	1:56P	2645	212	4
48	2465	53	7	23	611	85	5	57	2825	231	3
49	2645	54	7	24	809	88	3	58	3005	220	5
50	2825	61	5	25	1007	132	2	59	3185	213	7
51	3005	68	3	26	1196	124	4	2:00P	3365	223	6
52	3185	75	2	27	1385	193	5	01	3545	227	6
53	3365	64	2	28	1565	207	7	02	3725	228	5
54	3545	53	4	29	1745	215	8	03	3905	227	5
55	3725	55	5	30	1925	215	5				
56	3905	55	5	31	2105	145	2	**2**	(55)		
57	4085	47	4	32	2285	100	4	2:48 P	395	103	3
58	4265	59	3	33	2465	109	3	49	611	54	3
59	4445	82	4	34	2645	136	1	50	809	35	5
2:00P	4625	90	5	35	2825	169	1	51	1007	29	5
01	4805	93	6	36	3005	208	2	52	1196	33	3
02	4985	100	7	37	3185	221	5	53	1385	47	2
03	5165	108	7	38	3365	226	5	54	1565	47	1
04	5345	113	7	39	3545	223	6	55	1745	55	2
05	5525	108	7	40	3725	227	6	56	1925	75	2
06	5705	106	7	41	3905	231	5	57	2105	99	3
07	5885	111	7	42	4085	212	5	58	2285	113	3
08	6065	113	7	43	4265	190	5	59	2465	128	3
09	6245	112	7	44	4445	178	5	3:00P	2645	129	3
10	6425	116	8	45	4625	174	5	01	2825	127	3
11	6605	115	9	46	4805	166	5	02	3005	129	3
12	6785	101	8	47	4985	166	5	03	3185	133	3
13	6965	90	8	48	5165	179	6	04	3365	137	3
14	7145	86	7	49	5345	179	8	05	3545	135	4
15	7325	84	7	50	5525	174	7	06	3725	126	3
16	7505	81	7	51	5705	171	6	07	3905	120	2
17	7685	82	7	52	5885	173	6	08	4085	96	1
18	7865	86	7	53	6065	181	6	09	4265	2	1
19	8045	91	8	54	6245	184	7	10	4445	324	1
20	8225	94	9					11	4625	299	1
21	8405	98	10	**Oct.**				12	4805	261	1
22	8585	100	11	**1**	(54)			13	4985	233	1
23	8765	96	11	1:44P	395	106	2	14	5165	230	2
24	8945	95	11	45	611	156	2	15	5345	233	1
25	9125	95	12	46	809	178	3	16	5525	225	2
26	9305	94	14	47	1007	189	4	17	5705	221	3
27	9485	88	13	48	1196	186	5	18	5885	229	3
28	9665	79	13	49	1385	183	4	19	6065	235	3
29	9845	79	14	50	1565	202	3	20	6245	230	4
30	10025	80	14	51	1745	195	6	21	6425	232	5
31	10205	78	14	52	1925	196	7	22	6605	224	5
32	10385	77	12	53	2105	209	5	23	6785	220	6
				54	2285	211	5	24	6965	220	7
				55	2465	207	5	25	7145	219	7

TABLE II (1927)

Date and time	Height in meters	Azimuth from; degrees	Velocity m/s	Date and time	Height in meters	Azimuth from; degrees	Velocity m/s	Date and time	Height in meters	Azimuth from; degrees	Velocity m/s
Oct.				**Oct.**				**Oct.**			
2				**4**				**8**	(59)		
3:26P	7325	226	6	2:03P	2645	190	15	1:18P	395	114	6
27	7505	216	7	04	2825	184	16	19	611	104	4
28	7685	207	9	05	3005	182	16	20	809	98	3
29	7865	210	7	06	3185	182	14	21	1007	119	3
30	8045	214	7	07	3365	184	14	22	1196	135	3
31	8225	217	7	08	3545	182	15	23	1385	132	6
32	8405	214	7	09	3725	175	16	24	1565	132	9
33	8585	206	7	10	3905	177	19	25	1745	139	10
34	8765	210	7	11	4085	170	21	26	1925	146	10
35	8945	215	7	12	4265	173	22	27	2105	152	8
36	9125	211	8	13	4445	173	20	28	2285	153	7
37	9305	214	9	14	4625	171	19	29	2465	158	6
38	9485	219	8	15	4805	170	19	30	2645	197	3
39	9665	221	6	16	4985	167	19	31	2825	247	2
40	9845	213	7	17	5165	167	22	32	3005	237	1
41	10025	214	8	18	5345	169	28	33	3185	232	1
42	10205	217	7	19	5525	172	30	34	3365	249	2
43	10385	216	6	20	5705	171	28	35	3545	257	2
								36	3725	253	2
3	(56)			**5**	(58)			37	3905	247	2
5:20P	395	36	3	2:36P	395	186	3	38	4085	215	2
21	611	36	4	37	611	192	4	39	4265	220	2
22	809	32	6	38	809	193	3	40	4445	239	3
23	1007	46	11	39	1007	203	4	41	4625	233	3
24	1196	60	12	40	1196	202	3	42	4805	242	3
25	1385	78	12	41	1385	218	3	43	4985	252	5
26	1565	96	12	42	1565	222	5	44	5265	250	7
27	1745	113	14	43	1745	228	5	45	5345	252	8
28	1925	120	17	44	1925	228	6	46	5525	262	8
29	2105	127	17	45	2105	218	7	47	5705	261	9
30	2285	133	19	46	2285	207	7	48	5885	266	10
31	2465	156	21	47	2465	199	7	49	6065	270	11
32	2645	172	25	48	2645	201	7	50	6245	274	12
				49	2825	201	6	51	6425	277	12
				50	3005	208	5	52	6605	276	14
4	(57)			51	3185	221	5	53	6785	278	15
1:51P	395	168	4	52	3365	223	5	54	6965	282	15
52	611	171	3	53	3545	243	5	55	7145	285	15
53	809	172	3	54	3725	268	4	56	7325	283	15
54	1007	165	3	55	3905	275	3	57	7505	281	16
55	1196	162	3	56	4085	269	3	58	7685	283	18
56	1385	165	3	57	4265	252	4	59	7865	280	20
57	1565	165	3	58	4445	249	4	2:00P	8045	279	20
58	1745	162	4	59	4625	252	3	01	8225	279	21
59	1925	162	6	3:00P	4805	251	4	02	8405	278	22
2:00P	2105	177	7	01	4985	250	3	03	8585	276	22
01	2285	191	9	02	5165	243	7				
02	2465	194	12	03	5345	240	7				

TABLE II (1927)

Date and time	Height in meters	Azimuth from; degrees	Velocity m/s	Date and time	Height in meters	Azimuth from; degrees	Velocity m/s	Date and time	Height in meters	Azimuth from; degrees	Velocity m/s
Oct.				**Oct.**				**Oct.**			
8	(60)			**9**				**10**			
4:50P	395	56	7	2:53P	611	60	2	2:17P	1196	148	6
51	611	86	8	54	809	88	2	18	1385	155	7
52	809	102	7	55	1007	70	2	19	1565	146	7
53	1007	112	6	56	1196	78	1	20	1745	144	7
54	1196	110	5	57	1385	118	1	21	1925	151	7
55	1385	108	5	58	1565	169	2	22	2105	162	6
56	1565	111	6	59	1745	185	3	23	2285	168	6
57	1745	110	8	3:00P	1925	200	3	24	2465	167	7
58	1925	113	8	01	2105	216	4	25	2645	169	8
59	2105	118	6	02	2285	219	5	26	2825	169	8
5:00P	2285	115	4	03	2465	220	5	27	3005	162	8
01	2465	113	3	04	2645	219	6	28	3185	153	8
02	2645	106	2	05	2825	218	7	29	3365	149	8
03	2825	92	2	06	3005	215	8	30	3545	148	8
04	3005	104	2	07	3185	217	8	31	3725	149	8
05	3185	135	2	08	3365	217	7	32	3905	147	8
06	3365	185	3	09	3545	214	6	33	4085	143	7
07	3545	204	3	10	3725	207	7	34	4265	134	5
08	3725	237	4	11	3905	203	8	35	4445	127	6
09	3905	256	6	12	4085	208	8	36	4625	127	9
10	4085	262	6	13	4265	211	8	37	4805	124	9
11	4265	269	7	14	4445	209	9	38	4985	123	9
12	4445	273	7	15	4625	209	9	39	5165	129	11
13	4625	268	8	16	4805	200	9	40	5345	132	12
14	4805	264	9	17	4985	197	9	41	5525	131	10
15	4985	257	9	18	5165	207	10	42	5705	136	10
16	5165	253	10	19	5345	207	11	43	5885	147	10
17	5345	256	11	20	5525	202	12	44	6065	153	11
18	5525	254	12	21	5705	198	12	45	6245	155	13
19	5705	254	13	22	5885	195	14	46	6425	155	14
20	5885	258	14	23	6065	196	14	47	6605	158	16
21	6065	262	14	24	6245	195	13	48	6785	160	17
22	6245	259	15	25	6425	193	14	49	6965	163	18
23	6425	258	15	26	6605	204	15	50	7145	165	19
24	6605	260	15	27	6785	213	16	51	7325	169	19
25	6785	255	16	28	6965	211	14	52	7505	164	23
26	6965	247	16	29	7145	216	13	53	7685	169	25
27	7145	245	16	30	7325	222	16	54	7865	175	23
28	7325	249	16	31	7505	219	18	55	8045	172	21
29	7505	250	16	32	7685	220	18	56	8225	172	21
30	7685	249	18	33	7865	220	19				
31	7865	247	20	34	8045	219	17	**11**	(63)		
32	8045	247	20					3:45P	395	103	4
33	8225	247	20	**10**	(62)			46	611	136	4
34	8405	247	21	2:13P	395	15	4	47	809	171	4
				14	611	54	4	48	1007	200	4
9	(61)			15	809	95	4	49	1196	220	4
2:52P	395	26	1	16	1007	121	5	50	1385	216	2

TABLE II (1927)

Date and time	Height in meters	Azimuth from; degrees	Velocity m/s	Date and time	Height in meters	Azimuth from; degrees	Velocity m/s	Date and time	Height in meters	Azimuth from; degrees	Velocity m/s
Oct.				**Oct.**				**Oct.**			
11				**12**				**12**			
3:51P	1565	236	1	9:11A	611	52	4	9:59A	9305	209	31
52	1745	289	1	12	809	30	5	10:00A	9485	209	35
53	1925	280	3	13	1007	24	7				
54	2105	279	5	14	1196	28	9	**12**	(65)		
55	2285	271	3	15	1385	34	9	2:27P	395	93	2
56	2465	238	3	16	1565	44	7	28	611	70	6
57	2645	219	4	17	1745	51	7	29	809	65	6
58	2825	219	6	18	1925	62	5	30	1007	73	6
59	3005	219	6	19	2105	116	3	31	1196	85	5
4:00P	3185	196	4	20	2285	134	6	32	1385	73	5
01	3365	170	5	21	2465	134	7	33	1565	72	4
02	3545	182	6	22	2645	138	6	34	1745	108	5
03	3725	184	7	23	2825	139	6	35	1925	124	6
04	3905	175	9	24	3005	145	6	36	2105	125	8
05	4085	169	11	25	3185	151	8	37	2285	122	9
06	4265	160	13	26	3365	153	9	38	2465	119	10
07	4445	161	13	27	3545	150	10	39	2645	121	11
08	4625	168	12	28	3725	143	10	40	2825	126	11
09	4805	171	12	29	3905	149	9	41	3005	131	11
10	4985	174	13	30	4085	164	10	42	3185	137	12
11	5165	172	13	31	4265	173	9	43	3365	143	13
12	5345	176	11	32	4445	183	9	44	3545	148	13
13	5525	188	10	33	4625	186	10	45	3725	151	11
14	5705	198	11	34	4805	185	9	46	3905	149	10
15	5885	197	13	35	4985	185	8	47	4085	149	9
16	6065	192	14	36	5165	176	8	48	4265	154	9
17	6245	189	16	37	5345	174	8	49	4445	161	10
18	6425	185	17	38	5525	180	9	50	4625	162	11
19	6605	184	17	39	5705	190	10	51	4805	155	13
20	6785	187	16	40	5885	197	12	52	4985	153	14
21	6965	183	15	41	6065	196	14	53	5165	153	14
22	7145	178	14	42	6245	200	15				
23	7325	174	14	43	6425	208	17	**17**	(66)		
24	7505	173	15	44	6605	213	18	3:30P	395	103	4
25	7685	173	15	45	6785	217	20	31	611	136	5
26	7865	175	19	46	6965	217	21	32	809	169	5
27	8045	178	20	47	7145	216	22	33	1007	199	4
28	8225	179	16	48	7325	215	23	34	1196	231	4
29	8405	178	16	49	7505	213	22	35	1385	241	4
30	8585	178	16	50	7685	209	22	36	1565	252	7
31	8765	177	15	51	7865	207	22	37	1745	253	9
32	8945	176	17	52	8045	208	22	38	1925	249	11
33	9125	176	16	53	8225	209	23	39	2105	251	11
34	9305	177	14	54	8405	210	22	40	2285	255	11
35	9485	179	14	55	8585	206	24				
				56	8765	204	28	**18**	(67)		
12	(64)			57	8945	206	29	9:50A	395	111	5
9:10A	395	80	2	58	9125	207	29	51	611	123	9

TABLE II (1927)

Date and time	Height in meters	Azimuth from; degrees	Velocity m/s	Date and time	Height in meters	Azimuth from; degrees	Velocity m/s	Date and time	Height in meters	Azimuth from; degrees	Velocity m/s
Oct.				**Oct.**				**Oct.**			
18				**18**				**19**			
9:52A	809	146	8	1:56P	5345	305	8	1:29P	6425	89	7
53	1007	158	5	57	5525	305	9	30	6605	88	8
54	1196	144	4	58	5705	301	11	31	6785	90	8
55	1385	142	4	59	5885	302	11	32	6965	99	8
56	1565	136	3	2:00P	6065	304	12	33	7145	96	9
57	1745	124	2	01	6245	301	12	34	7325	90	9
58	1925	121	3	02	6425	298	12	35	7505	90	9
59	2105	126	5	03	6605	300	13	36	7685	92	10
10:00A	2285	122	6	04	6785	300	13	37	7865	100	11
01	2465	147	7	05	6965	298	14	38	8045	104	12
02	2645	153	7	06	7145	293	17				
03	2825	153	7	07	7325	295	20	**19**	(70)		
04	3005	167	4	08	7505	299	21	2:26P	395	93	2
05	3185	182	3					27	611	92	1
06	3365	185	3	**19**	(69)			28	809	300	1
07	3545	216	3	0:56P	395	66	2	29	1007	297	2
08	3725	239	4	57	611	61	2	30	1196	328	5
09	3905	244	4	58	809	24	2	31	1385	352	8
10	4085	244	4	59	1007	350	4	32	1565	351	9
				1:00P	1196	358	7	33	1745	3	9
18	(68)			01	1385	3	9	34	1925	14	9
1:29P	395	102	5	02	1565	12	10	35	2105	15	8
30	611	109	6	03	1745	25	10	36	2285	12	5
31	809	105	2	04	1925	24	9	37	2465	356	4
32	1007	94	1	05	2105	5	6	38	2645	354	5
33	1196	115	1	06	2285	354	5	39	2825	3	5
34	1385	113	3	07	2465	358	4	40	3005	7	5
35	1565	104	4	08	2645	12	4	41	3185	19	4
36	1745	108	5	09	2825	35	4	42	3365	38	5
37	1925	117	5	10	3005	51	5	43	3545	50	5
38	2105	125	5	11	3185	53	4	44	3725	67	5
39	2285	149	3	12	3365	62	5	45	3905	61	4
40	2465	161	3	13	3545	78	5	46	4085	48	4
41	2645	150	3	14	3725	79	5	47	4265	47	5
42	2825	147	2	15	3905	77	5	48	4445	46	4
43	3005	187	2	16	4085	80	5	49	4625	46	4
44	3185	236	2	17	4265	77	7	50	4805	55	5
45	3365	222	2	18	4445	69	8	51	4985	59	6
46	3545	209	3	19	4625	67	8	52	5165	53	8
47	3725	222	1	20	4805	69	8	53	5345	49	9
48	3905	225	3	21	4985	72	9	54	5525	48	11
49	4085	234	3	22	5165	74	10	55	5705	49	11
50	4265	255	3	23	5345	76	11	56	5885	57	10
51	4445	277	2	24	5525	76	9	57	6065	64	8
52	4625	278	2	25	5705	77	7	58	6245	64	7
53	4805	281	3	26	5885	82	6	59	6425	62	7
54	4985	291	4	27	6065	89	6	3:00P	6605	67	7
55	5165	300	7	28	6245	89	6	01	6785	71	8

TABLE II (1927)

Date and time	Height in meters	Azimuth from; degrees	Velocity m/s	Date and time	Height in meters	Azimuth from; degrees	Velocity m/s	Date and time	Height in meters	Azimuth from; degrees	Velocity m/s
Oct.				**Oct.**				**Oct.**			
19				**20**				**21**			
3:02P	6965	73	9	3:36P	8225	45	24	4:07P	4445	349	14
03	7145	73	9	37	8405	51	22	08	4625	345	15
				38	8585	55	24	09	4805	341	16
				39	8765	49	25	10	4985	341	16
20	(71)			40	8945	48	27	11	5165	341	18
2:53P	395	65	1	41	9125	48	25	12	5345	341	21
54	611	139	2	42	9305	48	25	13	5525	340	20
55	809	170	2	43	9485	47	29	14	5705	335	20
56	1007	105	1	44	9665	48	31	15	5885	333	21
57	1196	52	2	45	9845	48	29	16	6065	332	21
58	1385	25	1	46	10025	46	27	17	6245	339	20
59	1565	26	1	47	10205	44	27	18	6425	342	22
3:00P	1745	65	1	48	10385	43	25	19	6605	342	21
01	1925	74	1	49	10565	43	26	20	6785	343	22
02	2105	65	2	50	10745	44	28	21	6965	343	21
03	2285	54	3	51	10925	49	25	22	7145	345	22
04	2465	49	4	52	11105	49	16	23	7325	345	26
05	2645	50	4	53	11285	34	11				
06	2825	53	5	54	11465	13	7	**22**	(73)		
07	3005	48	6	55	11645	12	12	10:00A	395	53	5
08	3185	42	6	56	11825	22	16	01	611	44	6
09	3365	43	5	57	12005	11	14	02	809	34	3
10	3545	52	6	58	12185	15	10	03	1007	31	3
11	3725	53	6	59	12365	19	9	04	1196	29	7
12	3905	52	7					05	1385	22	10
13	4085	54	8	**21**	(72)			06	1565	22	9
14	4265	56	9	3:45P	395	60	3	07	1745	22	7
15	4445	62	12	46	611	57	4	08	1925	8	6
16	4625	63	14	47	809	41	4	09	2105	352	6
17	4805	62	14	48	1007	38	3	10	2285	349	7
18	4985	60	16	49	1196	42	3	11	2465	346	6
19	5165	60	17	50	1385	35	4	12	2645	344	7
20	5345	59	18	51	1565	29	4	13	2825	344	8
21	5525	56	18	52	1745	33	5	14	3005	341	8
22	5705	58	18	53	1925	27	5	15	3185	343	9
23	5885	58	19	54	2105	21	6	16	3365	348	9
24	6065	57	20	55	2285	23	7	17	3545	355	9
25	6245	57	21	56	2465	22	7	18	3725	358	9
26	6425	58	21	57	2645	19	7	19	3905	2	9
27	6605	59	21	58	2825	12	6	20	4085	7	9
28	6785	59	21	59	3005	1	6	21	4265	4	13
29	6965	61	22	4:00P	3185	354	7	22	4445	356	14
30	7145	58	24	01	3365	355	9	23	4625	354	15
31	7325	56	24	02	3545	0	12	24	4805	357	15
32	7505	53	25	03	3725	2	13	25	4985	1	14
33	7685	53	22	04	3905	358	12	26	5165	1	15
34	7865	56	24	05	4085	354	13	27	5345	1	15
35	8045	54	27	06	4265	351	14	28	5525	2	18

TABLE II (1927)

Date and time	Height in meters	Azimuth from; degrees	Velocity m/s	Date and time	Height in meters	Azimuth from; degrees	Velocity m/s	Date and time	Height in meters	Azimuth from; degrees	Velocity m/s
Oct.				**Oct.**				**Oct.**			
22				**22**				**24**			
10:29A	5705	359	19	2:24P	7865	13	33	2:12P	2645	98	9
30	5885	351	21	25	8045	14	35	13	2825	93	9
31	6065	356	21	26	8225	14	35	14	3005	86	8
32	6245	354	20	27	8405	13	33	15	3185	76	7
33	6425	351	20	28	8585	11	36	16	3365	66	7
				29	8765	13	39	17	3545	63	8
22	(74)			30	8945	13	36	18	3725	70	9
1:43P	395	40	5	31	9125	11	33	19	3905	72	11
44	611	50	6	32	9305	8	37	20	4085	69	11
45	809	43	5					21	4265	69	11
46	1007	33	3	**23**	(75)			22	4445	69	10
47	1196	37	4	1:42P	395	97	11	23	4625	66	8
48	1385	37	5	43	611	117	9	24	4805	66	7
49	1565	8	4	44	809	140	10	25	4985	65	7
50	1745	335	3	45	1007	148	10	26	5165	62	8
51	1925	322	4	46	1196	161	10	27	5345	61	7
52	2105	316	4	47	1385	166	11	28	5525	57	7
53	2285	320	3	48	1565	160	10	29	5705	58	8
54	2465	335	2	49	1745	150	7	30	5885	53	7
55	2645	359	2	50	1925	132	7	31	6065	37	7
56	2825	344	4	51	2105	124	10	32	6245	29	8
57	3005	348	6	52	2285	119	11	33	6425	33	8
58	3185	2	7	53	2465	110	10	34	6605	45	8
59	3365	1	8	54	2645	109	11	35	6785	60	7
2:00P	3545	10	7	55	2825	108	13	36	6965	59	5
01	3725	19	7	56	3005	112	15	37	7145	43	3
02	3905	18	8	57	3185	119	16				
03	4085	12	9	58	3365	123	16	**25**	(77)		
04	4265	5	9	59	3545	119	18	1:00P	395	85	5
05	4445	1	9	2:00P	3725	134	18	01	611	98	7
06	4625	4	9	01	3905	141	17	02	809	116	4
07	4805	8	10	02	4085	140	18	03	1007	148	1
08	4985	14	12	03	4265	134	18	04	1196	91	1
09	5165	19	14	04	4445	131	18	05	1385	115	3
10	5345	18	13					06	1565	125	7
11	5525	19	13	**24**	(76)			07	1745	130	10
12	5705	19	13	2:00P	395	88	5	08	1925	131	12
13	5885	19	13	01	611	111	7	09	2105	131	13
14	6065	18	13	02	809	118	5	10	2285	135	12
15	6245	17	14	03	1007	78	2	11	2465	142	9
16	6425	16	16	04	1196	47	2	12	2645	141	10
17	6605	13	19	05	1385	72	2	13	2825	138	11
18	6785	13	20	06	1565	108	3	14	3005	137	13
19	6965	15	23	07	1745	127	5	15	3185	137	13
20	7145	15	26	08	1925	123	6	16	3365	139	9
21	7325	16	27	09	2105	101	6	17	3545	137	6
22	7505	14	30	10	2285	93	6	18	3725	122	4
23	7685	13	33	11	2465	98	8	19	3905	86	3

TABLE II (1927)

Date and time	Height in meters	Azimuth from; degrees	Velocity m/s	Date and time	Height in meters	Azimuth from; degrees	Velocity m/s	Date and time	Height in meters	Azimuth from; degrees	Velocity m/s
Oct. 25				**Oct. 26**				**Oct. 26**			
1:20P	4085	65	4	9:13A	1007	196	3	10:01A	9665	86	18
21	4265	63	5	14	1196	197	3	02	9845	86	19
22	4445	71	7	15	1385	221	1	03	10025	83	20
23	4625	78	10	16	1565	288	1	04	10205	80	22
24	4805	88	12	17	1745	112	1	05	10385	77	22
25	4985	93	12	18	1925	138	2	06	10565	78	18
26	5165	101	10	19	2105	122	4	07	10745	80	17
27	5345	107	9	20	2285	102	4	08	10925	80	17
28	5525	104	9	21	2465	80	4	09	11105	77	12
29	5705	100	8	22	2645	76	4	10	11285	74	9
30	5885	88	7	23	2825	81	4				
31	6065	82	6	24	3005	74	4	**26**	(79)		
32	6245	82	6	25	3185	69	5	2:12P	395	89	3
33	6425	86	5	26	3365	68	5	13	611	141	2
34	6605	94	5	27	3545	76	5	14	809	202	3
35	6785	95	7	28	3725	87	6	15	1007	224	4
36	6965	97	9	29	3905	89	6	16	1196	234	5
37	7145	107	9	30	4085	86	7	17	1385	242	4
38	7325	123	6	31	4265	88	10	18	1565	242	3
39	7505	140	4	32	4445	85	11	19	1745	214	2
40	7685	132	3	33	4625	79	10	20	1925	152	2
41	7865	98	3	34	4805	75	11	21	2105	106	2
42	8045	76	4	35	4985	77	11	22	2285	85	3
43	8225	74	5	36	5165	82	11	23	2465	80	3
44	8405	85	6	37	5345	86	11	24	2645	75	3
45	8585	91	6	38	5525	87	12	25	2825	62	2
46	8765	92	5	39	5705	89	13	26	3005	53	1
47	8945	94	4	40	5885	91	13	27	3185	73	2
48	9125	96	5	41	6065	86	13	28	3365	57	4
49	9305	108	4	42	6245	83	12	29	3545	57	5
50	9485	113	3	43	6425	83	13	30	3725	75	4
51	9665	91	3	44	6605	83	12	31	3905	77	4
52	9845	58	4	45	6785	82	12	32	4085	76	6
53	10025	58	7	46	6965	81	12	33	4265	80	8
54	10205	58	8	47	7145	79	13	34	4445	79	8
55	10385	68	8	48	7325	77	14	35	4625	74	8
56	10565	77	7	49	7505	78	15	36	4805	69	9
57	10745	88	7	50	7685	80	16	37	4985	71	9
58	10925	87	6	51	7865	80	17	38	5165	74	9
59	11105	63	7	52	8045	80	18	39	5345	75	10
2:00P	11285	59	9	53	8225	79	17	40	5525	77	10
01	11465	62	9	54	8405	78	15	41	5705	80	11
02	11645	62	9	55	8585	82	17	42	5885	83	12
				56	8765	86	18	43	6065	85	12
26	(78)			57	8945	83	17	44	6245	84	11
9:10A	395	91	2	58	9125	84	20	45	6425	84	14
11	611	123	4	59	9305	83	20	46	6605	83	14
12	809	174	3	10:00A	9485	81	20	47	6785	80	14

TABLE II (1927)

Date and time	Height in meters	Azimuth from; degrees	Velocity m/s	Date and time	Height in meters	Azimuth from; degrees	Velocity m/s	Date and time	Height in meters	Azimuth from; degrees	Velocity m/s
Oct.				**Oct.**				**Oct.**			
26				**27**				**28**			
2:48P	6965	83	15	2:22P	4445	76	14	1:34P	3005	148	9
49	7145	84	16	23	4625	76	15	35	3185	141	9
50	7325	84	16	24	4805	76	15	36	3365	133	11
51	7505	87	17	25	4985	75	16	37	3545	130	12
52	7685	91	19	26	5165	74	16	38	3725	128	11
53	7865	91	21	27	5345	76	17	39	3905	125	11
54	8045	90	22	28	5525	76	17	40	4085	121	11
55	8225	88	24	29	5705	75	18	41	4265	119	12
56	8405	87	23	30	5885	75	18	42	4445	123	13
57	8585	86	23	31	6065	74	18	43	4625	125	13
58	8765	85	25	32	6245	77	18				
59	8945	87	27	33	6425	77	19	**28**	(82)		
3:00P	9125	88	29	34	6605	73	21	2:03P	395	99	3
01	9305	89	32	35	6785	72	23	04	611	126	4
02	9485	91	31	36	6965	70	22	05	809	169	4
03	9665	89	29	37	7145	72	21	06	1007	197	5
04	9845	88	29	38	7325	73	20	07	1196	210	6
05	10025	86	28	39	7505	70	21	08	1385	206	7
06	10205	87	22	40	7685	68	21	09	1565	199	8
07	10385	86	22	41	7865	67	20	10	1745	196	7
08	10565	85	25	42	8045	65	18	11	1925	198	7
09	10745	84	22	43	8225	64	20	12	2105	196	7
10	10925	84	19	44	8405	65	20	13	2285	185	7
11	11105	84	19	45	8585	67	20	14	2465	170	6
				46	8765	70	20	15	2645	164	7
27	(80)			47	8945	72	18	16	2825	164	8
2:00P	395	88	2	48	9125	73	19	17	3005	150	9
01	611	113	3	49	9305	75	17	18	3185	145	10
02	809	131	2	50	9485	75	17	19	3365	138	10
03	1007	144	1	51	9665	74	17	20	3545	130	12
04	1196	283	1	52	9845	73	17	21	3725	127	12
05	1385	316	1	53	10025	74	20	22	3905	123	12
06	1565	45	2					23	4085	123	11
07	1745	78	4	**28**	(81)			24	4265	122	12
08	1925	72	6	1:20P	395	99	3	25	4445	120	13
09	2105	64	7	21	611	122	5	26	4625	122	14
10	2285	74	8	22	809	164	5	27	4805	119	15
11	2465	82	8	23	1007	201	6	28	4985	119	16
12	2645	74	8	24	1196	207	8	29	5165	124	17
13	2825	71	8	25	1385	195	7	30	5345	124	17
14	3005	65	9	26	1565	195	6	31	5525	117	17
15	3185	67	9	27	1745	207	7	32	5705	115	17
16	3365	74	9	28	1925	203	7	33	5885	108	16
17	3545	77	10	29	2105	187	6	34	6065	111	17
18	3725	81	11	30	2285	173	7	35	6245	111	18
19	3905	83	11	31	2465	170	8	36	6425	107	19
20	4085	79	11	32	2645	163	7	37	6605	105	20
21	4265	77	13	33	2825	154	8	38	6785	106	20

TABLE II (1927)

Date and time	Height in meters	Azimuth from; degrees	Velocity m/s	Date and time	Height in meters	Azimuth from; degrees	Velocity m/s	Date and time	Height in meters	Azimuth from; degrees	Velocity m/s
Oct.				**Oct.**				**Oct.**			
28				**29**				**30**			
2:39P	6965	101	19	2:49P	8405	178	19	2:08P	7325	224	21
40	7145	96	16	50	8585	176	16	09	7505	221	22
				51	8765	178	14	10	7685	220	23
29	(83)			52	8945	172	11	11	7865	216	22
2:05P	395	65	3	53	9125	157	10	12	8045	214	20
06	611	80	5	54	9305	156	11	13	8225	213	17
07	809	88	3	55	9485	165	11	14	8405	209	17
08	1007	90	2	56	9665	172	11	15	8585	203	17
09	1196	86	2					16	8765	196	14
10	1385	60	3	**30**	(84)			17	8945	190	12
11	1565	40	3	1:30P	395	55	4	18	9125	187	12
12	1745	32	4	31	611	61	3				
13	1925	27	4	32	809	136	1	**31**	(85)		
14	2105	26	2	33	1007	166	3	1:00P	395	76	5
15	2285	8	1	34	1196	183	3	01	611	117	3
16	2465	359	1	35	1385	178	3	02	809	190	3
17	2645	210	1	36	1565	183	3	03	1007	234	2
18	2825	197	3	37	1745	197	2	04	1196	318	2
19	3005	205	5	38	1925	190	2	05	1385	311	2
20	3185	207	6	39	2105	202	1	06	1565	256	1
21	3365	213	7	40	2285	188	1				
22	3545	219	7	41	2465	148	1	**31**	(86)		
23	3725	206	8	42	2645	20	1	1:18P	395	69	4
24	3905	199	9	43	2825	346	1	19	611	101	3
25	4085	200	10	44	3005	298	1	20	809	187	3
26	4265	197	11	45	3185	289	1	21	1007	224	3
27	4445	195	11	46	3365	279	2	22	1196	277	2
28	4625	197	12	47	3545	266	4	23	1385	319	2
29	4805	196	11	48	3725	262	4	24	1565	308	3
30	4985	192	11	49	3905	251	6	25	1745	294	6
31	5165	196	12	50	4085	247	7	26	1925	285	8
32	5345	198	14	51	4265	245	8	27	2105	279	7
33	5525	196	15	52	4445	239	9	28	2285	267	6
34	5705	194	16	53	4625	238	9	29	2465	248	5
35	5885	193	16	54	4805	237	11	30	2645	237	5
36	6065	195	18	55	4985	239	11	31	2825	221	5
37	6245	195	18	56	5165	239	12	32	3005	224	5
38	6425	195	16	57	5345	240	13	33	3185	227	6
39	6605	191	16	58	5525	239	14	34	3365	215	5
40	6785	191	16	59	5705	240	15	35	3545	195	5
41	6965	190	16	2:00P	5885	243	16	36	3725	173	6
42	7145	191	17	01	6065	244	17	37	3905	162	7
43	7325	187	17	02	6245	241	18	38	4085	161	8
44	7505	184	17	03	6425	239	19	39	4265	168	9
45	7685	179	15	04	6605	235	19	40	4445	169	10
46	7865	173	17	05	6785	231	21	41	4625	169	11
47	8045	173	17	06	6965	229	20	42	4805	166	11
48	8225	175	18	07	7145	227	20	43	4985	164	11

TABLE II (1927)

Date and time	Height in meters	Azimuth from; degrees	Velocity m/s
Oct. 31			
1:44P	5165	159	11
45	5345	153	11
46	5525	153	11
47	5705	155	11
48	5885	158	11
49	6065	159	11
Nov. 1	(87)		
2:02P	395	63	4
03	611	80	4
04	809	164	4
05	1007	182	5
06	1196	178	4
07	1385	174	4
08	1565	160	5
09	1745	144	6
10	1925	150	8
11	2105	165	10
12	2285	174	10
13	2465	183	10
14	2645	191	10
15	2825	193	10
16	3005	192	10
17	3185	194	9
18	3365	195	9
2	(88)		
9:45A	395	95	3
46	611	140	8
47	809	160	11
48	1007	167	9
49	1196	168	7
50	1385	170	8
51	1565	174	9
52	1745	173	10
53	1925	173	10
54	2105	170	9
55	2285	169	10
56	2465	167	10
57	2645	164	10
58	2825	162	9
59	3005	170	8
10:00A	3185	189	8
01	3365	200	9
02	3545	198	11
03	3725	201	12
04	3905	205	11
05	4085	208	12

Date and time	Height in meters	Azimuth from; degrees	Velocity m/s
Nov. 2			
10:06A	4265	209	12
07	4445	208	12
08	4625	208	11
2	(89)		
1:18P	395	109	5
19	611	147	8
20	809	165	12
21	1007	170	13
22	1196	183	13
23	1385	201	13
24	1565	215	14
25	1745	218	14
26	1925	214	15
27	2105	213	17
28	2285	212	18
29	2465	212	17
30	2645	213	14
31	2825	205	11
32	3005	191	12
33	3185	185	13
34	3365	184	15
35	3545	185	17
36	3725	185	19
37	3905	182	19
38	4085	179	19
39	4265	178	18
3	(90)		
10:03A	395	125	15
04	611	142	23
05	809	155	28
06	1007	162	25
07	1196	181	13
08	1385	210	9
09	1565	186	20
10	1745	183	27
11	1925	194	24
12	2105	202	26
13	2285	207	27
14	2465	210	27
3	(91)		
0:06P	395	171	19
07	611	179	18
08	809	187	16
09	1007	193	14
10	1196	198	13
11	1385	201	14

Date and time	Height in meters	Azimuth from; degrees	Velocity m/s
Nov. 3			
0:12P	1565	201	18
13	1745	200	21
3	(92)		
1:20P	395	181	19
21	611	180	28
22	809	190	17
23	1007	196	17
24	1196	203	16
25	1385	208	18
26	1565	206	18
27	1745	206	19
28	1925	210	19
29	2105	209	20
30	2285	208	21
4	(93)		
1:15P	395	228	10
16	611	234	20
17	809	237	17
18	1007	232	21
19	1196	226	22
20	1385	227	18
21	1565	227	21
22	1745	225	21
23	1925	225	19
24	2105	231	18
25	2285	235	19
26	2465	234	21
4	(94)		
1:30P	395	236	10
31	611	230	14
32	809	227	13
33	1007	225	13
34	1196	227	15
35	1385	228	16
36	1565	230	15
37	1745	233	14
38	1925	237	13
39	2105	241	13
40	2285	246	13
41	2465	250	14
6	(95)		
9:00A	395	153	7
01	611	165	9
02	809	178	10
03	1007	185	10
04	1196	190	9

TABLE II (1927)

Date and time	Height in meters	Azimuth from; degrees	Velocity m/s	Date and time	Height in meters	Azimuth from; degrees	Velocity m/s	Date and time	Height in meters	Azimuth from; degrees	Velocity m/s
Nov.				**Nov.**				**Nov.**			
6	(96)			**7**	(99)			**8**			
9:14A	395	119	7	1:30P	395	36	1	2:23P	4625	214	18
15	611	162	7	31	611	112	4	24	4805	213	18
16	809	180	8	32	809	150	7	25	4985	211	17
17	1007	205	8	33	1007	179	7	26	5165	209	16
18	1196	227	12	34	1196	203	7				
19	1385	230	15	35	1385	214	7	**9**	(101)		
20	1565	236	18	36	1565	216	8	9:49A	395	89	2
21	1745	239	22	37	1745	229	9	50	611	125	6
22	1925	240	24	38	1925	244	11	51	809	158	9
23	2105	243	25	39	2105	250	12	52	1007	179	11
24	2285	244	26	40	2285	245	10	53	1196	191	11
25	2465	249	25	41	2465	242	10	54	1385	201	11
26	2645	254	24	42	2645	242	9	55	1565	202	12
				43	2825	245	10	56	1745	199	16
				44	3005	250	11	57	1925	200	21
6	(97)			45	3185	253	12	58	2105	202	22
1:15P	395	110	7	46	3365	251	12	59	2285	198	21
16	611	152	7	47	3545	246	13	10:00A	2465	190	21
17	809	171	12	48	3725	238	13	01	2645	185	23
18	1007	180	13	49	3905	233	14	02	2825	184	25
19	1196	192	13	50	4085	236	14				
20	1385	207	13	51	4265	242	14	**9**	(102)		
21	1565	227	14	52	4445	245	14	1:10P	395	128	4
22	1745	231	17					11	611	142	6
23	1925	234	19	**8**	(100)			12	809	177	6
24	2105	238	22	2:00P	395	72	5	13	1007	182	6
25	2285	238	23	01	611	139	4	14	1196	184	9
				02	809	191	6	15	1385	188	10
				03	1007	212	7	16	1565	192	11
6	(98)			04	1196	230	8	17	1745	197	13
2:10P	395	130	7	05	1385	238	9	18	1925	193	15
11	611	147	11	06	1565	237	8	19	2105	192	17
12	809	172	12	07	1745	231	7	20	2285	194	21
13	1007	188	13	08	1925	223	6	21	2465	190	23
14	1196	202	14	09	2105	219	6	22	2645	187	22
15	1385	213	14	10	2285	218	7				
16	1565	225	15	11	2465	217	8	**10**	(103)		
17	1745	227	18	12	2645	206	8	2:20P	395	207	5
18	1925	232	20	13	2825	193	7	21	611	172	13
19	2105	244	21	14	3005	198	7	22	809	174	24
20	2285	252	23	15	3185	204	9	23	1007	176	32
21	2465	252	24	16	3365	205	12	24	1196	177	26
22	2645	251	23	17	3545	204	14	25	1385	180	21
23	2825	255	23	18	3725	205	15	26	1565	180	29
24	3005	258	21	19	3905	207	16	27	1745	184	33
25	3185	260	17	20	4085	210	16	28	1925	203	27
26	3365	271	16	21	4265	212	16	29	2105	216	21
27	3545	278	15	22	4445	214	17	30	2285	221	17

TABLE II (1927)

Date and time	Height in meters	Azimuth from; degrees	Velocity m/s	Date and time	Height in meters	Azimuth from; degrees	Velocity m/s	Date and time	Height in meters	Azimuth from; degrees	Velocity m/s
Nov.				**Nov.**				**Nov.**			
10				**13**				**19**			
2:31P	2465	228	16	9:35A	1007	98	4	9:56A	3725	280	6
32	2645	227	16	36	1196	114	5	57	3905	282	6
33	2825	227	16	37	1385	111	7	58	4085	284	7
34	3005	228	16	38	1565	111	8	59	4265	284	9
35	3185	232	14	39	1745	107	9	10:00A	4445	277	10
36	3365	235	11	40	1925	107	10	01	4625	276	10
37	3545	215	8	41	2105	114	10	02	4805	277	11
38	3725	198	13	42	2285	116	9	03	4985	279	13
39	3905	201	18	43	2465	118	7	04	5165	279	11
				44	2645	120	6	05	5345	277	11
11	(104)			45	2825	125	5	06	5525	276	11
1:30P	395	82	6	46	3005	134	6	07	5705	278	12
31	611	73	6	47	3185	140	7	08	5885	280	11
32	809	60	5	48	3365	145	7	09	6065	282	11
33	1007	90	5	49	3545	152	7	10	6245	281	13
34	1196	114	6	50	3725	160	8	11	6425	280	13
35	1385	119	7	51	3905	168	10	12	6605	280	12
36	1565	119	7	52	4085	176	13	13	6785	280	13
37	1745	117	9	53	4265	182	14	14	6965	279	13
38	1925	116	10	54	4445	179	13	15	7145	283	11
39	2105	123	9	55	4625	177	13	16	7325	285	12
40	2285	137	8	56	4805	181	12	17	7505	282	14
41	2465	141	8	57	4985	188	13	18	7685	281	15
42	2645	132	9	58	5165	182	14	19	7865	282	16
43	2825	118	11	59	5345	178	14	20	8045	281	17
44	3005	116	11	10:00A	5525	173	16	21	8225	281	18
45	3185	124	11	01	5705	176	16	22	8405	281	17
46	3365	124	9	02	5885	180	15	23	8585	277	17
47	3545	117	5					24	8765	272	20
48	3725	136	4	**19**	(106)			25	8945	269	21
49	3905	150	5	9:38A	395	66	3	26	9125	268	20
50	4085	163	8	39	611	65	2	27	9305	269	19
51	4265	169	6	40	809	177	1	28	9485	269	18
52	4445	192	4	41	1007	183	1				
53	4625	199	7	42	1196	172	3	**20**	(107)		
54	4805	199	9	43	1385	144	4	0:10P	395	75	4
55	4985	204	9	44	1565	133	4	11	611	67	5
56	5165	206	11	45	1745	150	3	12	809	37	3
57	5345	217	15	46	1925	175	2	13	1007	46	3
58	5525	227	19	47	2105	199	2	14	1196	48	3
59	5705	230	21	48	2285	204	3	15	1385	34	7
2:00P	5885	227	22	49	2465	215	5	16	1565	31	9
01	6065	225	23	50	2645	225	6	17	1745	32	9
				51	2825	226	5	18	1925	32	10
13	(105)			52	3005	216	6	19	2105	36	9
9:32A	395	65	4	53	3185	217	7	20	2285	38	10
33	611	82	11	54	3365	237	6	21	2465	33	9
34	809	85	10	55	3545	267	6	22	2645	30	7

TABLE II (1927)

Date and time	Height in meters	Azimuth from; degrees	Velocity m/s	Date and time	Height in meters	Azimuth from; degrees	Velocity m/s	Date and time	Height in meters	Azimuth from; degrees	Velocity m/s
Nov.				**Nov.**				**Nov.**			
20				**21**				**21**			
0:23P	2825	29	5	1:02P	809	39	6	1:50P	9485	203	9
24	3005	33	5	03	1007	19	5	51	9665	208	9
25	3185	26	5	04	1196	23	7	52	9845	214	11
26	3365	359	6	05	1385	24	10	53	10025	216	11
27	3545	348	7	06	1565	21	10				
28	3725	355	7	07	1745	22	7	**22**	(109)		
29	3905	354	5	08	1925	31	4	1:10P	395	46	1
30	4085	358	7	09	2105	40	5	11	611	58	2
31	4265	358	8	10	2285	56	5	12	809	320	1
32	4445	349	11	11	2465	67	4	13	1007	242	1
33	4625	343	11	12	2645	67	2	14	1196	246	1
34	4805	339	11	13	2825	71	2	15	1385	311	1
35	4985	339	12	14	3005	52	2	16	1565	240	1
36	5165	339	11	15	3185	51	2	17	1745	232	2
37	5345	342	12	16	3365	5	1	18	1925	249	3
38	5525	345	13	17	3545	308	1	19	2105	234	2
39	5705	342	13	18	3725	259	3	20	2285	209	3
40	5885	337	13	19	3905	246	4	21	2465	211	5
41	6065	341	12	20	4085	238	3	22	2645	212	5
42	6245	341	10	21	4265	233	5	23	2825	211	4
43	6425	337	9	22	4445	240	5	24	3005	227	3
44	6605	324	9	23	4625	250	6	25	3185	231	5
45	6785	314	10	24	4805	247	5	26	3365	233	6
46	6965	313	10	25	4985	246	7	27	3545	232	5
47	7145	308	10	26	5165	259	7	28	3725	231	5
48	7325	305	10	27	5345	260	7	29	3905	233	5
49	7505	305	12	28	5525	247	7	30	4085	232	5
50	7685	289	11	29	5705	240	7	31	4265	231	5
51	7865	275	10	30	5885	245	7	32	4445	233	5
52	8045	282	11	31	6065	246	6	33	4625	226	5
53	8225	287	10	32	6245	239	5	34	4805	221	5
54	8405	287	11	33	6425	230	6	35	4985	225	5
55	8585	284	13	34	6605	217	6	36	5165	212	4
56	8765	283	13	35	6785	205	7	37	5345	197	4
57	8945	284	12	36	6965	204	6	38	5525	195	5
58	9125	287	13	37	7145	217	6	39	5705	201	4
59	9305	285	14	38	7325	226	6	40	5885	198	4
1:00P	9485	282	13	39	7505	218	6	41	6065	191	4
01	9665	280	14	40	7685	210	6	42	6245	185	5
02	9845	275	13	41	7865	209	6	43	6425	180	5
03	10025	263	12	42	8045	204	5	44	6605	184	6
04	10205	254	14	43	8225	201	6	45	6785	188	5
05	10385	253	16	44	8405	195	6	46	6965	183	5
06	10565	253	15	45	8585	196	6	47	7145	179	5
				46	8765	201	6	48	7345	176	5
21	(108)			47	8945	193	5	49	7505	175	4
1:00P	395	61	4	48	9125	197	6	50	7685	179	5
01	611	63	4	49	9305	199	9	51	7865	185	5

TABLE II (1927)

Date and time	Height in meters	Azimuth from; degrees	Velocity m/s
Nov.			
22			
1:52P	8045	190	5
53	8225	186	6
54	8405	178	6
25	(110)		
1:32P	395	265	2
33	611	327	3
34	809	5	5
35	1007	17	9
36	1196	15	12
37	1385	14	13
38	1565	17	14
39	1745	20	16
26	(111)		
10:20A	395	78	5
21	611	124	5
22	809	162	4
23	1007	173	3
24	1196	200	3
25	1385	234	4
26	1565	265	5
27	1745	285	6
28	1925	301	5
29	2105	302	4
30	2285	309	4
31	2465	323	5
32	2645	321	5
33	2825	318	6
34	3005	307	6
35	3185	292	6
36	3365	280	4
37	3545	235	4
38	3725	213	4
39	3905	193	5
40	4085	190	6
41	4265	189	7
42	4445	187	8
43	4625	183	10
44	4805	173	11
45	4985	174	12
46	5165	177	12
27	(112)		
1:30P	395	91	2
31	611	139	3
32	809	166	3
33	1007	149	3

Date and time	Height in meters	Azimuth from; degrees	Velocity m/s
Nov.			
27			
1:34P	1196	145	3
35	1385	158	5
36	1565	165	6
37	1745	177	6
38	1925	198	5
39	2105	204	6
40	2285	205	6
41	2465	217	6
42	2645	221	7
43	2825	229	9
44	3005	234	10
45	3185	230	11
46	3365	229	12
28	(113)		
1:18P	395	346	7
19	611	10	17
20	809	34	18
21	1007	48	15
22	1196	60	9
23	1385	46	7
24	1565	39	9
25	1745	51	10
26	1925	65	11
27	2105	69	11
28	2285	75	10
29	2465	83	8
30	2645	86	9
31	2825	94	8
32	3005	103	7
33	3185	147	6
34	3365	183	8
35	3545	175	8
36	3725	173	8
37	3905	181	7
38	4085	187	8
39	4265	185	9
40	4445	187	11
41	4625	189	12
42	4805	195	12
43	4985	197	12
44	5165	198	12
45	5345	196	14
46	5525	195	16
Dec.			
1	(114)		
1:05P	395	65	3

Date and time	Height in meters	Azimuth from; degrees	Velocity m/s
Dec.			
1			
1:06P	611	90	4
07	809	115	5
08	1007	124	7
09	1196	147	7
10	1385	160	9
11	1565	160	9
12	1745	165	9
13	1925	166	8
14	2105	161	7
15	2285	157	8
16	2465	149	8
17	2645	144	8
18	2825	138	8
19	3005	129	9
20	3185	124	9
21	3365	129	8
22	3545	122	9
23	3725	121	9
24	3905	119	7
25	4085	123	6
26	4265	131	5
27	4445	146	5
28	4625	156	5
29	4805	167	4
30	4985	359	5
31	5165	5	5
32	5345	21	7
33	5525	22	8
34	5705	25	9
35	5885	28	9
36	6065	24	10
37	6245	27	9
2	(115)		
11:56A	395	156	12
57	611	157	13
58	809	171	12
59	1007	171	15
12:00M	1196	169	23
0:01P	1385	174	25
02	1565	182	19
03	1745	185	17
2	(116)		
2:12P	395	122	9
13	611	130	12
14	809	138	11

TABLE II (1927)

Date and time	Height in meters	Azimuth from; degrees	Velocity m/s
Dec. 2			
2:15P	1007	155	9
16	1196	169	9
17	1385	177	10
18	1565	181	15
19	1745	186	18
20	1925	186	19
21	2105	184	19
22	2285	187	17
4	(117)		
0:52P	395	296	4
53	611	281	8
54	809	245	4
55	1007	220	4
56	1196	227	5
5	(118)		
2:10P	395	341	1
11	611	72	1
12	809	143	1
13	1007	79	1
14	1196	2	1
15	1385	333	1
16	1565	348	1
17	1745	355	1
18	1925	353	2
6	(119)		
1:45P	395	93	4
46	611	132	3
47	809	230	1
48	1007	351	5
49	1196	1	8
50	1385	5	9
7	(120)		
10:42A	395	84	1
43	611	56	3
44	809	47	3
45	1007	41	5
46	1196	43	6
47	1385	50	5
48	1565	58	4
49	1745	75	4
50	1925	88	4
51	2105	91	3
52	2285	94	2
53	2465	121	2
54	2645	151	4

Date and time	Height in meters	Azimuth from; degrees	Velocity m/s
Dec. 7			
10:55A	2825	161	5
56	3005	163	5
57	3185	163	8
58	3365	165	10
59	3545	168	10
11:00A	3725	166	10
01	3905	168	9
02	4085	174	9
7	(121)		
1:42P	395	65	2
43	611	56	3
44	809	39	5
45	1007	35	8
46	1196	38	9
47	1385	42	6
48	1565	53	3
49	1745	71	3
50	1925	81	4
51	2105	101	5
52	2285	105	5
53	2465	110	5
54	2645	114	5
55	2825	130	4
56	3005	147	6
57	3185	168	7
58	3365	178	8
59	3545	176	10
2:00P	3725	173	11
01	3905	168	12
02	4085	163	11
03	4265	161	10
04	4445	160	10
05	4625	162	10
06	4805	161	11
07	4985	170	11
08	5165	177	11
09	5345	175	13
10	5525	172	14
11	5705	170	13
12	5885	167	12
13	6065	166	11
14	6245	166	11
8	(122)		
1:28P	395	235	3
29	611	288	3
30	809	278	4

Date and time	Height in meters	Azimuth from; degrees	Velocity m/s
Dec. 8			
1:31P	1007	265	5
32	1196	240	3
33	1385	181	4
34	1565	175	5
35	1745	166	3
36	1925	202	4
37	2105	210	6
38	2285	210	8
39	2465	205	8
40	2645	196	8
41	2825	186	8
42	3005	187	9
43	3185	185	11
44	3365	177	12
45	3545	171	13
46	3725	173	14
47	3905	171	16
48	4085	172	16
49	4265	174	16
50	4445	169	17
51	4625	168	19
52	4805	168	20
53	4985	166	19
54	5165	165	20
9	(123)		
0:35P	395	69	3
36	611	83	6
37	801	99	5
38	1007	121	6
39	1196	137	7
40	1385	152	7
41	1565	157	8
42	1745	159	8
43	1925	165	8
44	2105	171	8
45	2285	175	8
46	2465	180	9
47	2645	179	9
48	2825	179	8
49	3005	174	8
50	3185	173	5
51	3365	210	2
52	3545	246	2
53	3725	285	4
54	3905	279	4
55	4085	286	4
56	4265	295	4

TABLE II (1927)

Date and time	Height in meters	Azimuth from; degrees	Velocity m/s	Date and time	Height in meters	Azimuth from; degrees	Velocity m/s	Date and time	Height in meters	Azimuth from; degrees	Velocity m/s
Dec.				**Dec.**				**Dec.**			
9				**10**				**11**			
0:57P	4445	293	3	2:18P	1925	182	7	2:03P	1925	170	20
58	4625	304	3	19	2105	182	9	04	2105	173	20
59	4805	323	4	20	2285	198	10	05	2285	179	19
1:00P	4985	344	9	21	2465	193	13				
01	5165	342	12	22	2645	192	15	**12**	(129)		
02	5345	333	11					11:05A	395	126	8
03	5525	331	10	**10**	(126)			06	611	130	13
04	5705	333	10	2:45P	395	110	11	07	809	140	10
05	5885	329	11	46	611	139	11	08	1007	150	7
06	6065	319	14	47	809	154	14	09	1196	184	6
07	6245	319	14	48	1007	157	18	10	1385	190	7
08	6425	320	13	49	1196	157	21	11	1565	189	8
09	6605	310	14	50	1385	156	17	12	1745	175	10
10	6785	307	15	51	1565	156	12	13	1925	173	11
11	6965	304	16	52	1745	160	11	14	2105	179	12
12	7145	301	17	53	1925	168	13	15	2285	186	15
13	7325	301	15	54	2105	176	17	16	2465	189	18
14	7505	300	19	55	2285	183	17	17	2645	186	19
15	7685	300	17	56	2465	189	16	18	2825	181	17
16	7865	299	17					19	3005	175	17
17	8045	300	17	**11**	(127)			20	3185	167	16
18	8225	297	18	11:16A	395	128	9	21	3365	158	16
19	8405	300	17	17	611	147	13	22	3545	152	17
20	8585	301	18	18	809	152	16	23	3725	149	19
21	8765	299	18	19	1007	155	17	24	3905	149	19
22	8945	299	19	20	1196	154	17	25	4085	149	19
				21	1385	155	19				
10	(124)			22	1565	161	20	**12**	(130)		
0:56P	395	133	8	23	1745	169	21	1:58P	395	107	8
57	611	151	14	24	1925	174	24	59	611	126	11
58	809	165	13	25	2105	175	25	2:00P	809	144	9
59	1007	178	11	26	2285	178	26	01	1007	163	7
1:00P	1196	183	12	27	2465	183	29	02	1196	185	5
01	1385	187	11	28	2645	185	29	03	1385	193	4
02	1565	194	13	29	2825	183	28	04	1565	177	4
03	1745	193	16	30	3005	182	29	05	1745	163	4
04	1925	193	18	31	3185	175	29	06	1925	149	4
05	2105	194	18	32	3365	173	30	07	2105	139	5
								08	2285	140	5
10	(125)			**11**	(128)			09	2465	151	6
2:10P	395	138	8	1:55P	395	149	10	10	2645	168	8
11	611	138	19	56	611	151	19	11	2825	175	12
12	809	145	20	57	809	152	22	12	3005	169	15
13	1007	153	25	58	1007	152	20	13	3185	161	16
14	1196	160	25	59	1196	153	17	14	3365	152	16
15	1385	163	17	2:00P	1385	158	17	15	3545	146	15
16	1565	169	7	01	1565	164	20	16	3725	149	13
17	1745	179	5	02	1745	169	19	17	3905	162	14

TABLE II (1927)

Date and time	Height in meters	Azimuth from; degrees	Velocity m/s	Date and time	Height in meters	Azimuth from; degrees	Velocity m/s	Date and time	Height in meters	Azimuth from; degrees	Velocity m/s
Dec. 12				**Dec. 14**				**Dec. 15**			
2:18P	4085	171	16	11:30A	6245	172	18	10:22A	809	152	11
19	4265	174	19	31	6425	174	17	23	1007	162	9
20	4445	177	22	32	6605	177	17	24	1196	168	10
13	(131)			33	6785	178	16	25	1385	166	16
1:49P	395	112	13	34	6965	179	16	26	1565	161	17
50	611	132	10	35	7145	181	15	27	1745	145	12
51	809	136	15	36	7325	185	13	28	1925	132	10
52	1007	150	14	37	7505	195	15	29	2105	127	9
53	1196	169	11	38	7685	200	16	30	2285	127	11
54	1385	165	13	39	7865	198	17	31	2465	132	12
55	1565	164	20	40	8045	195	18	32	2645	139	12
56	1745	171	23	**14**	(133)			33	2825	134	11
57	1925	177	22	2:00P	395	94	3	34	3005	131	11
58	2105	179	21	01	611	105	4	**15**	(135)		
14	(132)			02	809	48	1	2:00P	395	128	9
10:58A	395	87	5	03	1007	152	1	01	611	141	13
59	611	115	5	04	1196	133	2	02	809	155	11
11:00A	809	134	2	05	1385	158	3	03	1007	162	14
01	1007	103	1	06	1565	151	3	04	1196	163	19
02	1196	117	1	07	1745	115	3	05	1385	167	24
03	1385	133	1	08	1925	117	4	06	1565	172	26
04	1565	126	1	09	2105	132	5	07	1745	177	22
05	1745	121	1	10	2285	137	4	08	1925	180	18
06	1925	306	1	11	2465	141	4	09	2105	180	13
07	2105	242	1	12	2645	142	5	10	2285	166	7
08	2285	131	2	13	2825	126	5	11	2465	126	7
09	2465	169	4	14	3005	118	3	12	2645	124	9
10	2645	183	5	15	3185	145	3	**16**	(136)		
11	2825	170	4	16	3365	166	5	0:48P	395	147	11
12	3005	171	3	17	3545	167	7	49	611	154	13
13	3185	209	3	18	3725	162	7	50	809	160	15
14	3365	221	6	19	3905	162	8	51	1007	170	21
15	3545	212	7	20	4085	159	9	52	1196	174	26
16	3725	198	7	21	4265	162	8	53	1385	177	22
17	3905	178	8	22	4445	167	9	54	1565	185	14
18	4085	164	9	23	4625	172	10	55	1745	202	8
19	4265	162	9	24	4805	180	10	56	1925	206	6
20	4445	165	9	25	4985	179	11	57	2105	191	11
21	4625	169	10	26	5165	174	12	58	2285	188	14
22	4805	170	12	27	5345	173	13	59	2465	187	15
23	4985	172	13	28	5525	176	13	1:00P	2645	186	17
24	5165	176	15	29	5705	177	14	01	2825	185	17
25	5345	175	15	30	5885	179	15	02	3005	185	15
26	5525	173	17	31	6065	182	15	03	3185	189	16
27	5705	170	17	**15**	(134)			04	3365	195	21
28	5885	170	17	10:20A	395	117	12	05	3545	198	23
29	6065	171	19	21	611	137	11				

TABLE II (1927)

Date and time	Height in meters	Azimuth from; degrees	Velocity m/s
Dec.			
17	(137)		
10:40A	395	86	4
41	611	105	7
42	809	118	7
43	1007	127	4
44	1196	84	3
45	1385	66	3
46	1565	67	3
47	1745	62	1
48	1925	155	2
49	2105	140	6
50	2285	139	8
17	(138)		
1:00P	395	97	5
01	611	115	6
02	809	151	6
03	1007	165	5
04	1196	149	3
05	1385	109	4
06	1565	88	5
07	1745	92	4
08	1925	128	4
09	2105	147	4
10	2285	162	3
11	2465	167	3
12	2645	187	2
13	2825	191	3
14	3005	177	3
15	3185	161	2
16	3365	111	2
17	3545	91	3
18	3725	83	6
19	3905	89	6
20	4085	104	4
21	4265	130	4
22	4445	141	5
23	4625	143	7
24	4805	143	7
25	4985	141	6
26	5165	148	6
27	5345	156	6
28	5525	164	6
29	5705	175	6
18	(139)		
10:48A	395	1	2
49	611	75	2
50	809	94	4
51	1007	94	5

Date and time	Height in meters	Azimuth from; degrees	Velocity m/s
Dec.			
18			
10:52A	1196	80	5
53	1385	91	4
54	1565	88	3
55	1745	79	3
56	1925	79	3
57	2105	78	3
58	2285	50	3
59	2465	10	3
11:00A	2645	346	1
01	2825	346	1
02	3005	306	1
03	3185	252	2
04	3365	238	3
05	3545	245	4
06	3725	252	5
07	3905	252	6
08	4085	250	6
09	4265	243	7
10	4445	234	8
11	4625	233	8
12	4805	231	8
13	4985	228	9
14	5165	237	9
15	5345	254	9
16	5525	256	11
17	5705	251	11
18	5885	241	14
19	6065	239	16
20	6245	239	19
21	6425	242	20
22	6605	247	19
23	6785	249	20
24	6965	249	23
25	7145	248	27
26	7325	247	25
27	7505	253	23
28	7685	253	26
29	7865	252	26
18	(140)		
1:40P	395	34	2
41	611	65	5
42	809	79	3
43	1007	108	3
44	1196	128	3
45	1385	155	2
46	1565	169	2
47	1745	173	2

Date and time	Height in meters	Azimuth from; degrees	Velocity m/s
Dec.			
18			
1:48P	1925	131	1
49	2105	64	2
50	2285	66	3
51	2465	66	3
52	2645	59	2
53	2825	51	2
54	3005	20	1
55	3185	247	1
56	3365	221	2
57	3545	220	2
58	3725	233	3
59	3905	246	3
2:00P	4085	254	4
01	4265	233	4
02	4445	229	4
03	4625	233	5
04	4805	230	5
05	4985	225	6
19	(141)		
10:48A	395	46	3
49	611	41	10
50	809	35	6
51	1007	85	1
52	1196	101	2
53	1385	71	5
54	1565	56	8
55	1745	54	10
56	1925	51	11
57	2105	47	10
58	2285	48	9
59	2465	58	8
11:00A	2645	65	7
01	2825	57	6
02	3005	45	5
03	3185	47	3
04	3365	52	2
05	3545	51	2
06	3725	43	3
07	3905	38	3
08	4085	38	3
09	4265	31	2
10	4445	352	2
11	4625	319	4
12	4805	310	6
13	4985	301	6
14	5165	293	5
15	5345	288	4

TABLE II (1927)

Date and time	Height in meters	Azimuth from; degrees	Velocity m/s	Date and time	Height in meters	Azimuth from; degrees	Velocity m/s	Date and time	Height in meters	Azimuth from; degrees	Velocity m/s
Dec.				**Dec.**				**Dec.**			
19				**19**				**20**			
11:16A	5525	279	4	2:03P	1925	74	6	11:18A	3185	325	3
17	5705	284	5	04	2105	71	5	19	3365	297	2
18	5885	284	5	05	2285	71	4	20	3545	251	2
19	6065	281	6	06	2465	72	4	21	3725	232	2
20	6245	282	6	07	2645	59	5	22	3905	232	2
21	6425	288	6	08	2825	49	4	23	4085	239	2
22	6605	290	6	09	3005	49	2	24	4265	235	2
23	6785	286	7	10	3185	61	2	25	4445	243	2
24	6965	290	7	11	3365	60	3	26	4625	257	3
25	7145	291	7	12	3545	34	3	27	4805	255	4
26	7325	293	7	13	3725	14	4	28	4985	247	5
27	7505	297	8	14	3905	7	4	29	5165	238	5
28	7685	299	9	15	4085	353	4	30	5345	234	5
29	7865	299	8	16	4265	335	4	31	5525	230	5
30	8045	294	8	17	4445	326	4	32	5705	220	6
31	8225	291	9	18	4625	326	3	33	5885	209	6
32	8405	291	9	19	4805	327	3	34	6065	220	7
33	8585	287	8	20	4985	329	3	35	6245	226	9
34	8765	292	9	21	5165	330	3	36	6425	220	9
35	8945	282	11	22	5345	323	3	37	6605	217	10
36	9125	274	11	23	5525	323	4	38	6785	214	10
37	9305	269	13	24	5705	318	4	39	6965	213	11
38	9485	274	13	25	5885	308	5	40	7145	213	11
39	9665	276	13	26	6065	308	5	41	7325	219	11
40	9845	279	13	27	6245	315	5	42	7505	226	11
41	10025	280	12	28	6425	318	6	43	7685	230	11
42	10205	278	12	29	6605	307	7	44	7865	234	11
43	10385	274	12	30	6785	297	7	45	8045	239	11
44	10565	270	13	31	6965	297	7	46	8225	240	11
45	10745	267	13	32	7145	295	7	47	8405	242	10
46	10925	262	13	33	7325	294	7	48	8585	240	10
47	11105	262	14					49	8765	240	9
48	11285	264	15	**20**	(143)			50	8945	245	9
49	11465	264	15	11:03A	395	Calm	Calm	51	9125	242	11
50	11645	260	15	04	611	161	1	52	9305	233	12
51	11825	258	17	05	809	182	1	53	9485	236	13
52	12005	261	17	06	1007	329	2	54	9665	237	13
53	12185	267	16	07	1196	319	6	55	9845	238	10
				08	1385	300	7	56	10025	237	8
19	(142)			09	1565	297	6				
1:55P	395	38	1	10	1745	318	5	**20**	(144)		
56	611	67	3	11	1925	347	2	2:00P	395	160	1
57	809	58	4	12	2105	20	2	01	611	195	2
58	1007	45	3	13	2285	26	4	02	809	242	3
59	1196	43	4	14	2465	18	5	03	1007	270	4
2:00P	1385	43	6	15	2645	7	3	04	1196	310	4
01	1565	56	6	16	2825	359	2	05	1385	331	5
02	1745	72	6	17	3005	338	2	06	1565	322	6

TABLE II (1927)

Date and time	Height in meters	Azimuth from; degrees	Velocity m/s
Dec. 20			
2:07P	1745	308	5
08	1925	299	3
09	2105	294	3
10	2285	309	2
11	2465	330	3
12	2645	333	4
13	2825	328	4
14	3005	309	4
15	3185	306	3
16	3365	311	3
17	3545	292	3
18	3725	278	4
19	3905	278	4
20	4085	269	4
21	4265	259	4
22	4445	252	3
23	4625	242	4
24	4805	239	5
25	4985	238	7
26	5165	234	7
27	5345	237	7
28	5525	239	9
29	5705	238	10
21	(145)		
10:30A	395	73	2
31	611	85	3
32	809	78	2
33	1007	70	2
34	1196	346	1
35	1385	302	2
36	1565	304	2
37	1745	281	2
38	1925	248	2
39	2105	241	3
40	2285	249	4
41	2465	237	7
42	2645	233	9
43	2825	251	8
44	3005	259	9
45	3185	258	10
21	(146)		
1:50P	395	45	1
51	611	38	4
52	809	34	4
53	1007	29	3
54	1196	14	2

Date and time	Height in meters	Azimuth from; degrees	Velocity m/s
Dec. 21			
1:55P	1385	282	2
56	1565	266	3
57	1745	264	4
58	1925	266	5
59	2105	275	5
2:00P	2285	285	6
01	2465	282	8
02	2645	278	9
24	(147)		
11:42A	395	168	5
43	611	176	10
44	809	184	12
45	1007	191	12
46	1196	199	8
47	1385	214	6
48	1565	205	9
49	1745	210	13
50	1925	202	17
51	2105	206	20
52	2285	204	23
53	2465	206	23
54	2645	208	22
24	(148)		
1:40P	395	118	3
41	611	187	10
42	809	193	17
43	1007	195	18
26	(149)		
1:43P	395	110	2
44	611	161	3
45	809	186	3
46	1007	183	2
47	1196	189	1
48	1385	198	1
49	1565	205	2
50	1745	213	3
51	1925	223	4
52	2105	228	4
53	2285	238	4
54	2465	250	5
55	2645	250	5
56	2825	244	5
57	3005	237	7
58	3185	235	8
59	3365	237	8

Date and time	Height in meters	Azimuth from; degrees	Velocity m/s
Dec. 26			
2:00P	3545	242	8
01	3725	242	11
02	3905	245	12
03	4085	245	13
04	4265	244	13
05	4445	250	14
06	4625	256	16
07	4805	257	17
08	4985	256	18
09	5165	257	18
27	(150)		
1:40P	395	35	1
41	611	24	3
42	809	17	5
43	1007	20	9
44	1196	23	11
45	1385	24	8
46	1565	32	6
47	1745	36	4
48	1925	57	2
49	2105	115	2
50	2285	161	3
51	2465	182	4
52	2645	187	6
53	2825	179	7
54	3005	174	7
55	3185	185	9
56	3365	188	9
57	3545	191	10
58	3725	192	13
59	3905	193	15
2:00P	4085	193	16
01	4265	192	18
02	4445	190	18
03	4625	188	18
04	4805	188	18
05	4985	190	20
06	5165	190	22
07	5345	191	24
08	5525	192	26
09	5705	191	27
28	(151)		
2:00P	395	51	5
01	611	48	12
02	809	46	19
03	1007	49	19

TABLE II (1927–1928)

Date and time	Height in meters	Azimuth from; degrees	Velocity m/s	Date and time	Height in meters	Azimuth from; degrees	Velocity m/s	Date and time	Height in meters	Azimuth from; degrees	Velocity m/s
Dec.				**Jan.**				**Jan.**			
28				**1**				**3**			
2:04P	1196	55	12	1:33P	1007	232	9	1:32P	809	260	3
05	1385	58	11	34	1196	226	8	33	1007	286	5
06	1565	69	14	35	1385	220	8	34	1196	299	6
07	1745	73	18	36	1565	220	9	35	1385	288	6
08	1925	65	18	37	1745	216	11	36	1565	267	8
09	2105	61	15	38	1925	218	14	37	1745	264	9
10	2285	74	13	39	2105	222	17	38	1925	261	9
11	2465	90	13	40	2285	225	19	39	2105	265	8
12	2645	97	12	41	2465	229	21	40	2285	270	7
13	2825	97	8	42	2645	232	19	41	2465	259	6
14	3005	88	6	43	2825	231	17	42	2645	244	5
15	3185	72	8	44	3005	229	16	43	2825	240	6
16	3365	68	11	45	3185	226	16	44	3005	229	5
17	3545	66	12	46	3365	222	16	45	3185	217	5
				47	3545	216	17	46	3365	207	6
31	(152)			48	3725	211	18	47	3545	209	7
11:40A	395	53	2	49	3905	216	18	48	3725	217	7
41	611	44	4					49	3905	220	8
42	809	29	6	**2**	(154)			50	4085	217	9
43	1007	30	10	1:40P	395	59	4				
44	1196	31	12	41	611	68	11	**4**	(157)		
45	1385	34	7	42	809	71	9	1:57P	395	279	13
46	1565	6	2	43	1007	76	7	58	611	274	25
47	1745	256	2	44	1196	94	9	59	809	173	5
48	1925	216	7	45	1385	111	13	2:00P	1007	170	5
49	2105	204	10	46	1565	116	15	01	1196	177	7
50	2285	201	9	47	1745	119	18	02	1385	199	9
51	2465	194	7	48	1925	125	18	03	1565	202	12
52	2645	187	7	49	2105	134	16	04	1745	202	14
53	2825	198	7	50	2285	135	16	05	1925	204	16
54	3005	215	9	51	2465	131	15	06	2105	211	19
55	3185	218	10	52	2645	136	14	07	2285	206	23
56	3365	226	14	53	2825	143	14	08	2465	206	27
57	3545	226	17	54	3005	143	17	09	2645	204	26
58	3725	220	19	55	3185	144	17	10	2825	202	25
59	3905	217	22	56	3365	144	17	11	3005	201	26
12:00M	4085	211	24	57	3545	145	18				
0:01P	4265	208	26	58	3725	143	18	**5**	(158)		
02	4445	207	27					1:58P	395	135	9
03	4625	206	27	**2**	(155)			59	611	149	13
04	4805	201	27	9:57P	395	89	5	2:00P	809	167	12
05	4985	198	28	58	611	143	9	01	1007	174	12
				59	809	159	12	02	1196	179	12
Jan.				10:00P	1007	167	13	03	1385	182	11
1, 1928	(153)							04	1565	184	11
1:30P	395	95	4	**3**	(156)			05	1745	192	10
31	611	185	7	1:30P	395	130	1	06	1925	199	12
32	809	220	8	31	611	250	1	07	2105	197	15

TABLE II (1928)

Date and time	Height in meters	Azimuth from; degrees	Velocity m/s
Jan. 5			
2:08P	2285	202	17
09	2465	200	19
10	2645	198	21
11	2825	200	21
12	3005	203	19
13	3185	205	19
14	3365	205	18
6	(159)		
1:55P	395	161	19
56	611	167	19
57	809	171	19
58	1007	173	16
59	1196	180	14
2:00P	1385	188	11
01	1565	195	10
02	1745	194	11
03	1925	188	12
04	2105	183	12
05	2285	183	12
06	2465	183	12
07	2645	183	13
7	(160)		
1:42P	395	28	1
43	611	39	7
44	809	45	9
45	1007	65	8
46	1196	88	8
47	1385	98	7
48	1565	86	7
49	1745	74	9
50	1925	74	11
51	2105	75	12
52	2285	72	13
53	2465	70	14
54	2645	70	13
55	2825	71	13
56	3005	76	14
57	3185	83	15
58	3365	90	15
59	3545	91	15
8	(161)		
1:30P	395	77	2
31	611	356	6
32	809	1	10
33	1007	7	13

Date and time	Height in meters	Azimuth from; degrees	Velocity m/s
Jan. 8			
1:34P	1196	4	14
35	1385	1	13
36	1565	11	13
37	1745	24	11
38	1925	33	9
39	2105	35	9
40	2285	39	7
41	2465	52	5
42	2645	56	4
43	2825	58	4
44	3005	99	4
45	3185	128	6
46	3365	132	6
47	3545	142	5
48	3725	156	4
49	3905	153	5
50	4085	137	5
51	4265	133	6
9	(162)		
1:22P	395	187	2
23	611	274	3
24	809	296	6
25	1007	308	6
26	1196	334	8
27	1385	349	10
28	1565	356	12
29	1745	1	10
30	1925	345	7
31	2105	328	8
32	2285	322	7
33	2465	331	5
34	2645	338	3
35	2825	243	2
36	3005	217	4
37	3185	206	5
38	3365	204	7
39	3545	198	10
40	3725	192	11
41	3905	187	11
10	(163)		
2:00P	395	108	1
01	611	103	2
02	809	19	1
03	1007	44	3
04	1196	32	7
05	1385	24	9

Date and time	Height in meters	Azimuth from; degrees	Velocity m/s
Jan. 10			
2:06P	1565	18	10
07	1745	42	4
08	1925	61	7
09	2105	57	9
10	2285	63	8
11	2465	69	8
12	2645	73	8
13	2825	72	7
14	3005	68	8
15	3185	66	9
16	3365	63	9
17	3545	57	11
18	3725	53	15
19	3905	49	19
20	4085	46	20
21	4265	44	18
22	4445	47	16
23	4625	50	15
24	4805	52	14
25	4985	54	14
26	5165	68	15
27	5345	80	16
28	5525	79	10
29	5705	85	8
30	5885	77	8
31	6065	75	4
32	6245	154	2
33	6425	167	3
34	6605	169	3
35	6785	127	3
36	6965	123	3
37	7145	127	4
38	7325	115	5
39	7505	85	4
40	7685	89	7
41	7865	117	8
42	8045	125	11
43	8225	125	13
44	8405	132	14
45	8585	130	16
46	8765	122	15
47	8945	123	14
48	9125	126	15
49	9305	132	17
50	9485	137	19
51	9665	140	21
52	9845	146	22
53	10025	154	22

TABLE II (1928)

Date and time	Height in meters	Azimuth from; degrees	Velocity m/s	Date and time	Height in meters	Azimuth from; degrees	Velocity m/s	Date and time	Height in meters	Azimuth from; degrees	Velocity m/s
Jan. 10				**Jan. 12**				**Jan. 13**			
2:54P	10205	154	21	1:47P	3005	256	6	2:28P	5525	213	12
55	10385	151	21	48	3185	257	5	29	5705	210	13
				49	3365	259	6	30	5885	213	12
11	(164)			50	3545	259	7	31	6065	213	12
2:00P	395	Calm	Calm	51	3725	264	7	32	6245	220	11
01	611	333	1	52	3905	267	8	33	6425	209	13
02	809	6	4	53	4085	270	7	34	6605	211	15
03	1007	18	6	54	4265	262	7	35	6785	211	16
04	1196	18	11	55	4445	258	8	36	6965	209	17
05	1385	19	11	56	4625	257	8	37	7145	205	17
06	1565	28	8	57	4805	256	9	38	7325	203	17
07	1745	20	6	58	4985	251	9				
08	1925	11	4	59	5165	244	9				
09	2105	32	3	2:00P	5345	234	9	**14**	(167)		
10	2285	147	1	01	5525	224	9	1:38P	395	78	6
11	2465	190	5	02	5705	225	12	39	611	118	7
12	2645	192	8	03	5885	216	12	40	809	160	7
13	2825	197	11	04	6065	205	10	41	1007	178	7
14	3005	199	15					42	1196	162	8
15	3185	194	17	**13**	(166)			43	1385	158	9
16	3365	190	15	2:00P	395	75	3	44	1565	174	10
17	3545	186	18	01	611	118	2	45	1745	186	10
18	3725	184	21	02	809	272	2	46	1925	185	10
19	3905	184	19	03	1007	302	3	47	2105	183	10
20	4085	186	20	04	1196	0	1	48	2285	188	10
21	4265	190	19	05	1385	155	2	49	2465	191	10
22	4445	194	19	06	1565	273	1	50	2645	196	12
23	4625	197	19	07	1745	308	3	51	2825	199	11
24	4805	199	21	08	1925	315	4	52	3005	200	10
25	4985	199	25	09	2105	280	2	53	3185	206	9
26	5165	199	23	10	2285	233	2	54	3365	211	7
27	5345	199	18	11	2465	194	3	55	3545	213	5
				12	2645	199	3	56	3725	213	5
12	(165)			13	2825	216	3	57	3905	209	5
1:33P	395	80	1	14	3005	236	4	58	4085	207	6
34	611	83	3	15	3185	241	5	59	4265	203	6
35	809	167	1	16	3365	241	4	2:00P	4445	203	7
36	1007	120	3	17	3545	246	5	01	4625	203	7
37	1196	109	4	18	3725	245	5	02	4805	196	6
38	1385	138	4	19	3905	242	6	03	4985	188	6
39	1565	156	1	20	4085	237	6	04	5165	194	8
40	1745	297	1	21	4265	244	6	05	5345	202	10
41	1925	182	1	22	4445	256	6	06	5525	204	12
42	2105	246	1	23	4625	262	7	07	5705	204	12
43	2285	250	2	24	4805	255	8	08	5885	204	14
44	2465	244	3	25	4985	242	7	09	6065	206	16
45	2645	269	4	26	5165	245	7	10	6245	209	14
46	2825	268	6	27	5345	234	9	11	6425	211	11

TABLE II (1928)

Date and time	Height in meters	Azimuth from; degrees	Velocity m/s	Date and time	Height in meters	Azimuth from; degrees	Velocity m/s	Date and time	Height in meters	Azimuth from; degrees	Velocity m/s
Jan.				**Jan.**				**Jan.**			
15	(168)			**21**				**25**			
1:48P	395	150	26	1:40P	3725	217	9	2:24P	3005	226	7
49	611	152	15	41	3905	212	9	25	3185	219	8
50	809	155	21	42	4085	213	9	26	3365	213	7
51	1007	157	22	43	4265	217	8	27	3545	212	7
52	1196	163	21	44	4445	225	9	28	3725	208	7
53	1385	163	25	45	4625	226	9				
				46	4805	222	9	**26**	(174)		
16	(169)			47	4985	221	9	1:10P	395	1	3
0:38P	395	163	13	48	5165	223	8	11	611	26	6
39	611	176	13	49	5345	226	7	12	809	22	8
40	809	185	13	50	5525	235	7	13	1007	14	9
41	1007	184	11	51	5705	237	8	14	1196	14	10
42	1196	185	8	52	5885	230	11	15	1385	18	11
43	1385	191	10	53	6065	225	14	16	1565	42	12
44	1565	188	17					17	1745	59	10
45	1745	182	21	**23**	(171)			18	1925	77	8
46	1925	180	21	1:11P	395	212	4	19	2105	84	6
47	2105	173	21	12	611	291	2	20	2285	96	7
48	2285	168	20	13	809	348	3	21	2465	105	9
49	2465	164	19	14	1007	359	3	22	2645	110	9
50	2645	159	16	15	1196	346	2	23	2825	109	9
51	2825	159	14	16	1385	320	3	24	3005	110	10
52	3005	157	13	17	1565	341	2	25	3185	117	9
53	3185	146	12	18	1745	87	3	26	3365	126	7
54	3365	161	13	19	1925	89	7	27	3545	129	7
55	3545	164	13	20	2105	89	7				
56	3725	158	13					**26**	(175)		
57	3905	155	13	**24**	(172)			2:05P	395	43	3
				1:28P	395	83	2	06	611	32	5
21	(170)			29	611	182	1	07	809	14	8
1:22P	395	70	4	30	809	218	2	09	1007	10	10
23	611	83	7	31	1007	213	3	09	1196	14	11
24	809	83	6					10	1385	23	11
25	1007	77	5	**25**	(173)			11	1565	31	12
26	1196	120	4	2:10P	395	76	1	12	1745	47	11
27	1385	145	6	11	611	150	1	13	1925	66	8
28	1565	152	7	12	809	215	2	14	2105	83	6
29	1745	161	5	13	1007	213	4	15	2285	90	7
30	1925	148	4	14	1196	194	4	16	2465	102	7
31	2105	138	6	15	1385	173	3	17	2645	116	7
32	2285	141	6	16	1565	164	4	18	2825	122	7
33	2465	150	4	17	1745	165	5	19	3005	119	8
34	2645	169	3	18	1925	174	5	20	3185	117	8
35	2825	183	3	19	2105	186	5	21	3365	122	7
36	3005	201	3	20	2285	207	5	22	3545	125	7
37	3185	210	5	21	2465	223	5	23	3725	124	6
38	3365	219	6	22	2645	229	5	24	3905	125	5
39	3545	222	8	23	2825	222	6	25	4085	118	5

TABLE II (1928)

Date and time	Height in meters	Azimuth from; degrees	Velocity m/s	Date and time	Height in meters	Azimuth from; degrees	Velocity m/s	Date and time	Height in meters	Azimuth from; degrees	Velocity m/s
Jan.				**Jan.**				**Jan.**			
26				**27**	(177)			**29**			
2:26P	4265	125	4	3:12P	395	121	11	1:45P	2825	247	10
27	4445	143	5	13	611	130	9	46	3005	238	8
28	4625	165	6	14	809	141	12	47	3185	226	7
29	4805	182	7	15	1007	149	13	48	3365	219	9
30	4985	202	7	16	1196	156	11	49	3545	217	9
31	5165	220	10	17	1385	162	10	50	3725	221	9
32	5345	226	13	18	1565	167	9	51	3905	221	9
33	5525	228	16	19	1745	166	9	52	4085	231	8
34	5705	225	17	20	1925	161	7	53	4265	235	7
35	5885	227	20	21	2105	143	6	54	4445	246	8
36	6065	234	24	22	2285	116	5	55	4625	248	8
37	6245	233	27	23	2465	78	3	56	4805	250	8
38	6425	229	29	24	2645	56	3	57	4985	248	9
39	6605	232	32	25	2825	56	4	58	5165	242	9
40	6785	232	35	26	3005	65	4	59	5345	245	11
41	6965	228	35	27	3185	77	5	2:00P	5525	247	13
42	7145	232	34	28	3365	84	7	01	5705	246	13
43	7325	236	35	29	3545	90	9				
44	7505	234	35	30	3725	96	10	**29**	(180)		
45	7685	231	34	31	3905	103	12	3:12P	395	173	3
46	7865	230	34	32	4085	111	13	13	611	183	6
47	8045	231	34	33	4265	119	13	14	809	193	7
48	8225	231	34	34	4445	123	13	15	1007	207	6
49	8405	228	33	35	4625	118	13	16	1196	222	5
50	8585	227	34	36	4805	115	12	17	1385	233	5
51	8765	229	34	37	4985	116	11	18	1565	231	5
								19	1745	227	6
27	(176)			**28**	(178)			20	1925	225	7
1:25P	395	105	8	1:20P	395	267	4	21	2105	222	7
26	611	122	10	21	611	244	7	22	2285	223	8
27	809	132	14	22	809	226	6	23	2465	222	8
28	1007	142	14	23	1007	232	5	24	2645	231	8
29	1196	152	11	24	1196	238	5	25	2825	238	8
30	1385	154	12					26	3005	233	7
31	1565	147	13	**29**	(179)			27	3185	216	8
32	1745	144	13	1:32P	395	205	6	28	3365	232	7
33	1925	149	13	33	611	193	9	29	3545	247	8
34	2105	156	11	34	809	192	10	30	3725	243	8
35	2285	148	11	35	1007	199	10	31	3905	240	7
36	2465	138	11	36	1196	206	9	32	4085	246	6
37	2645	135	8	37	1385	211	7	33	4265	246	6
38	2825	132	4	38	1565	216	6	34	4445	242	7
39	3005	79	2	39	1745	227	6	35	4625	239	9
40	3185	50	2	40	1925	235	5	36	4805	237	9
41	3365	106	3	41	2105	238	5	37	4985	238	11
42	3545	121	5	42	2285	245	4	38	5165	239	12
43	3725	123	11	43	2465	247	5	39	5345	237	13
44	3905	124	11	44	2645	245	9	40	5525	236	14

TABLE II (1928)

Date and time	Height in meters	Azimuth from; degrees	Velocity m/s	Date and time	Height in meters	Azimuth from; degrees	Velocity m/s	Date and time	Height in meters	Azimuth from; degrees	Velocity m/s
Jan. 29				**Jan. 30**				**Jan. 31**			
3:41P	5705	234	14	2:19P	2105	242	5	2:13P	5705	198	8
42	5885	230	15	20	2285	261	3	14	5885	202	8
43	6065	228	16	21	2465	291	2	15	6065	202	7
44	6245	228	18	22	2645	326	1	16	6245	201	7
45	6425	224	19	23	2825	251	1	17	6425	199	8
				24	3005	253	1	18	6605	198	9
30	(181)			25	3185	247	1				
1:28P	395	79	4	26	3365	253	2				
29	611	93	3	27	3545	283	1	**Feb.**			
30	809	126	1	28	3725	96	1	**1**	(184)		
31	1007	124	2	29	3905	127	2	2:00P	395	73	4
32	1196	148	2	30	4085	137	2	01	611	73	4
33	1385	170	2	31	4265	123	4	02	809	86	2
34	1565	210	3	32	4445	116	5	03	1007	72	2
35	1745	225	5	33	4625	125	7	04	1196	46	2
36	1925	224	5	34	4805	140	9	05	1385	33	3
37	2105	227	5	35	4985	143	9	06	1565	37	3
38	2285	238	5					07	1745	43	4
39	2465	272	3	**31**	(183)			08	1925	43	5
40	2645	321	2	1:44P	395	72	5	09	2105	34	6
41	2825	347	1	45	611	86	4	10	2285	20	5
42	3005	199	1	46	809	116	2	11	2465	9	4
43	3185	227	1	47	1007	96	2	12	2645	344	3
44	3365	248	2	48	1196	80	3	13	2825	294	2
45	3545	258	2	49	1385	87	2	14	3005	269	3
46	3725	270	1	50	1565	116	2	15	3185	273	3
47	3905	128	2	51	1745	183	1	16	3365	247	3
48	4085	139	2	52	1925	244	2	17	3545	207	4
49	4265	136	3	53	2105	215	2	18	3725	191	5
50	4445	117	5	54	2285	239	3	19	3905	199	6
51	4625	109	5	55	2465	267	3	20	4085	187	8
52	4805	128	7	56	2645	273	4	21	4265	183	10
53	4985	139	9	57	2825	255	5	22	4445	190	10
54	5165	140	10	58	3005	245	5	23	4625	192	11
55	5345	134	11	59	3185	234	5	24	4805	196	11
56	5525	133	12	2:00P	3365	223	6	25	4985	199	13
57	5705	133	13	01	3545	221	6	26	5165	200	13
				02	3725	213	6	27	5345	201	13
30	(182)			03	3905	211	8	28	5525	201	13
2:10P	395	70	4	04	4085	215	8	29	5705	201	13
11	611	81	2	05	4265	225	7	30	5885	201	13
12	809	92	1	06	4445	233	7	31	6065	200	14
13	1007	97	2	07	4625	231	7	32	6245	201	14
14	1196	122	1	08	4805	221	7	33	6425	203	14
15	1385	212	1	09	4985	210	6	34	6605	201	14
16	1565	238	3	10	5165	208	6	35	6785	197	13
17	1745	238	5	11	5345	202	6	36	6965	194	14
18	1925	237	5	12	5525	196	7	37	7145	194	14

TABLE II (1928)

Date and time	Height in meters	Azimuth from; degrees	Velocity m/s	Date and time	Height in meters	Azimuth from; degrees	Velocity m/s	Date and time	Height in meters	Azimuth from; degrees	Velocity m/s
Feb.				**Feb.**				**Feb.**			
2	(185)			**3**				**3**			
11:10A	395	151	1	11:05A	1007	45	4	2:09P	2105	104	2
11	611	161	4	06	1196	31	6	10	2285	89	2
12	809	171	6	07	1385	23	5	11	2465	119	3
13	1007	176	8	08	1565	15	4	12	2645	130	3
14	1196	180	7	09	1745	11	4	13	2825	144	4
15	1385	181	7	10	1925	17	3	14	3005	147	5
16	1565	181	6	11	2105	36	3	15	3185	147	7
17	1745	190	6	12	2285	45	4	16	3365	158	8
18	1925	197	7	13	2465	54	3	17	3545	169	8
19	2105	204	9	14	2645	78	2	18	3725	169	7
20	2285	203	13	15	2825	122	1	19	3905	164	8
21	2465	196	15	16	3005	143	2	20	4085	168	10
22	2645	197	15	17	3185	184	3	21	4265	173	10
23	2825	202	14	18	3365	176	4	22	4445	173	10
24	3005	200	13	19	3545	165	5	23	4625	172	11
25	3185	198	11	20	3725	165	7	24	4805	177	12
26	3365	185	10	21	3905	168	8	25	4985	173	13
27	3545	175	10	22	4085	177	7	26	5165	172	13
28	3725	173	11	23	4265	175	9	27	5345	179	15
				24	4445	177	10	28	5525	183	17
2	(186)			25	4625	180	10	29	5705	183	19
2:22P	395	97	2	26	4805	178	10	30	5885	183	20
23	611	115	5	27	4985	177	10	31	6065	184	20
24	809	147	5	28	5165	173	11	32	6245	187	21
25	1007	178	5	29	5345	178	11	33	6425	189	26
26	1196	174	4	30	5525	177	12	34	6605	191	31
27	1385	161	5	31	5705	174	13	35	6785	192	34
28	1565	169	6	32	5885	177	13	36	6965	192	35
29	1745	175	7	33	6065	179	14	37	7145	189	34
30	1925	179	7	34	6245	180	15	38	7325	189	34
31	2105	182	7	35	6425	180	16	39	7505	191	33
32	2285	175	7	36	6605	181	16	40	7685	192	32
33	2465	164	7	37	6785	181	16				
34	2645	161	7	38	6965	180	17	**4**	(189)		
35	2825	165	7	39	7145	180	17	10:43A	395	Calm	Calm
36	3005	168	7	40	7325	179	16	44	611	62	2
37	3185	169	7	41	7505	179	16	45	809	36	4
38	3365	162	9					46	1007	11	6
39	3545	160	11	**3**	(188)			47	1196	11	7
40	3725	169	11	2:00P	395	44	2	48	1385	36	5
41	3905	177	11	01	611	33	2	49	1565	63	4
42	4085	176	13	02	809	2	3	50	1745	111	4
43	4265	176	14	03	1007	5	5	51	1925	147	6
				04	1196	18	6	52	2105	151	7
3	(187)			05	1385	25	5	53	2285	145	7
11:02A	395	78	2	06	1565	28	4	54	2465	151	8
03	611	106	2	07	1745	31	5	55	2645	151	8
04	809	85	2	08	1925	56	3	56	2825	151	11

TABLE II (1928)

Date and time	Height in meters	Azimuth from; degrees	Velocity m/s	Date and time	Height in meters	Azimuth from; degrees	Velocity m/s	Date and time	Height in meters	Azimuth from; degrees	Velocity m/s
Feb. 4				Feb. 6				Feb. 7			
10:57A	3005	157	14	10:51A	4265	134	17	2:04P	1196	142	5
58	3185	157	14	52	4445	128	20	05	1385	130	6
59	3365	150	12	53	4625	134	21	06	1565	106	5
11:00A	3545	144	13	54	4805	136	20	07	1745	99	5
01	3725	141	13	55	4985	137	19	08	1925	104	4
02	3905	141	13					09	2105	93	5
03	4085	143	13	6	(192)			10	2285	91	6
04	4265	143	15	1:42P	395	333	2	11	2465	101	6
05	4445	137	17	43	611	152	3	12	2645	114	7
06	4625	130	20	44	809	158	9	13	2825	118	8
07	4805	129	20	45	1007	165	12	14	3005	126	10
08	4985	133	22	46	1196	173	12	15	3185	132	10
09	5165	138	23	47	1385	177	12	16	3365	138	8
10	5345	146	21	48	1565	170	11	17	3545	140	6
11	5525	157	21	49	1745	162	9	18	3725	121	5
12	5705	162	20	50	1925	164	9	19	3905	102	6
				51	2105	175	9	7	(194)		
4	(190)			52	2285	187	10	3:00P	395	20	1
2:00P	395	212	2	53	2465	188	13	01	611	244	1
01	611	130	1	54	2645	185	14	02	809	176	1
02	809	357	5	55	2825	186	14	03	1007	131	3
03	1007	226	7	56	3005	187	14	04	1196	157	5
04	1196	205	12	57	3185	188	13	05	1385	151	6
05	1385	220	4	58	3365	182	11	06	1565	142	6
				59	3545	177	10	07	1745	143	7
6	(191)			2:00P	3725	177	9	08	1925	137	7
10:30A	395	57	3	01	3905	165	10	09	2105	125	7
31	611	97	5	02	4085	157	11	10	2285	117	8
32	809	136	11	03	4265	153	13	11	2465	115	7
33	1007	142	15	04	4445	148	15	12	2645	116	7
34	1196	145	14	05	4625	145	15	13	2825	122	7
35	1385	141	11	06	4805	146	17	14	3005	127	8
36	1565	140	9	07	4985	158	16	15	3185	129	9
37	1745	149	8	08	5165	169	16	16	3365	131	9
38	1925	161	9	09	5345	174	17	17	3545	133	8
39	2105	169	10	10	5525	178	16	18	3725	131	6
40	2285	178	11	11	5705	183	15	19	3905	116	5
41	2465	184	12	12	5885	185	13	20	4085	99	6
42	2645	184	14	13	6065	186	11	21	4265	95	9
43	2825	179	17	14	6245	186	9	22	4445	95	12
44	3005	175	16	15	6425	194	9	23	4625	99	14
45	3185	172	14	16	6605	198	10	24	4805	101	13
46	3365	166	15					25	4985	100	13
47	3545	161	14	7	(193)			26	5165	101	14
48	3725	154	13	2:00P	395	32	1	8	(195)		
49	3905	146	13	01	611	61	1	11:30A	395	25	2
50	4085	144	14	02	809	193	1	31	611	33	3
				03	1007	145	2				

TABLE II (1928)

Date and time	Height in meters	Azimuth from; degrees	Velocity m/s	Date and time	Height in meters	Azimuth from; degrees	Velocity m/s	Date and time	Height in meters	Azimuth from; degrees	Velocity m/s
Feb. 8				**Feb. 8**				**Feb. 8**			
11:32A	809	30	2	0:20P	9485	161	4	1:31P	2465	206	5
33	1007	1	1	21	9665	163	4	32	2645	196	6
34	1196	224	1	22	9845	170	5	33	2825	189	5
35	1385	227	1	23	10025	170	5	34	3005	192	3
36	1565	288	2	24	10205	171	4	35	3185	208	2
37	1745	231	2	25	10385	162	3	36	3365	291	1
38	1925	201	3	26	10565	168	3	37	3545	344	3
39	2105	187	4	27	10745	172	8	38	3725	344	3
40	2285	183	5	28	10925	168	11	39	3905	320	1
41	2465	173	5	29	11105	170	11	40	4085	310	2
42	2645	175	5	30	11285	172	10	41	4265	318	3
43	2825	178	4	31	11465	165	9	42	4445	322	2
44	3005	178	4	32	11645	164	10	43	4625	321	1
45	3185	201	3	33	11825	169	12	44	4805	321	1
46	3365	228	2	34	12005	162	15	45	4985	298	1
47	3545	285	2	35	12185	152	18	46	5165	267	1
48	3725	315	2	36	12365	140	14	47	5345	201	1
49	3905	325	1	37	12545	139	13	48	5525	182	2
50	4085	242	1	38	12725	135	16	49	5705	167	1
51	4265	251	2	39	12905	138	17	50	5885	126	1
52	4445	262	2	40	13085	159	18	51	6065	132	2
53	4625	267	2	41	13265	162	21	52	6245	148	2
54	4805	310	1	42	13445	163	26	53	6425	165	3
55	4985	104	1	43	13625	162	24	54	6605	174	2
56	5165	156	2	44	13805	159	22	55	6785	216	1
57	5345	158	3	45	13985	157	21	56	6965	190	1
58	5525	157	4	46	14165	161	23	57	7145	150	1
59	5705	161	4	47	14345	170	23	58	7325	70	1
12:00M	5885	153	3	48	14525	180	21	59	7505	52	1
0:01P	6065	122	3	49	14705	179	24	2:00P	7685	70	1
02	6245	110	5	50	14885	170	27	01	7865	197	1
03	6425	111	5	51	15065	167	27	02	8045	210	1
04	6605	119	4	52	15245	168	26	03	8225	202	1
05	6785	133	3	53	15425	169	37	04	8405	209	1
06	6965	153	3	54	15605	169	50	05	8585	211	3
07	7145	155	3					06	8765	232	3
08	7325	160	4	**8**	(196)			07	8945	264	3
09	7505	176	3	1:20P	395	37	2	08	9125	280	5
10	7685	251	1	21	611	47	6	09	9305	302	6
11	7865	285	1	22	809	44	5	10	9485	328	5
12	8045	161	1	23	1007	28	2	11	9665	340	5
13	8225	152	1	24	1196	266	2	12	9845	340	5
14	8405	156	1	25	1385	249	2	13	10025	327	6
15	8585	130	2	26	1565	257	2	14	10205	315	7
16	8765	127	5	27	1745	247	2	15	10385	120	8
17	8945	138	5	28	1925	226	3	16	10565	290	8
18	9125	153	4	29	2105	213	4	17	10745	281	6
19	9305	163	4	30	2285	212	5	18	10925	276	6

TABLE II (1928)

Date and time	Height in meters	Azimuth from; degrees	Velocity m/s	Date and time	Height in meters	Azimuth from; degrees	Velocity m/s	Date and time	Height in meters	Azimuth from; degrees	Velocity m/s
Feb.				**Feb.**				**Feb.**			
8				**10**				**12**	(200)		
2:19P	11105	268	8	3:15P	1385	241	2	1:28P	395	67	2
20	11285	254	9	16	1565	231	2	29	611	71	3
21	11465	245	7					30	809	87	1
22	11645	227	6	**11**	(199)			31	1007	122	1
23	11825	230	4	1:35P	395	79	1	32	1196	101	1
24	12005	243	5	36	611	78	3	33	1385	39	2
25	12185	246	6	37	809	85	2	34	1565	20	2
26	12365	254	7	38	1007	107	1	35	1745	15	2
27	12545	239	5	39	1196	233	4	36	1925	5	3
28	12725	207	1	40	1385	275	3	37	2105	351	5
29	12905	1	1	41	1565	309	3	38	2285	346	7
30	13085	346	1	42	1745	270	3	39	2465	331	5
31	13265	262	5	43	1925	290	3	40	2645	349	8
32	13445	254	7	44	2105	287	3	41	2825	353	12
33	13625	253	6	45	2285	294	5	42	3005	347	11
34	13805	260	8	46	2465	301	6	43	3185	347	12
35	13985	253	11	47	2645	301	7	44	3365	344	13
36	14165	240	11	48	2825	294	7	45	3545	346	13
37	14345	236	10	49	3005	289	8	46	3725	349	15
38	14525	235	10	50	3185	286	8	47	3905	347	17
39	14705	231	11	51	3365	281	8	48	4085	343	19
40	14885	237	9	52	3545	274	9	49	4265	341	19
41	15065	239	9	53	3725	270	9	50	4445	339	21
42	15245	250	12	54	3905	271	9	51	4625	341	21
43	15425	259	12	55	4085	277	9	52	4805	340	21
44	15605	266	12	56	4265	279	9	53	4985	338	24
45	15785	265	14	57	4445	278	11	54	5165	338	24
46	15965	261	14	58	4625	278	11	55	5345	339	23
47	16145	277	16	59	4805	272	11	56	5525	337	25
48	16325	288	19	2:00P	4985	266	10	57	5705	340	26
49	16505	287	24	01	5165	265	11	58	5885	340	25
50	16685	281	29	02	5345	264	11	59	6065	340	26
51	16865	278	36	03	5525	260	11	2:00P	6245	344	28
52	17045	278	38	04	5705	260	12	01	6425	346	29
53	17225	280	34	05	5885	260	13	02	6605	345	29
				06	6065	257	14	03	6785	342	29
9	(197)			07	6245	255	13	04	6965	342	29
2:30P	395	133	1	08	6425	255	13	05	7145	342	31
31	611	199	3	09	6605	260	14	06	7325	344	33
32	809	249	3	10	6785	262	14				
33	1007	280	4	11	6965	263	13	**13**	(201)		
34	1196	286	4	12	7145	267	12	2:18P	395	98	5
10	(198)			13	7325	272	11	19	611	127	6
3:10P	395	333	2	14	7505	278	12	20	809	168	7
11	611	337	2	15	7685	281	13	21	1007	188	7
12	809	247	1	16	7865	281	13	22	1196	201	6
13	1007	218	2	17	8045	281	13	23	1385	198	5
14	1196	236	2	19	8225	281	13	24	1565	195	4

TABLE II (1928)

Date and time	Height in meters	Azimuth from; degrees	Velocity m/s
Feb. 13			
2:25P	1745	197	3
26	1925	210	2
27	2105	200	4
28	2285	188	5
29	2465	202	5
30	2645	218	6
31	2825	226	8
32	3005	233	7
33	3185	242	5
34	3365	228	5
35	3545	235	6
36	3725	263	8
37	3905	279	10
38	4085	279	13
39	4265	284	14
40	4445	294	15
41	4625	299	17
42	4805	296	17
43	4985	292	17
44	5165	292	18
45	5345	293	18
46	5525	293	18
47	5705	292	17
48	5885	294	16
49	6065	292	16
50	6245	290	16
14	(202)		
1:39P	395	99	10
40	611	115	10
41	809	153	9
42	1007	182	11
43	1385	177	13
44	1565	173	13
45	1745	171	12
46	1925	174	12
47	2105	173	13
48	2285	172	12
49	2465	167	11
50	2645	164	10
51	2825	160	10
52	3005	155	10
15	(203)		
9:48A	395	270	10
49	611	28	10
50	809	132	7
51	1007	182	8
52	1196	184	9

Date and time	Height in meters	Azimuth from; degrees	Velocity m/s
Feb. 15			
9:53A	1385	207	9
54	1565	213	10
55	1745	213	11
56	1925	208	10
57	2105	208	9
16	(204)		
11:00A	395	225	2
01	611	209	2
02	809	190	3
03	1007	188	5
04	1196	195	6
05	1385	200	7
16	(205)		
3:20P	395	279	5
21	611	241	4
22	809	217	4
23	1007	215	5
24	1196	209	7
25	1385	199	8
26	1565	193	9
27	1745	196	10
28	1925	200	12
29	2105	200	15
30	2285	203	17
31	2465	204	15
32	2645	182	12
33	2825	165	14
34	3005	168	16
35	3185	172	18
36	3365	173	19
17	(206)		
10:55A	395	78	3
56	611	79	3
57	809	39	2
58	1007	20	3
59	1196	32	5
11:00A	1385	48	6
01	1565	47	6
02	1745	44	5
03	1925	49	4
04	2105	44	3
05	2285	27	2
06	2465	14	2
07	2645	193	1
08	2825	193	3
09	3005	197	5

Date and time	Height in meters	Azimuth from; degrees	Velocity m/s
Feb. 17			
11:10A	3185	197	5
11	3365	198	5
12	3545	216	6
13	3725	219	7
14	3905	218	7
15	4085	221	6
16	4265	225	6
17	4445	229	7
18	4625	229	8
19	4805	232	7
20	4985	239	7
21	5165	237	8
22	5345	226	9
23	5525	219	9
24	5705	213	9
25	5885	209	10
26	6065	209	12
27	6245	211	13
28	6425	208	16
29	6605	208	17
30	6785	210	17
31	6965	211	17
32	7145	213	16
33	7325	216	16
34	7505	215	15
35	7685	212	15
36	7865	215	17
37	8045	219	18
38	8225	217	17
39	8405	214	17
40	8585	217	19
41	8765	220	23
42	8945	221	26
17	(207)		
2:02P	395	79	2
03	611	79	2
04	809	58	3
05	1007	47	5
06	1196	39	7
07	1385	30	6
08	1565	24	5
09	1745	20	4
10	1925	2	3
11	2105	336	2
12	2285	326	2
13	2465	340	1
14	2645	147	1

TABLE II (1928)

Date and time	Height in meters	Azimuth from; degrees	Velocity m/s
Feb. 17			
2:15P	2825	238	2
16	3005	242	3
17	3185	236	4
18	3365	242	6
19	3545	232	6
20	3725	214	6
21	3905	205	7
22	4085	212	7
23	4265	217	7
24	4445	221	8
25	4625	222	10
26	4805	217	9
27	4985	217	9
28	5165	216	10
29	5345	215	11
30	5525	215	11
31	5705	211	11
32	5885	208	9
33	6065	211	11
34	6245	212	13
35	6425	212	13
36	6605	212	13
37	6785	212	13
38	6965	216	14
39	7145	219	16
40	7325	217	17
41	7505	219	17
42	7685	218	17
43	7865	215	17
44	8045	219	17
45	8225	221	17
46	8405	221	18
47	8585	220	18
48	8765	218	19
49	8945	219	20
50	9125	223	20
51	9305	223	19
52	9485	225	19
53	9665	226	22
54	9845	223	22
55	10025	223	22
56	10205	220	22
18	(208)		
10:30A	395	93	3
31	611	134	4
32	809	185	5
33	1007	203	4

Date and time	Height in meters	Azimuth from; degrees	Velocity m/s
Feb. 18			
10:34A	1196	218	2
35	1385	159	1
36	1565	103	1
37	1745	152	2
38	1925	174	4
39	2105	177	5
40	2285	187	7
41	2465	197	7
42	2645	198	8
43	2825	206	9
44	3005	211	11
45	3185	215	11
46	3365	219	12
47	3545	216	12
48	3725	215	12
49	3905	217	12
50	4085	220	12
51	4265	221	13
52	4445	219	14
53	4625	222	15
54	4805	223	17
55	4985	223	17
56	5165	223	17
57	5345	224	19
58	5525	225	21
59	5705	228	22
11:00A	5885	229	22
01	6065	227	24
02	6245	227	25
18	(209)		
1:45P	395	72	3
46	611	77	3
47	809	110	1
48	1007	116	2
49	1196	113	2
50	1385	159	1
51	1565	181	2
52	1745	179	4
53	1925	189	6
54	2105	190	8
55	2285	194	8
56	2465	198	9
57	2645	199	10
58	2825	199	11
59	3005	206	12
2:00P	3185	211	12
01	3365	210	13

Date and time	Height in meters	Azimuth from; degrees	Velocity m/s
Feb. 18			
2:02P	3545	211	13
03	3725	213	14
04	3905	214	15
05	4085	218	16
06	4265	220	18
07	4445	220	19
08	4625	221	21
09	4805	223	23
10	4985	225	24
11	5165	224	25
12	5345	225	25
13	5525	225	27
14	5705	226	29
15	5885	227	29
16	6065	227	29
20	(210)		
10:56A	395	325	2
57	611	1	4
58	809	5	5
59	1007	357	6
11:00A	1196	354	10
01	1385	2	14
02	1565	12	15
03	1745	18	12
04	1925	11	10
05	2105	10	8
06	2285	18	6
20	(211)		
2:04P	395	245	5
05	611	297	5
06	809	342	7
07	1007	340	12
08	1196	343	15
09	1385	355	12
10	1565	359	13
11	1745	356	13
21	(212)		
10:03A	395	36	2
04	611	31	3
05	809	26	3
06	1007	26	3
07	1196	34	5
08	1385	46	6
09	1565	46	7
10	1745	43	9

TABLE II (1928)

Date and time	Height in meters	Azimuth from; degrees	Velocity m/s
Feb. 21			
10:11A	1925	31	11
12	2105	23	13
13	2285	23	15
14	2465	24	15
15	2645	25	14
16	2825	21	13
17	3005	15	12
18	3185	18	11
19	3365	16	11
20	3545	13	12
21	3725	9	12
22	3905	16	11
23	4085	24	10
24	4265	28	10
25	4445	30	9
26	4625	28	7
27	4805	30	8
28	4985	34	9
29	5165	45	9
30	5345	56	8
31	5525	73	6
32	5705	91	5
33	5885	97	4
21	(213)		
2:11P	395	334	3
12	611	357	6
13	809	14	6
14	1007	23	7
15	1196	25	7
16	1385	28	8
17	1565	31	7
18	1745	30	10
19	1925	35	13
20	2105	34	15
21	2285	33	17
22	2465	33	16
23	2645	40	15
24	2825	44	14
25	3005	43	13
26	3185	39	13
27	3365	42	12
28	3545	42	12
29	3725	41	11
30	3905	41	10
31	4085	39	11
32	4265	39	11
33	4445	41	10

Date and time	Height in meters	Azimuth from; degrees	Velocity m/s
Feb. 21			
2:34P	4625	42	11
35	4805	48	11
36	4985	49	10
37	5165	70	11
38	5345	90	11
39	5525	92	9
22	(214)		
0:05P	395	27	1
06	611	143	3
07	809	185	4
08	1007	199	5
09	1196	206	6
10	1385	202	7
11	1565	193	7
12	1745	191	7
13	1925	189	7
14	2105	183	10
15	2285	183	12
16	2465	177	13
17	2645	165	14
18	2825	158	14
19	3005	157	14
22	(215)		
2:06P	395	53	2
07	611	153	4
08	809	178	4
09	1007	203	6
10	1196	206	6
11	1385	185	4
12	1565	174	5
13	1745	179	7
14	1925	186	9
15	2105	190	11
16	2285	192	12
17	2465	190	11
18	2645	184	10
19	2825	177	10
20	3005	163	12
21	3185	161	13
22	3365	160	12
23	3545	158	12
24	3725	156	11
25	3905	151	11
26	4085	145	11
27	4265	141	11

Date and time	Height in meters	Azimuth from; degrees	Velocity m/s
Feb. 23	(216)		
10:00A	395	188	3
01	611	187	4
02	809	193	5
03	1007	198	6
04	1196	196	7
05	1385	200	7
06	1565	205	8
07	1745	204	10
08	1925	196	11
09	2105	180	11
10	2285	172	14
11	2465	171	15
12	2645	165	14
13	2825	160	14
23	(217)		
2:02P	395	215	1
03	611	207	2
04	809	206	3
05	1007	203	7
06	1196	205	10
07	1385	209	8
08	1565	215	6
09	1745	215	5
10	1925	216	6
11	2105	215	7
12	2285	209	7
13	2465	205	7
14	2645	204	10
15	2825	199	12
16	3005	197	13
24	(218)		
11:01A	395	130	5
02	611	160	8
03	809	186	11
04	1007	202	11
05	1196	211	9
06	1385	212	9
07	1565	216	8
08	1745	220	8
09	1925	220	8
10	2105	221	8
11	2285	228	8
12	2465	231	7
13	2645	236	6
14	2825	236	7
15	3005	225	7

TABLE II (1928)

Date and time	Height in meters	Azimuth from; degrees	Velocity m/s	Date and time	Height in meters	Azimuth from; degrees	Velocity m/s	Date and time	Height in meters	Azimuth from; degrees	Velocity m/s
Feb.				**Feb.**				**Feb.**			
24				**24**				**25**			
11:16A	3185	211	5	2:32P	5705	218	9	1:59P	3005	157	13
17	3365	207	6	33	5885	205	9	2:00P	3185	171	14
18	3545	209	9	34	6065	202	11	01	3365	185	14
19	3725	206	10	35	6245	215	13	02	3545	186	17
20	3905	203	10	36	6425	221	14	03	3725	193	22
21	4085	204	11	37	6605	223	14	04	3905	196	24
22	4265	205	9	38	6785	222	15	05	4085	195	25
23	4445	205	8	39	6965	222	16	06	4265	193	25
24	4625	201	8	40	7145	216	16	07	4445	189	26
25	4805	187	8	41	7325	224	18	08	4625	180	29
26	4985	183	8	42	7505	244	20	09	4805	175	32
27	5165	188	8	43	7685	240	20				
28	5345	188	8	44	7865	233	20	**25**	(222)		
29	5525	186	8	45	8045	235	21	3:40P	395	106	6
30	5705	194	9	46	8225	235	22	41	611	123	14
31	5885	201	8					42	809	139	17
32	6065	206	9	**25**	(220)			43	1007	153	18
				11:01A	395	56	3	44	1196	163	19
24	(219)			02	611	97	6	45	1385	172	19
2:03P	395	97	4	03	809	147	8	46	1565	176	18
04	611	123	8	04	1007	167	12				
05	809	161	8	05	1196	171	13	**26**	(223)		
06	1007	196	7	06	1385	170	11	10:05A	395	203	15
07	1196	212	7	07	1565	164	8	06	611	196	7
08	1385	215	5	08	1745	161	8	07	809	162	5
09	1565	210	4	09	1925	163	10	08	1007	141	10
10	1745	219	5	10	2105	168	11	09	1196	149	13
11	1925	234	5	11	2285	176	13	10	1385	151	11
12	2105	233	4	12	2465	180	13	11	1565	141	21
13	2285	211	3	13	2645	178	14	12	1745	148	35
14	2465	188	2	14	2825	177	13	13	1925	156	33
15	2645	187	1	15	3005	178	12	14	2105	169	18
16	2825	185	2					15	2285	176	15
17	3005	188	5	**25**	(221)			16	2465	170	21
18	3185	200	7	1:45P	395	58	4				
19	3365	210	8	46	611	98	8	**26**	(224)		
20	3545	214	8	47	809	140	10	0:05P	395	126	7
21	3725	219	9	48	1007	157	14	06	611	145	8
22	3905	223	9	49	1196	166	15	07	809	143	14
23	4085	227	9	50	1385	165	14	08	1007	135	19
24	4265	210	11	51	1565	149	12	09	1196	145	18
25	4445	213	11	52	1745	135	13	10	1385	153	18
26	4625	219	10	53	1925	121	13	11	1565	155	19
27	4805	213	8	54	2105	114	11	12	1745	160	21
28	4985	208	8	55	2285	118	11	13	1925	161	26
29	5165	211	8	56	2465	130	11	14	2105	161	32
30	5345	212	8	57	2645	141	12	15	2285	163	32
31	5525	216	9	58	2825	147	12	16	2465	165	31

TABLE II (1928)

Date and time	Height in meters	Azimuth from; degrees	Velocity m/s
Feb.			
26	(225)		
1:58P	395	132	26
59	611	137	17
2:00P	809	143	16
27	(226)		
10:03A	395	78	2
04	611	76	6
05	809	79	8
06	1007	96	7
07	1196	112	8
08	1385	121	10
09	1565	127	12
10	1745	126	14
11	1925	119	14
12	2105	117	16
13	2285	119	18
14	2465	119	18
15	2645	118	18
27	(227)		
1:30P	395	149	18
31	611	155	11
32	809	160	13
33	1007	164	13
34	1196	172	11
35	1385	179	15
36	1565	178	19
37	1745	172	20
38	1925	165	18
39	2105	156	17
40	2285	154	19
27	(228)		
3:03P	395	176	20
04	611	186	26
05	809	196	24
06	1007	198	17
07	1196	202	13
28	(229)		
10:15A	395	123	4
16	611	151	5
17	809	194	4
18	1007	208	5
19	1196	209	6
20	1385	215	5
21	1565	218	3
22	1745	218	1
23	1925	133	1

Date and time	Height in meters	Azimuth from; degrees	Velocity m/s
Feb.			
28			
10:24A	2105	165	2
25	2285	182	3
26	2465	177	5
27	2645	170	5
28	2825	178	5
29	3005	187	5
30	3185	187	6
31	3365	189	6
32	3545	192	6
33	3725	183	6
34	3905	170	7
35	4085	172	9
36	4265	178	11
37	4445	173	13
38	4625	169	16
39	4805	173	18
40	4985	176	19
28	(230)		
2:07P	395	53	2
08	611	105	1
09	809	110	1
10	1007	66	3
11	1196	48	5
12	1385	43	6
13	1565	43	7
14	1745	47	7
15	1925	57	6
16	2105	69	6
17	2285	85	5
18	2465	99	5
19	2645	95	4
20	2825	72	3
21	3005	49	2
22	3185	24	2
23	3365	15	2
24	3545	18	2
25	3725	318	1
26	3905	202	1
27	4085	169	1
28	4265	162	1
29	4445	159	2
29	(231)		
10:10A	395	329	8
11	611	341	4
12	809	159	4
13	1007	155	13

Date and time	Height in meters	Azimuth from; degrees	Velocity m/s
Feb.			
29			
10:14A	1196	148	17
15	1385	144	19
16	1565	140	19
17	1745	139	20
18	1925	136	23
19	2105	138	29
20	2285	140	36
21	2465	142	34
22	2645	144	30
23	2825	145	29
29	(232)		
2:00P	395	67	5
01	611	83	5
02	809	88	5
03	1007	87	8
04	1196	112	9
05	1385	121	16
06	1565	110	28
07	1745	116	23
08	1925	140	16
09	2105	138	21
10	2285	133	23
11	2465	138	24
12	2645	139	25
13	2825	139	27
14	3005	139	27
15	3185	143	24
16	3365	147	23
March			
1	(233)		
9:48A	395	173	15
49	611	178	13
50	809	185	13
51	1007	187	15
52	1196	191	19
53	1385	191	24
54	1565	189	31
55	1745	190	39
56	1925	192	35
57	2105	196	20
58	2285	203	12
59	2465	203	12
10:00A	2645	199	16
01	2825	193	25
02	3005	189	32
03	3185	187	34

TABLE II (1928)

Date and time	Height in meters	Azimuth from; degrees	Velocity m/s
March			
1	(234)		
2:11P	395	165	13
12	611	176	20
13	809	181	21
14	1007	186	20
15	1196	192	17
16	1385	191	19
17	1565	190	20
18	1745	194	18
19	1925	203	12
20	2105	216	10
21	2285	210	15
22	2465	203	23
23	2645	204	30
24	2825	208	34
25	3005	211	35
26	3185	211	29
27	3365	210	20
2	(235)		
10:13A	395	89	7
14	611	114	7
15	809	150	8
16	1007	171	9
17	1196	183	11
18	1385	185	10
19	1565	188	9
20	1745	187	8
21	1925	184	8
22	2105	179	9
23	2285	174	9
24	2465	174	7
25	2645	183	6
26	2825	189	7
27	3005	189	8
28	3185	197	8
29	3365	213	9
30	3545	208	10
31	3725	184	9
32	3905	166	9
33	4085	162	8
2	(236)		
2:04P	395	99	5
05	611	118	8
06	809	167	5
07	1007	213	6
08	1196	220	7
09	1385	216	8

Date and time	Height in meters	Azimuth from; degrees	Velocity m/s
March			
2			
2:10P	1565	204	8
11	1745	197	9
12	1925	191	8
13	2105	180	7
14	2285	173	7
15	2465	175	8
16	2645	175	7
17	2825	175	7
18	3005	184	7
19	3185	202	5
20	3365	209	5
21	3545	204	5
3	(237)		
10:00A	395	89	6
01	611	107	8
02	809	131	10
03	1007	156	11
04	1196	173	13
05	1385	179	12
06	1565	184	9
07	1745	182	9
08	1925	170	9
09	2105	159	9
10	2285	160	11
11	2465	164	13
12	2645	160	14
13	2825	156	14
14	3005	161	16
15	3185	169	15
16	3365	169	16
17	3545	166	17
18	3725	169	17
19	3905	174	17
20	4085	176	17
21	4265	174	19
22	4445	169	23
23	4625	165	23
24	4805	164	19
3	(238)		
2:05P	395	145	8
06	611	161	10
07	809	171	14
08	1007	177	16
09	1196	184	15
10	1385	186	12
11	1565	185	8

Date and time	Height in meters	Azimuth from; degrees	Velocity m/s
March			
3			
2:12P	1745	186	10
13	1925	198	9
14	2105	170	8
15	2285	172	11
16	2465	187	13
17	2645	192	15
18	2825	187	18
19	3005	180	17
20	3185	171	15
21	3365	173	17
22	3545	176	27
23	3725	173	35
24	3905	172	38
4	(239)		
10:03A	395	121	5
04	611	129	7
05	809	145	9
06	1007	153	12
07	1196	161	12
08	1385	172	11
09	1565	183	10
10	1745	187	8
11	1925	198	6
12	2105	196	6
13	2285	189	8
14	2465	189	11
15	2645	187	15
16	2825	181	16
17	3005	180	19
18	3185	180	21
19	3365	177	23
20	3545	175	21
21	3725	174	20
22	3905	170	22
23	4085	169	25
24	4265	166	26
4	(240)		
2:05P	395	69	4
06	611	109	4
07	809	121	9
08	1007	128	13
09	1196	135	15
10	1385	145	15
11	1565	142	14
12	1745	135	11
13	1925	151	7

TABLE II (1928)

Date and time	Height in meters	Azimuth from; degrees	Velocity m/s
March 4			
2:14P	2105	179	6
15	2285	192	7
16	2465	194	10
17	2645	191	13
18	2825	187	16
19	3005	187	15
20	3185	183	12
21	3365	181	12
22	3545	182	15
23	3725	172	18
24	3905	166	22
25	4085	163	23
26	4265	161	22
27	4445	159	21
5	(241)		
10:06A	395	136	5
07	611	148	8
08	809	158	9
09	1007	168	8
10	1196	176	9
11	1385	180	13
12	1565	183	16
13	1745	186	15
14	1925	185	13
15	2105	184	13
16	2285	184	15
17	2465	187	17
18	2645	180	17
19	2825	171	16
20	3005	176	17
21	3185	178	19
22	3365	178	22
23	3545	179	25
24	3725	179	25
25	3905	180	24
26	4085	181	22
5	(242)		
2:36P	395	129	12
37	611	138	17
38	809	145	16
39	1007	145	12
40	1196	161	7
41	1385	205	6
42	1565	211	10
43	1745	202	13
44	1925	195	15

Date and time	Height in meters	Azimuth from; degrees	Velocity m/s
March 5			
2:45P	2105	191	16
46	2285	189	15
47	2465	187	17
48	2645	183	19
49	2825	184	17
50	3005	192	16
51	3185	192	16
52	3365	189	17
6	(243)		
10:05A	395	141	8
06	611	158	11
07	809	167	15
09	1007	171	14
10	1196	175	10
10	1385	185	7
11	1565	199	7
12	1745	196	8
13	1925	187	12
14	2105	183	17
15	2285	187	19
16	2465	192	19
17	2645	191	17
18	2825	189	12
19	3005	200	9
20	3185	231	4
21	3365	248	4
22	3545	220	7
23	3725	210	12
24	3905	209	20
25	4085	205	30
26	4265	206	35
27	4445	207	34
6	(244)		
1:58P	395	73	4
59	611	92	5
2:00P	809	141	3
01	1007	138	3
02	1196	126	5
03	1385	128	5
04	1565	131	5
05	1745	147	5
06	1925	175	6
07	2105	189	10
08	2285	191	15
09	2465	194	16
10	2645	193	17

Date and time	Height in meters	Azimuth from; degrees	Velocity m/s
March 6			
2:11P	2825	188	21
12	3005	190	19
13	3185	193	16
14	3365	194	17
15	3545	196	19
16	3725	194	22
17	3905	189	22
18	4085	189	22
19	4265	189	23
7	(245)		
9:51A	395	124	11
52	611	140	10
53	809	150	10
54	1007	157	10
55	1196	165	12
56	1385	168	16
57	1565	167	21
58	1745	166	23
59	1925	170	23
10:00A	2105	170	22
01	2285	120	21
02	2465	174	19
03	2645	174	16
04	2825	176	13
05	3005	171	11
06	3185	160	11
07	3365	148	11
08	3545	138	9
7	(246)		
1:52P	395	125	7
53	611	130	9
54	809	147	9
55	1007	161	10
56	1196	169	12
57	1385	172	14
58	1565	173	17
59	1745	175	19
2:00P	1925	177	20
01	2105	177	21
02	2285	175	22
03	2465	175	23
04	2645	176	23
05	2825	176	21
06	3005	176	20

TABLE II (1928)

Date and time	Height in meters	Azimuth from; degrees	Velocity m/s	Date and time	Height in meters	Azimuth from; degrees	Velocity m/s	Date and time	Height in meters	Azimuth from; degrees	Velocity m/s
March				**March**				**March**			
8	(247)			**9**	(249)			**10**			
10:02A	395	101	5	10:02A	395	90	5	10:09A	2465	126	6
03	611	135	6	03	611	172	5	10	2645	145	7
04	809	167	7	04	809	202	7	11	2825	176	8
05	1007	187	8	05	1007	218	7	12	3005	184	10
06	1196	193	8	06	1196	232	6	13	3185	178	11
07	1385	190	7	07	1385	249	5	14	3365	175	11
08	1565	184	6	08	1565	242	5	15	3545	174	13
09	1745	180	6	09	1745	237	6	16	3725	177	13
10	1925	177	6	10	1925	235	6				
11	2105	165	7	11	2105	237	8	**10**	(252)		
12	2285	151	7	12	2285	237	9	2:02P	395	70	3
13	2465	146	8	13	2465	231	10	03	611	104	6
14	2645	159	10	14	2645	227	12	04	809	139	7
15	2825	172	13	15	2825	227	13	05	1007	165	8
16	3005	175	15	16	3005	227	12	06	1196	177	12
17	3185	177	14	17	3185	227	10	07	1385	185	13
18	3365	178	14	18	3365	224	9	08	1565	190	14
19	3545	179	15	19	3545	218	8	09	1745	187	13
20	3725	178	17					10	1925	185	14
21	3905	177	19	**9**	(250)			11	2105	179	13
22	4085	179	19	1:35P	395	113	5	12	2285	172	12
23	4265	177	19	36	611	157	5	13	2465	169	13
24	4445	173	17	37	809	182	9	14	2645	176	14
25	4625	166	18	38	1007	192	9	15	2825	177	15
26	4805	166	20	39	1196	198	9	16	3005	177	15
27	4985	169	22	40	1385	201	11	17	3185	181	15
28	5165	169	23	41	1565	208	11	18	3365	180	15
				42	1745	217	9	19	3545	177	17
				43	1925	226	8	20	3725	190	19
8	(248)			44	2105	232	7	21	3905	182	22
1:58P	395	187	5	45	2285	225	8	22	4085	169	23
59	611	172	7	46	2465	218	9	23	4265	175	23
2:00P	809	168	8	47	2645	214	9	24	4445	178	25
01	1007	173	10	48	2825	209	10	25	4625	179	24
02	1196	176	11	49	3005	208	10	26	4805	179	22
03	1385	179	10					27	4985	173	24
04	1565	187	9	**10**	(251)			28	5165	173	25
05	1745	199	8	9:58A	395	14	5	29	5345	174	25
06	1925	214	7	59	611	119	2	30	5525	175	26
07	2105	216	8	10:00A	809	125	4				
08	2285	209	9	01	1007	127	5	**11**	(253)		
09	2465	201	10	02	1196	137	6	9:56A	395	108	16
10	2645	196	11	03	1385	141	7	57	611	126	15
11	2825	195	14	04	1565	144	7	58	809	137	22
12	3005	192	19	05	1745	144	7	59	1007	144	25
13	3185	186	22	06	1925	147	7	10:00A	1196	151	23
14	3365	181	22	07	2105	139	7	01	1385	154	23
15	3545	178	20	08	2285	124	7	02	1565	160	23

TABLE II (1928)

Date and time	Height in meters	Azimuth from; degrees	Velocity m/s
March 11			
10:03A	1745	169	23
04	1925	176	23
05	2105	181	22
06	2285	180	21
11	(254)		
2:03P	395	136	16
04	611	135	15
05	809	137	10
06	1007	147	10
07	1196	148	30
08	1385	328	45
09	1565	329	39
10	1745	333	36
11	1925	159	35
12	2105	163	30
13	2285	164	25
14	2465	168	24
15	2645	170	31
16	2825	170	34
12	(255)		
10:00A	395	143	16
01	611	145	18
02	809	148	14
03	1007	157	9
04	1196	169	11
05	1385	175	15
06	1565	182	18
07	1745	187	21
08	1925	189	23
09	2105	191	22
10	2285	194	22
11	2465	194	25
12	2645	192	26
13	2825	188	31
14	3005	186	37
12	(256)		
1:34P	395	144	14
35	611	142	15
36	809	144	9
37	1007	158	8
38	1196	168	9
39	1385	176	9
40	1565	183	9
41	1745	189	11
42	1925	193	13

Date and time	Height in meters	Azimuth from; degrees	Velocity m/s
March 12			
1:43P	2105	192	15
44	2285	190	19
45	2465	188	26
46	2645	187	29
13	(257)		
10:00A	395	308	4
01	611	311	4
02	809	303	1
03	1007	190	3
04	1196	184	6
05	1385	191	6
06	1565	200	7
07	1745	198	11
08	1925	197	12
09	2105	199	13
10	2285	198	15
11	2465	198	18
12	2645	200	21
13	2825	201	22
13	(258)		
2:00P	395	340	1
01	611	348	1
02	809	318	1
03	1007	105	2
04	1196	135	4
05	1385	180	5
06	1565	197	7
07	1745	200	10
08	1925	200	12
09	2105	200	15
10	2285	200	17
14	(259)		
10:50A	395	258	2
51	611	252	3
52	809	250	3
53	1007	251	3
14	(260)		
5:06P	395	286	5
07	611	287	7
08	809	279	5
09	1007	253	1
10	1196	218	2
11	1385	208	3
12	1565	226	5
13	1745	230	7

Date and time	Height in meters	Azimuth from; degrees	Velocity m/s
March 14			
5:14P	1925	226	7
15	2105	228	9
15	(261)		
10:08A	395	303	1
09	611	326	1
10	809	117	1
11	1007	146	1
16	(262)		
10:54A	395	4	5
55	611	9	5
56	809	12	5
57	1007	10	5
16	(263)		
2:20P	395	24	5
21	611	15	4
22	809	0	3
23	1007	1	5
24	1196	7	5
25	1385	17	7
26	1565	19	6
27	1745	19	6
17	(264)		
10:10A	395	83	4
11	611	37	2
12	809	15	3
13	1007	41	4
14	1196	43	6
15	1385	37	8
16	1565	33	8
17	1745	25	9
18	1925	16	10
19	2105	6	12
20	2285	2	13
21	2465	2	13
22	2645	1	13
23	2825	351	14
24	3005	358	13
25	3185	359	13
26	3365	358	14
27	3545	357	20
28	3725	347	18
29	3905	341	13
30	4085	354	15
31	4265	343	17

TABLE II (1928)

Date and time	Height in meters	Azimuth from; degrees	Velocity m/s	Date and time	Height in meters	Azimuth from; degrees	Velocity m/s	Date and time	Height in meters	Azimuth from; degrees	Velocity m/s
March 17				**March 18**				**March 20**			
10:32A	4445	349	19	10:05A	3185	252	7	10:35A	4085	321	22
33	4625	2	24	06	3365	250	9	36	4265	320	24
				07	3545	256	8	37	4445	322	25
17	(265)			08	3725	263	10	38	4625	321	24
2:15P	395	74	5					39	4805	320	26
16	611	74	3	**19**	(267)			40	4985	320	27
17	809	52	1	2:22P	395	47	5	41	5165	320	27
18	1007	58	1	23	611	38	4	42	5345	320	26
19	1196	40	1	24	809	45	2	43	5525	323	27
20	1385	4	3	25	1007	22	1	44	5705	322	27
21	1565	351	5	26	1196	292	2	45	5885	321	27
22	1745	342	5	27	1385	298	4	46	6065	322	29
23	1925	325	4	28	1565	308	4	47	6245	322	30
24	2105	312	6	29	1745	324	3	48	6425	323	32
25	2285	315	6	30	1925	324	4	49	6605	320	31
26	2465	321	2	31	2105	313	4	50	6785	320	28
27	2645	331	10	32	2285	306	5	51	6965	321	31
28	2825	339	14	33	2465	304	7	52	7145	321	34
29	3005	340	15	34	2645	303	7	53	7325	321	33
30	3185	343	14	35	2825	304	6	54	7505	322	34
31	3365	347	15	36	3005	304	7	55	7685	325	33
32	3545	347	13	37	3185	304	7	56	7865	325	34
33	3725	343	12	38	3365	304	7	57	8045	323	34
34	3905	336	14	39	3545	309	8				
35	4085	332	15	40	3725	310	8	**20**	(269)		
36	4265	338	15	41	3905	309	9	2:06P	395	60	1
37	4445	344	14					07	611	103	2
38	4625	342	15	**20**	(268)			08	809	150	3
39	4805	337	17	10:15A	395	68	2	09	1007	165	3
40	4985	331	19	16	611	92	3	10	1196	184	3
41	5165	330	20	17	809	135	2	11	1385	197	3
				18	1007	172	2	12	1565	219	2
18	(266)			19	1196	202	2	13	1745	240	2
9:50A	395	67	5	20	1385	215	2	14	1925	249	3
51	611	101	3	21	1565	239	2	15	2105	260	4
52	809	207	2	22	1745	248	3	16	2285	274	4
53	1007	199	3	23	1925	254	4	17	2465	296	4
54	1196	154	4	24	2105	267	4	18	2645	309	7
55	1385	143	5	25	2285	278	4	19	2825	308	10
56	1565	117	3	26	2465	303	4	20	3005	305	11
57	1745	41	2	27	2645	316	5	21	3185	309	12
58	1925	343	3	28	2825	314	8	22	3365	313	14
59	2105	316	5	29	3005	316	10	23	3545	315	15
10:00A	2285	318	7	30	3185	314	11	24	3725	316	16
01	2465	288	5	31	3365	313	12	25	3905	315	17
02	2645	240	5	32	3545	317	15	26	4085	314	18
03	2825	247	5	33	3725	319	17	27	4265	316	19
04	3005	248	6	34	3905	320	19	28	4445	317	18

TABLE II (1928)

Date and time	Height in meters	Azimuth from; degrees	Velocity m/s	Date and time	Height in meters	Azimuth from; degrees	Velocity m/s	Date and time	Height in meters	Azimuth from; degrees	Velocity m/s
March				**March**				**March**			
20				**21**	(271)			**22**			
2:29P	4625	314	20	2:02P	395	86	3	10:19A	4265	170	14
30	4805	312	19	03	611	90	4	20	4445	170	14
31	4985	314	18	04	809	94	3	21	4625	178	13
32	5165	314	19	05	1007	105	3	22	4805	183	13
33	5345	314	20	06	·1196	110	3	23	4985	183	14
34	5525	313	20	07	1385	99	3	24	5165	175	14
35	5705	315	21	08	1565	148	1	25	5345	171	14
36	5885	316	21	09	1745	202	1	26	5525	173	14
37	6065	314	20	10	1925	255	2				
38	6245	313	22	11	2105	275	3	**22**	(273)		
39	6425	315	21	12	2285	277	1	1:46P	395	80	4
40	6605	313	21	13	2465	273	2	47	611	96	2
41	6785	313	21	14	2645	265	3	48	809	152	2
42	6965	317	22	15	2825	264	3	49	1007	148	2
43	7145	320	22	16	3005	264	3	50	1196	93	1
				17	3185	275	3	51	1385	48	1
				18	3365	268	5	52	1565	72	1
21	(270)			19	3545	257	6	53	1745	108	1
9:58A	395	66	2	20	3725	253	5	54	1925	119	3
59	611	97	2	21	3905	249	7	55	2105	120	3
10:00A	809	140	3	22	4085	248	7	56	2285	130	4
01	1007	151	3	23	4265	253	9	57	2465	140	4
02	1196	148	3	24	4445	258	10	58	2645	148	6
03	1385	145	1	25	4625	255	11	59	2825	154	7
04	1565	264	1	26	4805	251	11	2:00P	3005	160	8
05	1745	237	2					01	3185	170	9
06	1925	219	2	**22**	(272)			02	3365	173	10
07	2105	233	3	9:58A	395	81	3	03	3545	169	12
08	2285	246	4	59	611	97	2	04	3725	169	13
09	2465	263	4	10:00A	809	143	1	05	3905	170	14
10	2645	277	5	01	1007	70	2	06	4085	168	14
11	2825	286	5	02	1196	48	2	07	4265	168	14
12	3005	284	5	03	1385	60	2	08	4445	167	15
13	3185	279	5	04	1565	89	2	09	4625	165	14
14	3365	280	6	05	1745	121	2	10	4805	162	14
15	3545	282	7	06	1925	126	3	11	4985	164	16
16	3725	281	8	07	2105	126	3	12	5165	168	16
17	3905	280	8	08	2285	137	3	13	5345	171	16
18	4085	277	9	09	2465	140	4	14	5525	171	18
19	4265	275	9	10	2645	144	5	15	5705	169	18
20	4445	275	9	11	2825	147	6	16	5885	167	16
21	4625	274	9	12	3005	143	8	17	6065	169	16
22	4805	269	10	13	3185	146	9	18	6245	171	16
23	4985	266	10	14	3365	157	10	19	6425	170	16
24	5165	266	10	15	3545	162	11				
25	5345	265	10	16	3725	165	12	**23**	(274)		
26	5525	267	11	17	3905	174	13	10:55A	395	128	6
27	5705	268	11	18	4085	175	13	56	611	150	12

TABLE II (1928)

Date and time	Height in meters	Azimuth from; degrees	Velocity m/s
March 23			
10:57A	809	173	11
58	1007	212	7
59	1196	230	7
11:00A	1385	237	5
01	1565	244	4
02	1745	248	4
03	1925	250	7
04	2105	251	8
05	2285	252	7
06	2465	255	5
07	2645	241	6
08	2825	215	6
09	3005	198	7
10	3185	190	7
11	3365	187	7
23	(275)		
1:46P	395	110	3
47	611	147	4
48	809	175	5
49	1007	187	5
50	1196	189	5
51	1385	200	3
52	1565	218	3
53	1745	224	4
54	1925	207	5
55	2105	179	6
56	2285	161	7
57	2465	155	8
58	2645	157	8
59	2825	157	8
2:00P	3005	157	8
01	3185	148	4
02	3365	115	4
03	3545	107	3
04	3725	138	3
05	3905	176	4
06	4085	192	5
07	4265	204	7
08	4445	196	9
09	4625	178	9
10	4805	173	9
11	4985	175	10
12	5165	171	13
13	5345	172	13
14	5525	177	13
15	5705	176	14
16	5885	177	15

Date and time	Height in meters	Azimuth from; degrees	Velocity m/s
March 23			
2:17P	6065	177	16
18	6245	175	17
19	6425	174	18
20	6605	174	18
21	6785	176	19
22	6965	175	18
23	7145	174	17
24	(276)		
10:02A	395	67	4
03	611	90	5
04	809	102	4
05	1007	92	3
06	1196	88	4
07	1385	85	4
08	1565	71	4
09	1745	73	4
10	1925	76	5
11	2105	75	5
12	2285	85	5
13	2465	98	5
14	2645	108	5
15	2825	116	5
16	3005	129	4
17	3185	151	4
18	3365	154	5
19	3545	158	6
20	3725	170	8
21	3905	177	9
22	4085	177	10
23	4265	179	11
24	4445	179	10
25	4625	175	11
26	4805	177	12
27	4985	177	13
28	5165	172	14
29	5345	171	14
30	5525	170	14
31	5705	173	14
32	5885	172	14
33	6065	171	15
34	6245	176	16
35	6425	176	16
36	6605	182	15
37	6785	181	16
38	6965	178	15
39	7145	177	15

Date and time	Height in meters	Azimuth from; degrees	Velocity m/s
March 24			
10:40A	7325	177	16
41	7505	177	16
24	(277)		
2:02P	395	62	2
03	611	120	2
04	809	156	4
05	1007	157	4
06	1196	166	3
07	1385	172	3
08	1565	178	4
09	1745	204	4
10	1925	213	5
11	2105	204	5
12	2285	204	6
13	2465	208	6
14	2645	214	6
15	2825	215	7
16	3005	214	8
17	3185	209	7
18	3365	207	9
19	3545	203	10
20	3725	200	11
21	3905	203	11
22	4085	205	11
23	4265	210	11
24	4445	209	12
25	4625	204	15
26	4805	204	15
27	4985	206	14
28	5165	206	15
29	5345	206	14
30	5525	204	15
31	5705	199	18
32	5885	196	19
33	6065	201	19
34	6245	205	20
35	6425	207	20
25	(278)		
10:15A	395	109	3
16	611	118	9
17	809	126	9
18	1007	138	12
19	1196	145	18
20	1385	147	27
21	1565	152	36
22	1745	162	43

TABLE II (1928)

Date and time	Height in meters	Azimuth from; degrees	Velocity m/s
March 25			
10:23A	1925	174	32
24	2105	199	13
25	2285	201	7
26	2465	166	12
27	2645	158	22
28	2825	158	28
29	3005	158	30
30	3185	158	30
26	(279)		
10:02A	395	140	11
03	611	144	12
04	809	154	10
05	1007	159	12
06	1196	159	16
07	1385	161	17
08	1565	162	18
09	1745	163	21
10	1925	163	24
26	(280)		
2:00P	395	157	19
01	611	157	12
02	809	157	13
03	1007	157	14
04	1196	155	16
05	1385	156	21
06	1565	157	21
07	1745	159	18
08	1925	165	17
09	2105	175	17
10	2285	183	18
27	(281)		
10:10A	395	229	2
11	611	247	2
12	809	267	3
13	1007	274	1
14	1196	80	2
15	1385	83	4
16	1565	101	6
17	1745	121	9
18	1925	124	13
19	2105	118	16
20	2285	113	15
21	2465	115	13
22	2645	125	10
23	2825	132	9
24	3005	131	8

Date and time	Height in meters	Azimuth from; degrees	Velocity m/s
March 27	(282)		
2:14P	395	233	5
15	611	237	2
16	809	195	1
17	1007	154	3
18	1196	155	6
19	1385	158	9
20	1565	162	11
21	1745	161	13
22	1925	161	13
23	2105	161	12
24	2285	161	11
25	2465	161	11
26	2645	160	12
27	2825	154	13
28	3005	143	15
29	3185	133	16
30	3365	127	15
31	3545	124	16
32	3725	129	17
33	3905	135	17
34	4085	135	15
35	4265	129	14
36	4445	125	16
37	4625	126	18
38	4805	126	19
39	4985	126	19
40	5165	126	21
41	5345	127	19
42	5525	129	21
43	5705	131	20
44	5885	130	19
45	6065	130	19
46	6245	130	22
47	6425	130	23
48	6605	132	21
49	6785	130	21
50	6965	131	21
51	7145	133	18
52	7325	133	20
53	7505	138	21
54	7685	140	20
55	7865	138	20
28	(283)		
10:56A	395	91	1
57	611	126	1
58	809	149	2
59	1007	163	1

Date and time	Height in meters	Azimuth from; degrees	Velocity m/s
March 28			
11:00A	1196	280	1
01	1385	242	1
02	1565	198	1
03	1745	157	1
04	1925	156	3
05	2105	166	4
06	2285	176	5
07	2465	181	4
08	2645	179	4
09	2825	187	5
10	3005	192	5
11	3185	190	4
12	3365	207	5
13	3545	212	6
14	3725	201	5
15	3905	176	6
16	4085	164	8
17	4265	160	9
18	4445	162	12
19	4625	165	13
20	4805	170	18
21	4985	170	27
22	5165	168	34
23	5345	167	36
24	5525	166	37
25	5705	166	38
28	(284)		
2:48P	395	90	1
49	611	69	2
50	809	48	2
51	1007	35	1
52	1196	13	1
53	1385	37	1
54	1565	151	2
55	1745	174	4
56	1925	177	5
57	2105	177	5
58	2285	184	5
59	2465	199	5
3:00P	2645	199	5
01	2825	194	5
02	3005	197	5
03	3185	197	5
04	3365	197	5
05	3545	180	6
06	3725	174	8
07	3905	173	10

TABLE II (1928)

Date and time	Height in meters	Azimuth from; degrees	Velocity m/s
March 28			
3:08P	4085	171	13
09	4265	176	14
10	4445	190	11
11	4625	194	10
12	4805	188	13
13	4985	189	17
14	5165	190	18
15	5345	186	19
16	5525	182	18
17	5705	182	16
18	5885	185	16
19	6065	188	16
20	6245	189	15
29	(285)		
10:15A	395	76	5
16	611	81	3
17	809	102	2
18	1007	89	3
19	1196	84	5
20	1385	77	6
21	1565	78	5
22	1745	78	3
23	1925	88	2
24	2105	109	1
25	2285	112	1
26	2465	106	1
27	2645	95	1
28	2825	103	2
29	3005	102	2
30	3185	78	2
31	3365	53	3
32	3545	267	1
33	3725	245	6
34	3905	239	10
35	4085	234	11
36	4265	240	11
37	4445	253	12
38	4625	251	15
39	4805	247	17
40	4985	244	19
41	5165	242	21
29	(286)		
2:00P	395	33	3
01	611	54	3
02	809	128	2
03	1007	175	3

Date and time	Height in meters	Azimuth from; degrees	Velocity m/s
March 29			
2:04P	1196	178	3
05	1385	168	3
06	1565	165	3
07	1745	156	3
08	1925	145	3
09	2105	147	4
10	2285	151	4
11	2465	143	5
12	2645	134	4
13	2825	147	5
14	3005	172	7
15	3185	183	9
30	(287)		
10:12A	395	16	2
13	611	26	3
14	809	39	3
15	1007	60	4
16	1196	76	6
17	1385	82	7
18	1565	85	7
19	1745	92	7
20	1925	112	6
21	2105	135	5
22	2285	142	4
30	(288)		
2:04P	395	211	3
05	611	78	2
06	809	74	3
07	1007	70	3
08	1196	74	3
09	1385	117	2
10	1565	171	3
11	1745	192	4
12	1925	196	5
13	2105	188	5
14	2285	175	6
15	2465	166	6
31	(289)		
2:00P	395	329	2
01	611	347	4
02	809	11	4
03	1007	46	4
04	1196	35	5
05	1385	30	5

Date and time	Height in meters	Azimuth from; degrees	Velocity m/s
April 3	(290)		
10:00A	395	131	1
01	611	168	3
02	809	209	4
03	1007	228	4
04	1196	221	5
05	1385	220	5
06	1565	213	7
07	1745	203	7
08	1925	198	5
09	2105	199	3
10	2285	220	4
11	2465	227	6
12	2645	234	7
13	2825	234	7
14	3005	241	8
15	3185	245	9
16	3365	237	8
17	3545	238	7
18	3725	234	7
19	3905	232	6
20	4085	235	7
21	4265	238	8
22	4445	247	8
23	4625	255	5
24	4805	256	5
25	4985	258	7
26	5165	262	9
27	5345	265	9
28	5525	262	9
29	5705	259	8
30	5885	255	7
31	6065	248	6
32	6245	230	5
33	6425	215	5
34	6605	199	6
35	6785	193	8
36	6965	191	8
37	7145	190	7
38	7325	191	8
39	7505	192	9
40	7685	191	9
41	7865	188	9
42	8045	173	10
43	8225	164	10
44	8405	170	10
45	8585	186	10
46	8765	197	11
47	8945	203	11

TABLE II (1928)

Date and time	Height in meters	Azimuth from; degrees	Velocity m/s	Date and time	Height in meters	Azimuth from; degrees	Velocity m/s	Date and time	Height in meters	Azimuth from; degrees	Velocity m/s
April				**April**				**April**			
3				**4**				**8**			
10:48A	9125	212	12	2:42P	8045	247	29	2:10P	1565	199	4
49	9305	209	11	43	8225	247	28	11	1745	199	5
50	9485	204	11					12	1925	219	5
				6	(292)			13	2105	224	5
				2:25P	395	347	3	14	2285	220	4
4	(291)			26	611	21	3	15	2465	202	4
2:00P	395	326	5	27	809	67	3	16	2645	203	5
01	611	10	5	28	1007	64	2	17	2825	203	7
02	809	51	7	29	1196	21	1	18	3005	204	8
03	1007	45	6	30	1385	2	1	19	3185	202	9
04	1196	39	5	31	1565	338	2	20	3365	197	9
05	1385	24	3	32	1745	310	3	21	3545	185	9
06	1565	24	1	33	1925	290	4	22	3725	181	9
07	1745	105	1	34	2105	252	3	23	3905	182	9
08	1925	199	3	35	2285	210	3	24	4085	173	11
09	2105	194	6	36	2465	217	4	25	4265	174	12
10	2285	183	7	37	2645	222	5	26	4445	177	11
11	2465	190	7					27	4625	177	12
12	2645	190	7	**7**	(293)			28	4805	175	11
13	2825	190	5	2:00P	395	62	1	29	4985	170	12
14	3005	192	5	01	611	66	2	30	5165	170	13
15	3185	195	5	02	809	83	1	31	5345	166	13
16	3365	193	5	03	1007	187	2	32	5525	166	12
17	3545	198	5	04	1196	200	3	33	5705	168	13
18	3725	205	5	05	1385	208	4	34	5885	169	15
19	3905	204	7	06	1565	221	4	35	6065	169	17
20	4085	199	7	07	1745	228	5	36	6245	170	16
21	4265	205	8	08	1925	225	7	37	6425	169	15
22	4445	212	7	09	2105	219	8	38	6605	168	15
23	4625	209	7	10	2285	210	9	39	6785	169	15
24	4805	211	8	11	2465	202	8	40	6965	173	14
25	4985	225	8	12	2645	198	8	41	7145	177	12
26	5165	229	8	13	2825	198	9				
27	5345	222	9	14	3005	196	9				
28	5525	217	10	15	3185	193	10	**9**	(295)		
29	5705	217	10	16	3365	190	11	1:58P	395	80	5
30	5885	217	10	17	3545	185	14	59	611	96	7
31	6065	216	10	18	3725	185	18	2:00P	809	136	7
32	6245	217	11	19	3905	188	19	01	1007	167	11
33	6425	218	13	20	4085	190	19	02	1196	178	11
34	6605	230	14					03	1385	175	10
35	6785	242	15	**8**	(294)			04	1565	177	10
36	6965	247	18	2:04P	395	103	4	05	1745	178	9
37	7145	252	22	05	611	110	4	06	1925	179	10
38	7325	254	27	06	809	121	4	07	2105	180	11
39	7505	254	33	07	1007	136	4	08	2285	182	12
40	7685	252	33	08	1196	164	4	09	2465	188	12
41	7865	247	30	09	1385	201	3	10	2645	193	12

TABLE II (1928)

Date and time	Height in meters	Azimuth from; degrees	Velocity m/s	Date and time	Height in meters	Azimuth from; degrees	Velocity m/s	Date and time	Height in meters	Azimuth from; degrees	Velocity m/s
April				**April**				**April**			
10	(296)			**13**				**15**			
2:10P	395	53	7	2:22P	4445	266	24	1:42P	3005	203	20
11	611	53	9	23	4625	265	25	43	3185	203	12
12	809	44	9	24	4805	264	27	44	3365	203	9
13	1007	53	5	25	4985	266	29				
14	1196	120	5	26	5165	269	27	**16**	(301)		
15	1385	146	8	27	5345	270	24	2:15P	395	165	1
16	1565	152	9					16	611	266	1
17	1745	158	7	**14**	(299)			17	809	6	1
18	1925	172	6	2:27P	395	27	1	18	1007	78	2
19	2105	185	8	28	611	64	3	19	1196	87	3
20	2285	184	10	29	809	99	3	20	1385	84	2
21	2465	191	11	30	1007	127	3	21	1565	77	4
22	2645	209	11	31	1196	156	3	22	1745	86	5
23	2825	213	13	32	1385	192	2	23	1925	105	5
24	3005	208	16	33	1565	225	2	24	2105	130	4
25	3185	199	21	34	1745	225	3	25	2285	154	5
26	3365	192	21	35	1925	235	4	26	2465	170	7
27	3545	186	17	36	2105	249	4	27	2645	181	8
28	3725	186	15	37	2285	245	5	28	2825	179	9
				38	2465	251	6	29	3005	180	10
12	(297)			39	2645	253	7	30	3185	189	11
2:04P	395	256	1	40	2825	246	8	31	3365	192	11
05	611	305	1	41	3005	256	9	32	3545	196	12
06	809	340	1	42	3185	259	11	33	3725	201	12
				43	3365	251	13	34	3905	203	12
13	(298)			44	3545	246	13	35	4085	202	14
2:00P	395	302	5	45	3725	245	13	36	4265	209	16
01	611	273	3	46	3905	249	15	37	4445	215	16
02	809	206	3	47	4085	259	17	38	4625	220	15
03	1007	196	5	48	4265	262	18	39	4805	226	16
04	1196	192	6	49	4445	261	19	40	4985	223	17
05	1385	189	7	50	4625	262	21	41	5165	222	18
06	1565	203	7					42	5345	220	19
07	1745	230	6	**15**	(300)			43	5525	221	18
08	1925	239	7	1:28P	395	167	5	44	5705	225	17
09	2105	245	9	29	611	174	9	45	5885	226	17
10	2285	254	11	30	809	178	10	46	6065	226	19
11	2465	257	12	31	1007	180	12	47	6245	225	20
12	2645	260	13	32	1196	184	10	48	6425	225	18
13	2825	263	15	33	1385	190	13	49	6605	223	20
14	3005	262	16	34	1565	200	15	50	6785	217	20
15	3185	262	17	35	1745	224	11				
16	3365	266	18	36	1925	235	9	**17**	(302)		
17	3545	273	21	37	2105	236	11	2:03P	395	72	2
18	3725	274	22	38	2285	214	14	04	611	39	2
19	3905	273	24	39	2465	206	19	05	809	30	4
20	4085	271	25	40	2645	202	28	06	1007	44	4
21	4265	267	25	41	2825	203	29	07	1196	54	5

TABLE II (1928)

Date and time	Height in meters	Azimuth from; degrees	Veloc-ity m/s	Date and time	Height in meters	Azimuth from; degrees	Veloc-ity m/s	Date and time	Height in meters	Azimuth from; degrees	Veloc-ity m/s
April 17				**April 18**				**April 19**			
2:08P	1385	58	6	2:24P	2285	304	7	1:45P	2285	150	5
09	1565	70	5	25	2465	297	8	46	2465	156	6
10	1745	85	4	26	2645	285	8	47	2645	155	7
11	1925	122	4	27	2825	283	7	48	2825	152	8
12	2105	152	6	28	3005	288	4	49	3005	149	11
13	2285	154	7	29	3185	266	3	50	3185	145	12
14	2465	158	7	30	3365	245	4	51	3365	141	13
15	2645	167	7	31	3545	234	4	52	3545	139	13
16	2825	173	7	32	3725	224	5	53	3725	137	13
17	3005	176	7	33	3905	196	4	54	3905	136	12
18	3185	179	7	34	4085	161	4	55	4085	135	11
19	3365	178	8	35	4265	162	6	56	4265	126	9
20	3545	176	8	36	4445	167	7	57	4445	100	7
21	3725	177	9	37	4625	153	6	58	4625	88	7
22	3905	181	8	38	4805	156	6	59	4805	87	8
23	4085	190	8	39	4985	161	8	2:00P	4985	80	10
24	4265	193	10	40	5165	164	9	01	5165	77	13
25	4445	194	11	41	5345	161	9	02	5345	69	11
26	4625	197	11	42	5525	161	9	03	5525	63	7
27	4805	198	11	43	5705	169	8	04	5705	67	5
28	4985	197	12	44	5885	175	7	05	5885	50	4
29	5165	196	13	45	6065	175	8	06	6065	54	3
30	5345	196	13	46	6245	176	9	07	6245	76	3
31	5525	194	14	47	6425	175	10	08	6425	88	2
32	5705	191	15	48	6605	181	11	09	6605	30	1
33	5885	191	15	49	6785	178	13	10	6785	20	2
34	6065	196	15	50	6965	171	13	11	6965	24	4
35	6245	197	16	51	7145	168	12	12	7145	21	5
36	6425	197	15	52	7325	166	12	13	7325	3	7
37	6605	199	15	53	7505	166	14	14	7505	347	6
38	6785	198	15	54	7685	160	14	15	7685	334	8
39	6965	196	15	55	7865	151	13	16	7865	342	9
40	7145	189	15	56	8045	149	15	17	8045	336	9
41	7325	192	16	57	8225	153	18	18	8225	335	9
42	7505	197	17	58	8405	155	18	19	8405	328	10
43	7685	198	17	59	8585	153	16	20	8585	331	11
								21	8765	343	14
18	(303)			**19**	(304)			22	8945	344	13
2:14P	395	Calm	Calm	1:35P	395	102	6	23	9125	356	11
15	611	252	1	36	611	115	7	24	9305	352	13
16	809	282	2	37	809	122	8	25	9485	342	13
17	1007	321	2	38	1007	127	9	26	9665	338	13
18	1196	16	3	39	1196	129	8	27	9845	335	13
19	1385	13	5	40	1385	136	7	28	10025	339	13
20	1565	351	6	41	1565	153	8	29	10205	348	12
21	1745	341	6	42	1745	154	9	30	10385	4	11
22	1925	334	6	43	1925	144	8	31	10565	12	10
23	2105	315	5	44	2105	143	7	32	10745	12	9

TABLE II (1928)

Date and time	Height in meters	Azimuth from; degrees	Velocity m/s	Date and time	Height in meters	Azimuth from; degrees	Velocity m/s	Date and time	Height in meters	Azimuth from; degrees	Velocity m/s
April 19				**April 20**				**April 21**			
2:33P	10925	12	7	2:42P	7865	213	18	2:58P	4445	159	5
34	11105	29	3	43	8045	210	20	59	4625	164	4
35	11285	7	2	44	8225	207	19	3:00P	4805	137	3
36	11465	355	2	45	8405	208	16	01	4985	101	4
37	11645	7	2	46	8585	207	15	02	5165	95	4
				47	8765	198	15	03	5345	95	5
20	(305)			48	8945	192	12	04	5525	87	5
2:01P	395	71	4	49	9125	193	8	05	5705	70	6
02	611	90	7	50	9305	172	5	06	5885	62	7
03	809	92	9	51	9485	110	3	07	6065	55	7
04	1007	111	9	52	9665	124	6	08	6245	57	7
05	1196	123	8	53	9845	125	10	09	6425	54	7
06	1385	129	7	54	10025	120	10	10	6605	54	7
07	1565	128	7	55	10205	119	11	11	6785	48	7
08	1745	118	5	56	10385	109	13	12	6965	37	9
09	1925	80	3	57	10565	107	15	13	7145	33	9
10	2105	91	3	58	10745	110	17	14	7325	41	7
11	2285	106	4	59	10925	113	16	15	7505	50	6
12	2465	123	4	3:00P	11105	112	13	16	7685	46	5
13	2645	152	3	01	11285	119	12	17	7865	37	5
14	2825	187	4	02	11465	114	10	18	8045	36	6
15	3005	196	7	03	11645	100	10	19	8225	46	7
16	3185	187	11	04	11825	108	12	20	8405	59	9
17	3365	183	14	05	12005	118	12	21	8585	61	11
18	3545	181	12					22	8765	60	9
19	3725	179	8	**21**	(306)			23	8945	59	9
20	3905	171	6	2:36P	395	52	4	24	9125	67	12
21	4085	151	5	37	611	53	3	25	9305	73	12
22	4265	133	6	38	809	44	5	26	9485	74	10
23	4445	141	4	39	1007	50	6	27	9665	76	12
24	4625	198	4	40	1196	71	5	28	9845	86	13
25	4805	215	5	41	1385	75	5	29	10025	92	10
26	4985	201	10	42	1565	74	5	30	10205	80	10
27	5165	194	14	43	1745	74	6	31	10385	74	12
28	5345	193	13	44	1925	86	5	32	10565	74	13
29	5525	200	13	45	2105	107	5	33	10745	79	15
30	5705	200	11	46	2285	111	5	34	10925	82	14
31	5885	189	11	47	2465	113	5	35	11105	82	13
32	6065	194	12	48	2645	135	3	36	11285	84	9
33	6245	218	12	49	2825	170	2	37	11465	85	10
34	6425	226	13	50	3005	154	3				
35	6605	221	12	51	3185	163	3	**22**	(307)		
36	6785	219	14	52	3365	177	3	2:02P	395	307	1
37	6965	218	16	53	3545	160	4	03	611	315	1
38	7145	212	15	54	3725	152	4	04	809	333	1
39	7325	221	18	55	3905	160	4	05	1007	249	1
40	7505	226	19	56	4085	144	4	06	1196	231	1
41	7685	221	15	57	4265	146	5	07	1385	18	2

TABLE II (1928)

Date and time	Height in meters	Azimuth from; degrees	Velocity m/s	Date and time	Height in meters	Azimuth from; degrees	Velocity m/s	Date and time	Height in meters	Azimuth from; degrees	Velocity m/s
April 22				**April 22**				**April 24**			
2:08P	1565	17	3	2:56P	10205	114	8	2:42P	1196	357	6
09	1745	25	2	57	10385	104	9	43	1385	358	5
10	1925	21	3	58	10565	98	8	44	1565	351	5
11	2105	23	3	59	10745	88	5	45	1745	339	4
12	2285	19	4	3:00P	10925	96	8	46	1925	339	4
13	2465	14	4	01	11105	84	3	47	2105	345	5
14	2645	11	5	02	11285	172	2	48	2285	340	6
15	2825	18	5	03	11465	142	5	49	2465	339	7
16	3005	36	5					50	2645	334	9
17	3185	43	6	**23**	(308)			51	2825	334	11
18	3365	42	5	2:04P	395	6	6	52	3005	333	11
19	3545	41	5	05	611	5	5	53	3185	333	13
20	3725	40	4	06	809	12	3	54	3365	329	13
21	3905	36	5	07	1007	28	2	55	3545	327	13
22	4085	38	5	08	1196	23	5	56	3725	328	13
23	4265	43	6	09	1385	29	7	57	3905	327	14
24	4445	47	6	10	1565	28	9	58	4085	328	17
25	4625	58	5	11	1745	21	10	59	4265	328	17
26	4805	60	6	12	1925	20	10	3:00P	4445	330	15
27	4985	59	7	13	2105	16	9	01	4625	325	16
28	5165	58	8	14	2285	14	9	02	4805	323	17
29	5345	56	8	15	2465	9	8	03	4985	323	18
30	5525	59	8	16	2645	10	9	04	5165	320	18
31	5705	68	7	17	2825	8	10	05	5345	319	20
32	5885	91	7	18	3005	4	11				
33	6065	96	8	19	3185	6	12	**26**	(310)		
34	6245	91	8	20	3365	6	13	2:25P	395	180	4
35	6425	87	9	21	3545	2	12	26	611	188	5
36	6605	87	10	22	3725	358	10	27	809	205	3
37	6785	89	10	23	3905	351	11	28	1007	245	3
38	6965	89	10	24	4085	347	11	29	1196	255	4
39	7145	94	11	25	4265	355	11	30	1385	255	3
40	7325	102	12	26	4445	3	12				
41	7505	106	12	27	4625	1	11	**26**	(311)		
42	7685	106	12	28	4805	357	12	3:05P	395	276	4
43	7865	103	12	29	4985	352	13	06	611	276	2
44	8045	100	13	30	5165	345	11	07	809	233	2
45	8225	108	13	31	5345	340	9				
46	8405	107	11	32	5525	347	11	**27**	(312)		
47	8585	107	10	33	5705	351	12	1:00P	395	106	3
48	8765	122	11	34	5885	351	13	01	611	107	4
49	8945	122	12	35	6065	351	13	02	809	108	6
50	9125	120	12					03	1007	120	5
51	9305	112	13	**24**	(309)			04	1196	145	6
52	9485	106	14	2:38P	395	27	3	05	1385	161	10
53	9665	104	11	39	611	13	2	06	1565	169	10
54	9845	104	10	40	809	3	3	07	1745	175	9
55	10025	109	11	41	1007	0	4	08	1925	188	10

TABLE II (1928)

Date and time	Height in meters	Azimuth from; degrees	Velocity m/s	Date and time	Height in meters	Azimuth from; degrees	Velocity m/s	Date and time	Height in meters	Azimuth from; degrees	Velocity m/s
April				**April**				**May**			
27				**29**				**1**			
1:09P	2105	177	11	2:29P	2105	189	3	2:22P	1565	180	4
10	2285	159	11	30	2285	175	3	23	1745	168	4
11	2645	158	11	31	2465	147	4	24	1925	163	5
12	2645	145	12	32	2645	129	4	25	2105	170	7
13	2825	149	12					26	2285	170	7
14	3005	163	12	**30**	(314)			27	2465	166	6
15	3185	157	12	1:38P	395	52	2	28	2645	174	7
16	3365	153	11	39	611	50	2	29	2825	179	7
17	3545	156	9	40	809	100	3	30	3005	176	7
18	3725	152	10	41	1007	114	3	31	3185	176	7
19	3905	145	11	42	1196	109	3	32	3365	176	7
20	4085	137	10	43	1385	120	3	33	3545	178	6
21	4265	134	10	44	1565	116	3	34	3725	185	6
22	4445	137	9	45	1745	128	2	35	3905	190	5
23	4625	137	9	46	1925	147	2	36	4085	202	5
24	4805	137	10	47	2105	204	1	37	4265	210	7
25	4985	137	11	48	2285	219	1	38	4445	202	8
26	5165	138	9	49	2465	185	3	39	4625	202	8
27	5345	146	8	50	2645	187	5	40	4805	208	8
28	5525	149	9	51	2825	184	6	41	4985	209	9
29	5705	155	8	52	3005	187	6	42	5165	192	13
30	5885	159	8	53	3185	184	8	43	5345	189	12
31	6065	163	9	54	3365	181	9	44	5525	192	14
32	6245	168	9	55	3545	181	9	45	5705	185	14
33	6425	175	11	56	3725	188	9	46	5885	178	14
34	6605	173	10	57	3905	194	9	47	6065	182	14
35	6785	170	10	58	4085	194	10	48	6245	184	14
36	6965	168	10	59	4265	197	10	49	6425	189	13
37	7145	184	8	2:00P	4445	204	9	50	6605	192	13
38	7325	198	12	01	4625	205	9				
39	7505	190	10	02	4805	201	8	**2**	(316)		
40	7685	192	8	03	4985	215	7	2:00P	395	21	6
41	7865	192	8	04	5165	228	7	01	611	17	4
42	8045	191	9	05	5345	228	5	02	809	16	3
43	8225	191	10	06	5525	251	6	03	1007	25	4
44	8405	190	11	07	5705	301	8	04	1196	33	4
45	8585	190	11	08	5885	314	12	05	1385	60	3
				09	6065	316	14	06	1565	99	2
29	(313)			10	6245	320	14	07	1745	125	1
2:20P	395	183	4					08	1925	185	1
21	611	199	3	**May**				09	2105	221	3
22	809	223	3	**1**	(315)			10	2285	219	5
23	1007	232	3	2:16P	395	185	4	11	2465	211	6
24	1196	221	4	17	611	225	2				
25	1385	215	5	18	809	299	1	**3**	(317)		
26	1565	214	3	19	1007	358	2	2:06P	395	35	5
27	1745	214	4	20	1196	46	1	07	611	33	7
28	1925	206	4	21	1385	165	2	08	809	33	6

TABLE II (1928)

Date and time	Height in meters	Azimuth from; degrees	Velocity m/s
May 3			
2:09P	1007	11	4
10	1196	10	2
11	1385	10	2
12	1565	341	2
13	1745	325	5
14	1925	319	6
15	2105	310	6
16	2285	298	5
17	2465	295	4
18	2645	304	5
19	2825	315	6
20	3005	317	7
21	3185	313	8
22	3365	305	8
23	3545	295	8
24	3725	295	9
25	3905	294	12
26	4085	291	13
27	4265	292	14
28	4445	296	15
29	4625	296	16
30	4805	288	17
31	4985	291	17
32	5165	289	19
33	5345	290	19
34	5525	291	18
35	5705	291	17
36	5885	291	17
37	6065	291	17
4	(318)		
2:26P	395	105	3
27	611	101	3
28	809	110	3
29	1007	136	2
30	1196	161	3
31	1385	177	3
32	1565	212	2
33	1745	236	2
34	1925	247	1
35	2105	243	3
36	2285	255	3
37	2465	274	4
38	2645	286	5
39	2825	286	7
40	3005	279	6
41	3185	281	6
42	3365	285	8

Date and time	Height in meters	Azimuth from; degrees	Velocity m/s
May 4			
2:43P	3545	289	10
44	3725	300	13
45	3905	298	13
46	4085	297	12
47	4265	291	12
48	4445	288	11
49	4625	291	10
50	4805	291	9
51	4985	292	10
52	5165	293	12
53	5345	292	14
54	5525	287	14
55	5705	279	15
56	5885	284	16
57	6065	286	16
58	6245	294	16
5	(319)		
2:02P	395	180	7
03	611	181	13
04	809	180	13
05	1007	180	11
06	1196	185	10
07	1385	192	8
08	1565	205	8
09	1745	206	11
10	1925	200	12
11	2105	196	14
12	2285	199	15
13	2465	206	14
7	(320)		
2:16P	395	94	2
17	611	44	2
18	809	56	2
19	1007	102	3
20	1196	139	5
21	1385	163	7
22	1565	184	8
23	1745	197	9
24	1925	208	9
25	2105	210	9
26	2285	210	8
27	2465	212	7
28	2645	219	7
29	2825	218	9
30	3005	203	11
31	3185	192	13

Date and time	Height in meters	Azimuth from; degrees	Velocity m/s
May 7			
2:32P	3365	189	15
33	3545	189	15
34	3725	190	16
35	3905	191	16
36	4085	191	17
37	4265	191	15
38	4445	186	15
39	4625	180	16
40	4805	183	17
41	4985	184	17
42	5165	184	18
43	5345	184	19
44	5525	188	17
45	5705	190	18
46	5885	190	19
8	(321)		
1:51P	395	87	5
52	611	99	5
53	809	105	5
54	1007	135	6
55	1196	154	7
56	1385	150	7
57	1565	153	7
58	1745	159	5
59	1925	159	4
2:00P	2105	147	4
01	2285	142	5
02	2465	147	6
03	2645	148	7
9	(322)		
1:56P	395	183	5
57	611	199	7
58	809	203	6
59	1007	183	6
2:00P	1196	169	7
01	1385	171	6
02	1565	169	7
03	1745	166	7
10	(323)		
2:07P	395	188	5
08	611	203	3
09	809	201	5
10	1007	198	5
11	1196	197	6
12	1385	194	7

TABLE II (1928)

Date and time	Height in meters	Azimuth from; degrees	Velocity m/s
May 10			
2:13P	1565	189	8
14	1745	188	9
15	1925	189	10
16	2105	194	12
17	2285	196	13
18	2465	199	14
19	2645	203	14
20	2825	207	13
21	3005	206	11
22	3185	202	10
23	3365	204	9
11	(324)		
1:58P	395	218	7
59	611	211	7
2:00P	809	195	7
01	1007	186	7
02	1196	181	7
03	1385	175	6
04	1565	166	7
05	1745	164	9
06	1925	164	9
07	2105	178	8
08	2285	197	7
09	2465	209	6
10	2645	211	5
11	2825	202	5
12	3005	204	5
13	3185	220	6
14	3365	234	7
15	3545	238	7
16	3725	234	9
17	3905	224	11
18	4085	221	12
19	4265	227	12
12	(325)		
2:05P	395	165	12
06	611	162	14
07	809	162	14
08	1007	162	14
13	(326)		
1:45P	395	223	12
46	611	216	9
47	809	209	8
48	1007	195	7
49	1196	193	7

Date and time	Height in meters	Azimuth from; degrees	Velocity m/s
May 13			
1:50P	1385	194	8
51	1565	199	9
52	1745	200	10
14	(327)		
2:11P	395	196	2
12	611	233	2
13	809	222	1
14	1007	207	1
15	1196	177	1
16	1385	170	2
17	1565	169	3
18	1745	165	4
19	1925	176	5
20	2105	178	7
21	2285	174	9
22	2465	171	9
23	2645	170	9
24	2825	170	8
25	3005	172	9
26	3185	179	12
27	3365	192	14
28	3545	199	15
29	3725	201	16
30	3905	203	17
31	4085	205	15
32	4265	210	15
33	4445	212	15
34	4625	214	17
35	4805	215	20
15	(328)		
1:52P	395	269	5
53	611	271	6
54	809	267	5
55	1007	227	3
56	1196	197	6
57	1385	195	11
58	1565	193	12
59	1745	190	12
2:00P	1925	194	12
01	2105	207	12
02	2285	221	12
03	2465	230	13
04	2645	236	14
05	2825	239	14
06	3005	237	17
07	3185	237	19

Date and time	Height in meters	Azimuth from; degrees	Velocity m/s
May 15			
2:08P	3365	232	20
09	3545	231	21
10	3725	231	21
16	(329)		
1:56P	395	165	11
57	611	175	16
58	809	177	16
59	1007	182	12
2:00P	1196	196	8
01	1385	229	7
02	1565	259	7
17	(330)		
1:52P	395	99	4
53	611	132	5
54	809	164	5
55	1007	184	5
56	1196	191	5
57	1385	189	4
58	1565	199	3
59	1745	201	5
2:00P	1925	195	5
01	2105	200	5
02	2285	198	5
03	2465	194	5
18	(331)		
1:50P	395	313	5
51	611	288	4
52	809	234	5
53	1007	203	8
54	1196	200	11
55	1385	198	11
56	1565	184	8
57	1745	176	9
58	1925	181	11
19	(332)		
2:00P	395	289	4
01	611	282	4
02	809	280	5
03	1007	298	4
04	1196	318	4
20	(333)		
6:30P	395	340	4
31	611	290	2
32	809	223	2
33	1007	166	2

TABLE II (1928)

Date and time	Height in meters	Azimuth from; degrees	Velocity m/s
May 20			
6:34P	1196	140	1
35	1385	156	1
36	1565	164	2
37	1745	172	3
38	1925	166	4
39	2105	149	4
40	2285	150	4
41	2465	160	5
42	2645	172	6
43	2825	168	6
44	3005	162	6
45	3185	162	6
46	3365	165	6
47	3545	189	5
48	3725	194	6
49	3905	177	7
50	4085	161	7
51	4265	152	7
52	4445	145	8
53	4625	148	8
54	4805	154	9
55	4985	161	10
56	5165	179	9
57	5345	226	7
58	5525	282	9
59	5705	299	13
7:00P	5885	302	17
01	6065	308	23
02	6245	307	27
03	6425	305	27
04	6605	302	28
05	6785	301	28
06	6965	308	31
07	7145	300	28
08	7325	291	27
09	7505	297	29
10	7685	296	31
11	7865	298	30
12	8045	298	29
13	8225	297	29
21	(334)		
2:21P	395	150	6
22	611	159	7
23	809	161	8
24	1007	166	10
25	1196	176	9
26	1385	185	9

Date and time	Height in meters	Azimuth from; degrees	Velocity m/s
May 21			
2:27P	1565	196	9
28	1745	217	8
29	1925	235	8
30	2105	239	9
22	(335)		
1:50P	395	164	11
51	611	166	16
52	809	172	16
53	1007	176	16
54	1196	173	11
55	1385	176	8
56	1565	197	9
57	1745	210	11
58	1925	219	15
59	2105	221	16
2:00P	2285	220	18
23	(336)		
1:55P	395	187	5
56	611	185	10
57	809	187	11
58	1007	192	10
59	1196	198	7
2:00P	1385	208	5
01	1565	220	4
02	1745	212	3
03	1925	171	3
04	2105	181	4
05	2285	198	7
06	2465	199	7
07	2645	196	7
08	2825	190	6
09	3005	183	5
10	3185	193	6
11	3365	193	6
12	3545	190	6
24	(337)		
2:18P	395	182	8
19	611	176	10
20	809	177	9
21	1007	182	8
22	1196	192	8
23	1385	199	8
24	1565	202	7

Date and time	Height in meters	Azimuth from; degrees	Velocity m/s
May 26	(338)		
1:38P	395	336	4
39	611	333	2
40	809	332	2
41	1007	334	3
42	1196	1	2
43	1385	13	1
44	1565	337	2
45	1745	8	3
46	1925	350	2
47	2105	306	3
48	2285	304	2
49	2465	255	1
50	2645	262	2
51	2825	292	2
52	3005	285	3
53	3185	276	3
54	3365	269	3
55	3545	273	3
56	3725	279	4
57	3905	279	3
58	4085	279	3
59	4265	296	3
2:00P	4445	311	5
01	4625	321	5
02	4805	326	5
03	4985	334	5
04	5165	340	6
05	5345	332	8
06	5525	328	9
07	5705	333	8
08	5885	329	9
09	6065	326	10
10	6245	330	9
11	6425	337	8
12	6605	344	9
13	6785	345	9
14	6965	343	11
15	7145	341	13
16	7325	339	13
17	7505	342	12
18	7685	347	11
19	7865	349	11
20	8045	341	12
21	8225	341	13
22	8405	346	14
23	8585	344	14
24	8765	344	14

TABLE II (1928)

Date and time	Height in meters	Azimuth from; degrees	Velocity m/s	Date and time	Height in meters	Azimuth from; degrees	Velocity m/s	Date and time	Height in meters	Azimuth from; degrees	Velocity m/s
May				**May**				**May**			
27	(339)			**29**	(341)			**29**			
1:55P	395	28	6	10:29A	395	325	2	2:09P	2465	290	4
56	611	27	5	30	611	219	1	10	2645	290	4
57	809	20	4	31	809	169	2	11	2825	281	5
58	1007	17	3	32	1007	172	1	12	3005	276	7
59	1196	17	5	33	1196	233	3	13	3185	279	7
2:00P	1385	17	7	34	1385	246	3	14	3365	283	9
01	1565	14	7	35	1565	246	6	15	3545	275	9
02	1745	10	7	36	1745	248	6	16	3725	273	8
03	1925	10	7	37	1925	266	5	17	3905	280	8
04	2105	10	8	38	2105	303	3	18	4085	284	8
05	2285	10	9	39	2285	304	5	19	4265	290	8
06	2465	5	10	40	2465	295	5	20	4445	291	9
07	2645	358	9	41	2645	289	6	21	4625	291	10
08	2825	354	7	42	2825	290	6	22	4805	291	10
09	3005	352	5	43	3005	287	6	23	4985	287	12
10	3185	324	7	44	3185	289	7	24	5165	286	14
11	3365	304	10	45	3365	296	8	25	5345	287	13
12	3545	308	11	46	3545	296	9	26	5525	287	12
13	3725	320	10	47	3725	304	9	27	5705	287	12
14	3905	313	10	48	3905	299	10	28	5885	288	12
15	4085	310	11	49	4085	293	10	29	6065	288	13
16	4265	320	11	50	4265	296	10	30	6245	290	14
17	4445	318	11	51	4445	297	10	31	6425	291	14
18	4625	311	11	52	4625	299	11	32	6605	290	15
19	4805	307	10	53	4805	301	11	33	6785	291	15
20	4985	302	9	54	4985	302	12	34	6965	292	15
21	5165	299	10	55	5165	298	13	35	7145	292	16
22	5345	297	11	56	5345	298	14	36	7325	292	17
23	5525	300	13	57	5525	298	14				
24	5705	301	14	58	5705	299	14	**30**	(343)		
25	5885	299	15	59	5885	302	15	9:59A	395	120	3
26	6065	298	15	11:00A	6065	303	15	10:00A	611	334	1
27	6245	301	14	01	6245	303	15	01	809	316	1
28	6425	301	14	02	6425	303	17	02	1007	299	1
29	6605	293	13	03	6605	303	17	03	1196	36	1
30	6785	287	13					04	1385	46	1
31	6965	285	14	**29**	(342)			05	1565	53	2
32	7145	289	14	1:58P	395	157	2	06	1745	68	2
33	7325	295	15	59	611	186	1	07	1925	93	2
				2:00P	809	251	1	08	2105	110	2
				01	1007	256	2	09	2285	122	3
28	(340)			02	1196	248	2	10	2465	131	3
1:48P	395	32	5	03	1385	244	2	11	2645	135	3
49	611	23	4	04	1565	241	4	12	2825	141	2
50	809	25	3	05	1745	246	5	13	3005	137	2
51	1007	30	4	06	1925	256	4	14	3185	145	3
52	1196	21	5	07	2105	260	7	15	3365	161	3
53	1385	12	7	08	2285	268	7	16	3545	174	3

TABLE II (1928)

Date and time	Height in meters	Azimuth from; degrees	Velocity m/s
May 30			
10:17A	3725	180	3
18	3905	182	3
19	4085	182	4
20	4265	184	4
21	4445	184	4
22	4625	193	4
23	4805	199	4
24	4985	202	4
25	5165	209	4
26	5345	217	4
27	5525	221	4
28	5705	232	3
29	5885	234	3
30	6065	245	3
31	6245	258	4
32	6425	263	3
33	6605	268	3
34	6785	270	4
35	6965	264	4
36	7145	266	5
37	7325	266	7
38	7505	251	8
39	7685	245	8
40	7865	245	9
41	8045	240	10
42	8225	240	10
30	(344)		
1:54P	395	143	4
55	611	102	3
56	809	71	3
57	1007	38	2
58	1196	343	3
59	1385	344	4
2:00P	1565	356	4
01	1745	3	3
02	1925	55	3
03	2105	92	1
04	2285	159	1
05	2465	164	1
06	2645	170	2
07	2825	172	3
08	3005	169	4
09	3185	171	5
10	3365	171	5
11	3545	170	5
12	3725	174	5
13	3905	178	5

Date and time	Height in meters	Azimuth from; degrees	Velocity m/s
May 30			
2:14P	4085	177	6
15	4265	177	6
16	4445	182	5
17	4625	191	5
18	4805	198	5
19	4985	207	5
20	5165	217	5
21	5345	225	5
22	5525	236	4
23	5705	253	4
24	5885	263	4
25	6065	265	5
26	6245	265	5
27	6425	249	5
28	6605	251	5
29	6785	260	5
30	6965	261	5
31	7145	261	5
32	7325	265	5
33	7505	265	5
34	7685	258	6
35	7865	265	6
36	8045	270	7
37	8225	257	7
38	8405	253	7
39	8585	254	8
40	8765	257	7
41	8945	259	8
42	9125	264	8
43	9305	269	9
44	9485	271	10
45	9665	269	10
46	9845	267	9
47	10025	255	8
48	10205	255	9
49	10385	250	8
50	10565	249	10
51	10745	250	11
31	(345)		
1:51P	395	148	4
52	611	131	3
53	809	133	2
54	1007	133	2
55	1196	106	1
56	1385	155	1
57	1565	208	1
58	1745	293	1

Date and time	Height in meters	Azimuth from; degrees	Velocity m/s
May 31			
1:59P	1925	329	1
2:00P	2105	272	1
01	2285	234	3
02	2465	202	4
03	2645	188	4
04	2825	195	4
05	3005	200	4
06	3185	196	4
07	3365	188	6
08	3545	185	6
09	3725	191	6
10	3905	193	6
11	4085	190	7
12	4265	190	6
13	4445	190	6
14	4625	194	6
15	4805	203	7
16	4985	206	8
17	5165	209	9
18	5345	213	10
19	5525	218	10
20	5705	229	10
21	5885	230	10
22	6065	231	10
23	6245	231	10
24	6425	238	10
25	6605	253	10
26	6785	258	9
27	6965	265	9
28	7145	269	10
29	7325	265	12
30	7505	263	11
31	7685	271	13
32	7865	280	16
33	8045	278	15
34	8225	279	16
35	8405	282	15
36	8585	279	16
37	8765	279	16
38	8945	280	15
39	9125	279	16
40	9305	280	17
41	9485	264	16
42	9665	265	17
43	9845	266	17
44	10025	267	17

TABLE II (1928)

Date and time	Height in meters	Azimuth from; degrees	Velocity m/s
June			
1	(346)		
2:01P	395	133	2
02	611	33	2
03	809	15	3
04	1007	22	3
05	1196	30	3
06	1385	28	3
07	1565	30	3
08	1745	30	2
09	1925	19	3
10	2105	22	3
11	2285	112	1
12	2465	158	4
13	2645	157	6
14	2825	147	8
15	3005	138	10
1	(347)		
4:31P	395	91	4
32	611	85	4
33	809	72	3
34	1007	52	3
35	1196	30	2
36	1385	22	2
37	1565	35	3
38	1745	49	4
39	1925	44	5
40	2105	52	6
41	2285	74	3
42	2465	165	3
43	2645	174	4
44	2825	160	5
45	3005	158	7
46	3185	158	8
47	3365	168	7
48	3545	195	6
49	3725	205	6
50	3905	209	6
51	4085	199	5
52	4265	174	6
53	4445	160	6
54	4625	149	5
55	4805	145	5
56	4985	145	5
57	5165	140	4
58	5345	148	4
59	5525	164	4
5:00P	5705	164	3
01	5885	162	4

Date and time	Height in meters	Azimuth from; degrees	Velocity m/s
June			
2	(348)		
9:58P	395	33	6
59	611	93	5
10:00P	809	111	6
01	1007	117	7
02	1196	125	7
03	1385	134	8
04	1565	130	8
05	1745	114	7
06	1925	108	6
07	2105	114	6
08	2285	126	9
09	2465	130	10
10	2645	137	11
11	2825	143	11
12	3005	145	10
13	3185	149	11
14	3365	160	10
15	3545	165	9
16	3725	160	8
17	3905	147	8
18	4085	138	9
19	4265	144	8
20	4445	152	10
21	4625	154	10
22	4805	155	9
23	4985	156	10
24	5165	160	13
25	5345	162	14
26	5525	160	16
27	5705	164	16
28	5885	172	14
29	6065	173	14
30	6245	175	12
31	6425	179	11
32	6605	180	12
33	6785	166	11
34	6965	169	10
35	7145	176	9
36	7325	183	9
37	7505	182	11
2	(349)		
1:58P	395	107	7
59	611	133	7
2:00P	809	143	11
01	1007	145	13
02	1196	147	13
03	1385	150	10

Date and time	Height in meters	Azimuth from; degrees	Velocity m/s
June			
2			
2:04P	1565	165	8
05	1745	186	7
06	1925	185	6
07	2105	162	7
08	2285	151	10
09	2465	139	10
10	2645	130	9
11	2825	130	8
12	3005	146	8
13	3185	167	6
14	3365	175	5
15	3545	152	5
16	3725	145	6
17	3905	140	7
18	4085	140	8
3	(350)		
9:53A	395	33	4
54	611	92	4
55	809	141	5
56	1007	160	6
57	1196	174	8
58	1385	168	8
59	1565	189	11
10:00A	1745	200	15
01	1925	194	14
3	(351)		
2:10P	395	196	7
11	611	194	9
12	809	193	13
13	1007	194	14
14	1196	196	13
15	1385	198	13
16	1565	198	14
17	1745	196	14
18	1925	193	12
19	2105	200	12
20	2285	207	11
21	2465	210	13
22	2645	213	15
23	2825	214	15
4	(352)		
9:51A	395	198	4
52	611	327	1
53	809	17	4
54	1007	22	6

TABLE II (1928)

Date and time	Height in meters	Azimuth from; degrees	Velocity m/s	Date and time	Height in meters	Azimuth from; degrees	Velocity m/s	Date and time	Height in meters	Azimuth from; degrees	Velocity m/s
June 4				**June 4**				**June 4**			
9:55A	1196	33	5	10:43A	9845	205	9	2:24P	5345	141	13
56	1385	48	4	44	10025	208	9	25	5525	140	14
57	1565	64	5	45	10205	209	9	26	5705	139	13
58	1745	81	4	46	10385	209	8	27	5885	135	15
59	1925	100	3	47	10565	201	7	28	6065	137	14
10:00A	2105	108	4	48	10745	200	10	29	6245	140	13
01	2285	116	5	49	10925	210	14	30	6425	140	15
02	2465	125	6	50	11105	212	12				
03	2645	129	7	51	11285	187	11	**5**	(354)		
04	2825	131	9	52	11465	196	13	9:55A	395	96	3
05	3005	131	10	53	11645	216	11	56	611	49	3
06	3185	134	9	54	11825	205	8	57	809	34	4
07	3365	134	9	55	12005	194	9	58	1007	20	5
08	3545	130	9	56	12185	195	10	59	1196	16	4
09	3725	129	8	57	12365	195	9	10:00A	1385	47	3
10	3905	132	8	58	12545	206	8	01	1565	55	3
11	4085	133	9	59	12725	218	6	02	1745	64	3
12	4265	134	9	11:00A	12905	221	4	03	1925	91	2
13	4445	138	8	01	13085	222	2	04	2105	109	2
14	4625	135	9					05	2285	105	2
15	4805	131	10	**4**	(353)			06	2465	121	1
16	4985	130	10	1:57P	395	172	4	07	2645	301	1
17	5165	127	10	58	611	144	1	08	2825	299	2
18	5345	126	10	59	809	13	2	09	3005	262	1
19	5525	135	10	2:00P	1007	0	4	10	3185	185	2
20	5705	138	10	01	1196	352	4	11	3365	177	4
21	5885	140	10	02	1385	359	4	12	3545	187	7
22	6065	141	9	03	1565	10	1	13	3725	198	7
23	6245	136	11	04	1745	184	1	14	3905	204	7
24	6425	131	12	05	1925	183	2	15	4085	203	7
25	6605	131	11	06	2105	181	3	16	4265	202	7
26	6785	136	10	07	2285	180	4	17	4445	200	8
27	6965	142	10	08	2465	169?	5?	18	4625	199	8
28	7145	144	10	09	2645	158?	6?				
29	7325	142	10	10	2825	147	7	**5**	(355)		
30	7505	147	9	11	3005	158	8	1:56P	395	100	5
31	7685	147	8	12	3185	165	11	57	611	77	4
32	7865	152	7	13	3365	161	13	58	809	39	4
33	8045	163	7	14	3545	156	13	59	1007	21	4
34	8225	173	5	15	3725	145	11	2:00P	1196	32	3
35	8405	190	5	16	3905	132	9	01	1385	51	3
36	8585	185	6	17	4085	139	9	02	1565	50	4
37	8765	180	7	18	4265	153	11	03	1745	42	5
38	8945	186	7	19	4445	156	14	04	1925	40	4
39	9125	189	7	20	4625	152	16	05	2105	50	3
40	9305	182	8	21	4805	150	14	06	2285	59	3
41	9485	183	9	22	4985	146	13	07	2465	49	4
42	9665	190	9	23	5165	145	14	08	2645	46	5

TABLE II (1928)

Date and time	Height in meters	Azimuth from; degrees	Velocity m/s	Date and time	Height in meters	Azimuth from; degrees	Velocity m/s	Date and time	Height in meters	Azimuth from; degrees	Velocity m/s
June 5				**June 5**				**June 6**			
2:09P	2825	58	4	2:57P	11465	149	6	10:31A	8225	112	4
10	3005	96	4	58	11645	147	13	32	8405	114	3
11	3185	131	4	59	11825	146	8	33	8585	130	3
12	3365	159	4					34	8765	144	3
13	3545	181	5	**6**	(356)			35	8945	147	3
14	3725	181	7	9:47A	395	95	4	36	9125	137	4
15	3905	177	8	48	611	68	3	37	9305	125	3
16	4085	176	9	49	809	31	6	38	9485	110	2
17	4265	174	8	50	1007	14	7	39	9665	114	3
18	4445	178	9	51	1196	21	6	40	9845	133	4
19	4625	180	9	52	1385	37	6	41	10025	146	4
20	4805	184	8	53	1565	40	5	42	10205	147	5
21	4985	182	7	54	1745	39	5	43	10385	135	6
22	5165	185	8	55	1925	28	6	44	10565	130	7
23	5345	194	8	56	2105	31	5	45	10745	141	7
24	5525	200	8	57	2285	30	5	46	10925	148	6
25	5705	201	9	58	2465	32	5	47	11105	136	3
26	5885	197	9	59	2645	31	4	48	11285	112	4
27	6065	192	9	10:00A	2825	52	3	49	11465	105	3
28	6245	193	8	01	3005	86	3	50	11645	127	3
29	6425	194	8	02	3185	79	2	51	11825	127	3
30	6605	196	8	03	3365	102	2	52	12005	133	2
31	6785	200	7	04	3545	132	3	53	12185	115	1
32	6965	208	7	05	3725	154	4	54	12365	114	3
33	7145	207	6	06	3905	166	4	55	12545	115	8
34	7325	208	5	07	4085	167	4	56	12725	120	11
35	7505	205	5	08	4265	167	4	57	12905	120	12
36	7685	203	4	09	4445	155	5	58	13085	115	13
37	7865	211	4	10	4625	155	5	59	13265	115	13
38	8045	194	3	11	4805	149	5	11:00A	13445	124	16
39	8225	176	3	12	4985	161	4	01	13625	127	19
40	8405	175	2	14	5165	154	4	02	13805	126	20
41	8585	162	2	15	5345	151	4	03	13985	126	21
42	8765	124	4	16	5525	137	3	04	14165	129	18
43	8945	114	6	17	5705	140	3	05	14345	134	17
44	9125	120	5	18	5885	144	4	06	14525	134	18
45	9305	120	5	19	6065	138	4				
46	9485	118	6	20	6245	135	4	**6**	(357)		
47	9665	122	4	21	6425	142	5	1:53P	395	46	3
48	9845	119	5	22	6605	144	5	54	611	34	2
49	10025	110	6	23	6785	138	4	55	809	8	3
50	10205	110	6	24	6965	129	4	56	1007	357	5
51	10385	101	8	25	7145	120	5	57	1196	355	5
52	10565	106	8	26	7325	116	5	58	1385	354	6
53	10745	112	8	27	7505	114	5	59	1565	355	6
54	10925	137	6	28	7685	114	4	2:00P	1745	357	5
55	11105	137	6	29	7865	114	3	01	1925	22	3
56	11285	138	5	30	8045	114	4	02	2105	34	3

TABLE II (1928)

Date and time	Height in meters	Azimuth from; degrees	Velocity m/s	Date and time	Height in meters	Azimuth from; degrees	Velocity m/s	Date and time	Height in meters	Azimuth from; degrees	Velocity m/s
June 6				**June 6**				**June 6**			
2:03P	2285	22	4	2:51P	10925	137	7	3:39P	19565	58	7
04	2465	22	4	52	11105	135	6	40	19745	58	6
05	2645	38	3	53	11285	120	6	41	19925	64	5
06	2825	80	3	54	11465	134	5	42	20105	71	10
07	3005	121	3	55	11645	171	3	43	20285	64	10
08	3185	131	3	56	11825	187	2	44	20465	57	10
09	3365	127	3	57	12005	202	2	45	20645	58	7
10	3545	143	3	58	12185	203	1	46	20825	66	5
11	3725	149	4	59	12365	105	1	47	21005	94	4
12	3905	149	4	3:00P	12545	118	2	48	21185	95	6
13	4085	148	4	01	12725	109	2	49	21365	115	10
14	4265	151	4	02	12905	115	1	50	21545	115	13
15	4445	169	3	03	13085	166	2	51	21725	110	15
16	4625	171	3	04	13265	185	2	52	21905	106	14
17	4805	168	3	05	13445	188	2	53	22085	103	13
18	4985	160	4	06	13625	186	2	54	22265	90	15
19	5165	152	5	07	13805	181	1	55	22445	81	15
20	5345	151	5	08	13985	170	1	56	22625	76	14
21	5525	150	4	09	14165	138	1	57	22805	75	14
22	5705	146	4	10	14345	102	2	58	22985	69	12
23	5885	132	3	11	14525	78	2	59	23165	68	11
24	6065	111	3	12	14705	37	2	4:00P	23345	67	11
25	6245	111	2	13	14885	34	4	01	23525	66	10
26	6425	120	3	14	15065	42	5	02	23705	67	9
27	6605	125	5	15	15245	64	7	03	23885	71	8
28	6785	132	5	16	15425	93	7	04	24065	74	9
29	6965	129	5	17	15605	115	7	05	24245	73	8
30	7145	127	3	18	15785	131	5	06	24425	77	8
31	7325	129	3	19	15965	126	3	07	24605	77	8
32	7505	135	4	20	16145	56	3	08	24785	73	7
33	7685	131	4	21	16325	53	6	09	24965	73	6
34	7865	129	4	22	16505	60	5	10	25145	73	4
35	8045	132	3	23	16685	50	3	11	25325	77	5
36	8225	129	2	24	16865	36	4	12	25505	78	6
37	8405	89	3	25	17045	27	4	13	25685	78	7
38	8585	59	3	26	17225	35	5	14	25865	60	7
39	8765	60	2	27	17405	47	6	15	26045	35	6
40	8945	71	3	28	17585	54	7	16	26225	25	5
41	9125	89	4	29	17765	56	7	17	26405	20	4
42	9305	116	4	30	17945	64	7	18	26585	45	5
43	9485	137	4	31	18125	73	8	19	26765	82	6
44	9665	139	3	32	18305	84	7	20	26945	85	7
45	9845	145	2	33	18485	84	8	21	27125	85	7
46	10025	163	3	34	18665	75	9	22	27305	112	6
47	10205	161	3	35	18845	71	9	23	27485	136	6
48	10385	159	4	36	19025	65	9	24	27665	124	7
49	10565	154	6	37	19205	60	7	25	27845	104	8
50	10745	138	7	38	19385	58	7	26	28025	104	9

TABLE II (1928)

Date and time	Height in meters	Azimuth from; degrees	Velocity m/s	Date and time	Height in meters	Azimuth from; degrees	Velocity m/s	Date and time	Height in meters	Azimuth from; degrees	Velocity m/s
June 6				**June 7**				**June 7**			
4:27P	28205	104	10	10:35A	7325	5	3	2:09P	3005	30	9
28	28385	103	12	36	7505	10	3	10	3185	24	8
29	28565	102	13	37	7685	18	4	11	3365	19	5
30	28745	103	13	38	7865	26	4	12	3545	29	5
31	28925	116	14	39	8045	24	4	13	3725	32	6
32	29105	124	13	40	8225	25	5	14	3905	29	5
33	29285	124	14	41	8405	34	5	15	4085	17	4
34	29465	123	12	42	8585	34	5	16	4265	334	3
				43	8765	29	6	17	4445	321	3
7	(358)			44	8945	29	6	18	4625	332	4
9:57A	395	62	3	45	9125	29	6	19	4805	352	4
58	611	36	2	46	9305	29	7	20	4985	346	4
59	809	9	5	47	9485	31	6	21	5165	342	4
10:00A	1007	17	7	48	9665	31	5	22	5345	340	4
01	1196	26	9	49	9845	35	4	23	5525	324	3
02	1385	31	9	50	10025	37	3	24	5705	322	3
03	1565	39	8	51	10205	31	2	25	5885	327	4
04	1745	42	9	52	10385	13	3	26	6065	327	5
05	1925	38	8	53	10565	11	2	27	6245	329	4
06	2105	35	8	54	10745	10	2	28	6425	333	3
07	2285	26	7	55	10925	6	2	29	6605	332	2
08	2465	17	6	56	11105	12	2	30	6785	323	3
09	2645	15	5	57	11285	5	1	31	6965	319	2
10	2825	15	5	58	11465	357	2	32	7145	337	2
11	3005	15	6	59	11645	358	2	33	7325	352	3
12	3185	19	5	11:00A	11825	343	2	34	7505	1	3
13	3365	14	3	01	12005	226	2	35	7685	14	3
14	3545	9	3	02	12185	330	2	36	7865	32	3
15	3725	335	2	03	12365	337	2	37	8045	37	3
16	3905	317	3	04	12545	340	2	38	8225	36	5
17	4085	317	3	05	12725	335	1	39	8405	42	4
18	4265	313	3	06	12905	354	2	40	8585	46	4
19	4445	321	2					41	8765	46	4
20	4625	318	2	**7**	(359)			42	8945	42	6
21	4805	329	3	1:55P	395	99	4	43	9125	44	5
22	4985	341	3	56	611	55	4	44	9305	45	7
23	5165	341	2	57	809	32	6	45	9485	43	8
24	5345	341	2	58	1007	23	6	46	9665	44	8
25	5525	350	1	59	1196	11	5	47	9845	42	8
26	5705	354	1	2:00P	1385	0	4	48	10025	37	7
27	5885	336	1	01	1565	1	4	49	10205	41	7
28	6065	328	1	02	1745	9	6	50	10385	43	6
29	6245	335	1	03	1925	15	7	51	10565	42	5
30	6425	319	1	04	2105	38	6	52	10745	26	4
31	6605	302	1	05	2285	60	5	53	10925	22	4
32	6785	315	2	06	2465	60	5	54	11105	20	2
33	6965	338	2	07	2645	45	7	55	11285	341	1
34	7145	356	2	08	2825	35	8	56	11465	330	1

TABLE II (1928)

Date and time	Height in meters	Azimuth from; degrees	Veloc-ity m/s	Date and time	Height in meters	Azimuth from; degrees	Veloc-ity m/s	Date and time	Height in meters	Azimuth from; degrees	Veloc-ity m/s
June				June				June			
7				8				8			
2:57P	11645	329	3	10:37A	7865	179	7	2:08P	3545	23	4
58	11825	318	1	38	8045	182	7	09	3725	15	3
59	12005	343	2	39	8225	177	7	10	3905	344	1
3:00P	12185	3	6	40	8405	179	7	11	4085	279	1
01	12365	6	10	41	8585	176	7	12	4265	212	2
				42	8765	183	8	13	4445	207	3
8	(360)			43	8945	182	8	14	4625	208	4
9:56A	395	108	5	44	9125	182	7	15	4805	209	5
57	611	57	5	45	9305	182	8	16	4985	204	5
58	809	33	7	46	9485	179	8	17	5165	205	5
59	1007	26	8	47	9665	174	9	18	5345	207	5
10:00A	1196	25	8	48	9845	174	9	19	5525	209	4
01	1385	37	7	49	10025	174	9	20	5705	213	4
02	1565	55	5	50	10205	179	9	21	5885	215	5
03	1745	61	4	51	10385	188	8	22	6065	222	6
04	1925	53	3	52	10565	198	7	23	6245	224	5
05	2105	45	5	53	10745	204	7	24	6425	233	5
06	2285	41	7	54	10925	205	7	25	6605	230	5
07	2465	26	5	55	11105	217	6	26	6785	224	4
08	2645	16	6	56	11285	240	6	27	6965	224	5
09	2825	16	7	57	11465	257	4	28	7145	224	5
10	3005	26	7	58	11645	271	3	29	7325	220	7
11	3185	36	7	59	11825	270	2	30	7505	212	6
12	3365	46	6	11:00A	12005	271	2	31	7685	213	6
13	3545	44	5	01	12185	272	4	32	7865	216	7
14	3725	38	2	02	12365	271	6	33	8045	209	7
15	3905	4	1	03	12545	264	5	34	8225	201	8
16	4085	18	1	04	12725	240	5	35	8405	195	9
17	4265	38	1	05	12905	229	7	36	8585	195	9
18	4445	46	2					37	8765	198	9
19	4625	69	2	8	(361)			38	8945	200	10
20	4805	85	1	1:51P	395	104	7	39	9125	205	10
21	4985	94	1	52	611	89	4	40	9305	207	10
22	5165	89	2	53	809	71	5	41	9485	209	10
23	5345	40	1	54	1007	52	7	42	9665	210	10
24	5525	223	3	55	1196	31	7	43	9845	211	10
25	5705	236	4	56	1385	26	8	44	10025	211	9
26	5885	246	3	57	1565	27	7	45	10205	206	9
27	6065	242	3	58	1745	31	6	46	10385	219	9
28	6245	239	3	59	1925	36	5	47	10565	232	9
29	6425	231	3	2:00P	2105	40	5	48	10745	230	10
30	6605	227	4	01	2285	50	5	49	10925	229	9
31	6785	224	4	02	2465	56	7	50	11105	225	8
32	6965	220	5	03	2645	50	7	51	11285	235	6
33	7145	210	4	04	2825	42	6	52	11465	251	4
34	7325	195	4	05	3005	35	6	53	11645	254	5
35	7505	195	5	06	3185	26	6	54	11825	257	5
36	7685	183	6	07	3365	25	5	55	12005	257	4

TABLE II (1928)

Date and time	Height in meters	Azimuth from; degrees	Velocity m/s	Date and time	Height in meters	Azimuth from; degrees	Velocity m/s	Date and time	Height in meters	Azimuth from; degrees	Velocity m/s
June				**June**				**June**			
8				**9**				**10**			
2:56P	12185	263	5	2:27P	6785	273	10	3:18P	4985	220	11
57	12365	270	5	28	6965	268	9	19	5165	222	12
58	12545	269	5	29	7145	270	11	20	5345	222	13
59	12725	282	5	30	7325	271	14				
3:00P	12905	293	6	31	7505	277	11	**11**	(364)		
01	13085	294	5	32	7685	289	8	1:56P	395	105	3
02	13265	297	4	33	7865	295	8	57	611	99	3
03	13445	293	3	34	8045	299	11	58	809	95	2
04	13625	265	3	35	8225	302	11	59	1007	339	1
05	13805	276	3	36	8405	300	11	2:00P	1196	317	2
06	13985	275	3	37	8585	293	11	01	1385	341	1
				38	8765	288	11	02	1565	314	1
9	(362)			39	8945	285	11	03	1745	105	2
1:52P	395	95	4	40	9125	286	11	04	1925	117	5
53	611	96	6	41	9305	289	13	05	2105	128	6
54	809	63	4	42	9485	291	14	06	2285	120	7
55	1007	16	4	43	9665	292	15	07	2465	127	7
56	1196	0	4	44	9845	292	16	08	2645	134	7
57	1385	345	5	45	10025	290	15	09	2825	143	7
58	1565	346	5	46	10205	289	15	10	3005	152	7
59	1745	291	4	47	10385	288	15	11	3185	166	6
2:00P	1925	266	1					12	3365	171	6
01	2105	223	1	**10**	(363)						
02	2285	229	1	2:53P	395	160	4	**12**	(365)		
03	2465	235	1	54	611	217	1	2:06P	395	110	4
04	2645	204	1	55	809	305	3	07	611	106	4
05	2825	205	3	56	1007	248	2	08	809	75	2
06	3005	219	3	57	1196	225	4	09	1007	45	2
07	3185	223	5	58	1385	228	4	10	1196	39	2
08	3365	219	5	59	1565	227	5	11	1385	3	3
09	3545	223	6	3:00P	1745	218	7	12	1565	16	5
10	3725	225	7	01	1925	211	7	13	1745	14	3
11	3905	—	8	02	2105	211	8	14	1925	114	1
12	4085	—	8	03	2285	210	9	15	2105	206	2
13	4265	248	9	04	2465	212	9	16	2285	222	2
14	4445	250	8	05	2645	194	8	17	2465	215	3
15	4625	251	8	06	2825	186	8	18	2645	216	2
16	4805	255	7	07	3005	190	7	19	2825	212	3
17	4985	255	8	08	3185	204	7	20	3005	217	3
18	5165	263	8	09	3365	209	8	21	3185	221	3
19	5345	270	8	10	3545	218	8	22	3365	231	3
20	5525	268	10	11	3725	218	8	23	3545	220	3
21	5705	270	10	12	3905	213	8	24	3725	211	3
22	5885	272	9	13	4085	209	9	25	3905	228	3
23	6065	272	8	14	4265	213	10	26	4085	218	4
24	6245	272	7	15	4445	213	9	27	4265	212	4
25	6425	270	8	16	4625	213	11	28	4445	217	3
26	6605	272	10	17	4805	217	11	29	4625	212	3

TABLE II (1928)

Date and time	Height in meters	Azimuth from; degrees	Veloc-ity m/s	Date and time	Height in meters	Azimuth from; degrees	Veloc-ity m/s	Date and time	Height in meters	Azimuth from; degrees	Veloc-ity m/s
June 12				**June 14**				**June 14**			
2:30P	4805	204	3	1:53P	611	88	5	2:41P	9305	21	7
31	4985	207	3	54	809	89	4	42	9485	24	7
32	5165	212	3	55	1007	97	3	43	9665	23	7
33	5345	220	3	56	1196	96	2	44	9845	29	9
34	5525	230	3	57	1385	59	2	45	10025	29	9
35	5705	232	4	58	1565	41	3	46	10205	30	11
36	5885	245	4	59	1745	47	3	47	10385	35	13
37	6065	255	5	2:00P	1925	37	2	48	10565	40	13
38	6245	250	4	01	2105	24	1	49	10745	36	12
39	6425	247	4	02	2285	8	1	50	10925	31	14
40	6605	261	5	03	2465	4	1	51	11105	35	16
41	6785	263	6	04	2645	13	1	52	11285	37	17
42	6965	255	6	05	2825	13	2	53	11465	43	16
43	7145	257	5	06	3005	22	1	54	11645	48	15
44	7325	254	4	07	3185	19	1	55	11825	39	11
45	7505	254	4	08	3365	6	1	56	12005	15	8
46	7685	254	3	09	3545	342	1	57	12185	14	9
47	7865	254	3	10	3725	308	2	58	12365	357	13
48	8045	254	3	11	3905	297	3	59	12545	357	16
49	8225	237	4	12	4085	291	3	3:00P	12725	6	14
50	8405	226	5	13	4265	284	4	01	12905	14	13
51	8585	213	7	14	4445	274	4	02	13085	15	14
52	8765	208	8	15	4625	264	4	03	13265	14	13
53	8945	206	9	16	4805	252	4	04	13445	15	10
54	9125	206	8	17	4985	245	5	05	13625	2	7
55	9305	207	9	18	5165	254	6	06	13805	0	8
56	9485	206	9	19	5345	264	6	07	13985	0	8
57	9665	208	8	20	5525	260	5	08	14165	0	9
58	9845	208	8	21	5705	264	4	09	14345	8	11
59	10025	211	9	22	5885	280	4	10	14525	9	12
3:00P	10205	213	9	23	6065	295	5				
01	10385	206	9	24	6245	311	5	**15**	(368)		
02	10565	207	9	25	6425	321	4	1:52P	395	156	3
03	10745	229	8	26	6605	316	5	53	611	127	1
04	10925	230	7	27	6785	318	5	54	809	355	2
05	11105	217	7	28	6965	331	5	55	1007	357	4
06	11285	217	7	29	7145	336	6	56	1196	359	5
				30	7325	346	6	57	1385	359	7
13	(366)			31	7505	358	6	58	1565	4	6
1:48P	395	104	4	32	7685	18	7	59	1745	17	3
49	611	74	3	33	7865	13	6	2:00P	1925	32	2
50	809	44	4	34	8045	15	5	01	2105	34	2
51	1007	32	4	35	8225	18	5	02	2285	42	2
52	1196	28	5	36	8405	21	4	03	2465	72	2
53	1385	16	5	37	8585	20	5	04	2645	103	3
				38	8765	22	6	05	2825	121	3
14	(367)			39	8945	22	6	06	3005	122	3
1:52P	395	88	5	40	9125	23	6	07	3185	125	4

TABLE II (1928)

Date and time	Height in meters	Azimuth from; degrees	Veloc-ity m/s	Date and time	Height in meters	Azimuth from; degrees	Veloc-ity m/s	Date and time	Height in meters	Azimuth from; degrees	Veloc-ity m/s
June 15				**June 15**				**June 15**			
2:08P	3365	127	3	2:56P	12005	90	4	3:48P	7145	79	10
09	3545	121	3	57	12185	70	4	49	7325	83	9
10	3725	118	4	58	12365	64	7	50	7505	87	10
11	3905	125	4	59	12545	65	9	51	7685	87	11
12	4085	131	4	3:00P	12725	61	9	52	7865	85	12
13	4265	126	3	01	12905	61	9	53	8045	84	13
14	4445	124	2	02	13085	71	9	54	8225	82	14
15	4625	150	1	03	13265	77	10	55	8405	81	15
16	4805	156	3	04	13445	77	6	56	8585	80	15
17	4985	124	3					57	8765	80	16
18	5165	115	5	**15**	(369)			58	8945	81	16
19	5345	115	6	3:10P	395	144	5	59	9125	81	16
20	5525	104	7	11	611	177	1	4:00P	9305	81	16
21	5705	93	8	12	809	302	1	01	9485	83	17
22	5885	86	7	13	1007	332	3	02	9665	81	18
23	6065	86	6	14	1196	336	4	03	9845	81	19
24	6245	85	7	15	1385	336	6	04	10025	82	20
25	6425	84	7	16	1565	337	5	05	10205	83	18
26	6605	85	7	17	1745	340	5	06	10385	85	17
27	6785	88	6	18	1925	348	5	07	10565	87	16
28	6965	83	7	19	2105	353	2	08	10745	86	15
29	7145	83	9	20	2285	39	1	09	10925	82	13
30	7325	89	11	21	2465	99	1	10	11105	75	14
31	7505	91	11	22	2645	125	2	11	11285	73	13
32	7685	94	11	23	2825	135	3	12	11465	75	13
33	7865	95	12	24	3005	130	3	13	11645	67	9
34	8045	94	13	25	3185	119	3	14	11825	32	7
35	8225	93	15	26	3365	99	3	15	12005	40	8
36	8405	91	16	27	3545	92	4				
37	8585	90	16	28	3725	98	4	**16**	(370)		
38	8765	88	17	29	3905	98	4	2:54P	395	98	3
39	8945	86	18	30	4085	98	3	55	611	75	4
40	9125	84	20	31	4265	95	3	56	809	22	3
41	9305	82	20	32	4445	77	2	57	1007	349	4
42	9485	89	19	33	4625	52	2	58	1196	331	4
43	9665	95	20	34	4805	59	1	59	1385	321	5
44	9845	79	19	36	4985	105	1	3:00P	1565	327	5
45	10025	80	21	37	5165	109	3	01	1745	308	4
46	10205	80	23	38	5345	97	5	02	1925	267	4
47	10385	80	21	39	5525	88	6	03	2105	260	4
48	10565	80	21	40	5705	81	6	04	2285	268	3
49	10745	80	19	41	5885	78	6	05	2465	302	2
50	10925	82	17	42	6065	78	5	06	2645	325	3
51	11105	84	16	43	6245	77	5	07	2825	333	4
52	11285	80	13	44	6425	75	6	08	3005	337	4
53	11465	71	17	45	6605	76	6	09	3185	347	4
54	11645	79	19	46	6785	82	6	10	3365	328	3
55	11825	83	9	47	6965	81	9	11	3545	302	4

TABLE II (1928)

Date and time	Height in meters	Azimuth from; degrees	Velocity m/s	Date and time	Height in meters	Azimuth from; degrees	Velocity m/s	Date and time	Height in meters	Azimuth from; degrees	Velocity m/s
June 16				**June 17**	(371)			**June 17**			
3:12P	3725	298	4	1:51P	395	107	6	2:39P	9125	326	12
13	3905	302	6	52	611	89	5	40	9305	323	12
14	4085	307	7	53	809	64	4	41	9485	322	11
15	4265	313	7	54	1007	32	3	42	9665	322	9
16	4445	319	6	55	1196	18	3	43	9845	321	9
17	4625	317	6	56	1385	11	5	44	10025	321	9
18	4805	325	5	57	1565	359	7	45	10205	323	9
19	4985	352	5	58	1745	355	7	46	10385	308	8
20	5165	0	6	59	1925	351	5	47	10565	300	10
21	5345	357	5	2:00P	2105	352	3	48	10745	300	11
22	5525	357	6	01	2285	356	3	49	10925	295	11
23	5705	350	6	02	2465	335	4	50	11105	295	9
24	5885	340	6	03	2645	1	4	51	11285	295	6
25	6065	333	7	04	2825	9	4	52	11465	295	6
26	6245	332	7	05	3005	12	4	53	11645	295	7
27	6425	329	8	06	3185	11	3	54	11825	295	7
28	6605	330	9	07	3365	15	3	55	12005	293	6
29	6785	331	9	08	3545	358	4	56	12185	294	5
30	6965	329	10	09	3725	352	6	57	12365	295	5
31	7145	331	11	10	3905	358	6	58	12545	298	5
32	7325	329	11	11	4085	3	4	59	12725	297	6
33	7505	329	10	12	4265	3	5	3:00P	12905	297	5
34	7685	328	10	13	4445	359	7	01	13085	307	6
35	7865	326	11	14	4625	357	6	02	13265	316	4
36	8045	323	11	15	4805	0	6	03	13445	299	4
37	8225	320	12	16	4985	1	6	04	13625	271	4
38	8405	322	12	17	5165	1	6	05	13805	249	4
39	8585	324	12	18	5345	0	6	06	13985	248	4
40	8765	323	11	19	5525	2	7				
41	8945	323	11	20	5705	2	6	**18**	(372)		
42	9125	323	11	21	5885	359	5	1:56P	395	76	5
43	9305	323	10	22	6065	359	7	57	611	57	5
44	9485	319	9	23	6245	359	8	58	809	24	4
45	9665	320	9	24	6425	357	10	59	1007	25	5
46	9845	322	9	25	6605	354	11	2:00P	1196	23	7
47	10025	322	8	26	6785	353	12	01	1385	15	8
48	10205	322	8	27	6965	351	13	02	1565	15	7
49	10385	324	8	28	7145	349	13	03	1745	32	5
50	10565	329	8	29	7325	349	13	04	1925	47	5
51	10745	333	7	30	7505	346	12	05	2105	44	5
52	10925	345	7	31	7685	339	11	06	2285	30	6
53	11105	347	6	32	7865	337	12	07	2465	15	6
54	11285	341	5	33	8045	338	12	08	2645	15	5
55	11465	345	5	34	8225	338	12	09	2825	26	6
56	11645	16	5	35	8405	335	13	10	3005	34	7
57	11825	50	3	36	8585	332	13	11	3185	41	7
58	12005	52	2	37	8765	329	13	12	3365	37	7
59	12185	56	2	38	8945	327	12	13	3545	36	8

TABLE II (1928)

Date and time	Height in meters	Azimuth from; degrees	Velocity m/s
June 18			
2:14P	3725	37	8
15	3905	36	8
16	4085	37	9
17	4265	39	9
18	4445	38	9
19	4625	33	8
20	4805	32	9
21	4985	32	10
22	5165	32	10
23	5345	31	10
24	5525	28	11
25	5705	22	13
26	5885	19	14
27	6065	15	15
28	6245	10	15
29	6425	11	15
30	6605	11	15
19	(373)		
1:52P	395	103	4
53	611	85	2
54	809	38	3
55	1007	20	5
56	1196	5	5
57	1385	353	5
58	1565	354	5
59	1745	357	5
2:00P	1925	6	3
01	2105	16	1
02	2285	111	1
03	2465	184	1
04	2645	176	1
05	2825	142	1
06	3005	139	2
07	3185	137	3
08	3365	145	3
09	3545	134	4
10	3725	117	4
11	3905	119	4
12	4085	137	4
13	4265	152	6
14	4445	143	7
15	4625	156	6
16	4805	165	7
17	4985	163	8
18	5165	160	8
19	5345	155	8
20	5525	153	7

Date and time	Height in meters	Azimuth from; degrees	Velocity m/s
June 19			
2:21P	5705	150	8
22	5885	145	8
23	6065	148	8
24	6245	154	8
25	6425	154	7
26	6605	154	7
27	6785	154	10
28	6965	156	10
29	7145	162	7
30	7325	165	7
20	(374)		
1:46P	395	246	5
47	611	267	4
48	809	279	4
49	1007	295	4
50	1196	304	3
51	1385	313	2
52	1565	227	1
53	1745	184	1
54	1925	220	3
55	2105	203	6
56	2285	176	7
57	2465	179	6
58	2645	198	6
59	2825	204	6
2:00P	3005	200	6
01	3185	196	6
21	(375)		
1:49P	395	281	5
50	611	282	7
51	809	261	5
52	1007	244	5
53	1196	222	5
54	1385	178	6
55	1565	170	6
21	(376)		
11:56P	395	84	4
57	611	138	6
58	809	173	7
59	1007	194	12
12:00P	1196	190	14
0:01A	1385	182	13
02	1565	178	13
03	1745	177	13
04	1925	178	13

Date and time	Height in meters	Azimuth from; degrees	Velocity m/s
June 22			
0:05A	2105	180	14
06	2285	184	15
07	2465	187	15
08	2645	192	16
09	2825	196	18
10	3005	197	16
11	3185	201	13
12	3365	202	11
22	(377)		
0:18A	395	96	2
19	611	139	6
20	809	154	7
21	1007	164	8
22	1196	174	9
23	1385	177	9
24	1565	175	8
25	1745	181	7
26	1925	187	10
27	2105	188	13
23	(378)		
1:54P	395	190	6
55	611	188	4
56	809	175	5
57	1007	176	6
58	1196	178	7
59	1385	177	7
2:00P	1565	173	7
01	1745	168	7
02	1925	163	8
03	2105	160	7
04	2285	158	8
05	2465	153	9
06	2645	151	8
07	2825	146	8
08	3005	145	9
23	(379)		
2:10P	395	260	8
11	611	258	7
12	809	254	6
13	1007	231	3
14	1196	187	2
15	1385	106	3
16	1565	136	4
17	1745	160	4
18	1925	128	5

TABLE II (1928)

Date and time	Height in meters	Azimuth from; degrees	Velocity m/s
June 23			
2:19P	2105	117	7
20	2285	127	5
21	2465	133	5
22	2645	149	4
23	2825	167	4
24	3005	166	4
24	(380)		
1:56P	395	80	5
57	611	102	3
58	809	133	3
59	1007	130	4
2:00P	1196	126	4
01	1385	137	3
02	1565	153	2
03	1745	153	3
04	1925	162	2
05	2105	165	2
06	2285	162	3
07	2465	165	5
08	2645	163	6
09	2825	167	5
10	3005	172	6
11	3185	170	8
12	3365	168	8
13	3545	168	9
14	3725	167	9
15	3905	167	10
16	4085	167	10
17	4265	170	9
18	4445	174	10
19	4625	169	10
20	4805	167	8
21	4985	160	6
22	5165	144	6
23	5345	139	8
24	5525	137	9
25	5705	136	10
26	5885	133	11
27	6065	133	12
28	6245	131	11
25	(381)		
1:56P	395	172	4
57	611	260	2
58	809	304	3
59	1007	359	3
2:00P	1196	19	4

Date and time	Height in meters	Azimuth from; degrees	Velocity m/s
June 25			
2:01P	1385	32	4
02	1565	41	5
03	1745	46	5
04	1925	48	6
05	2105	48	5
06	2285	48	5
07	2465	48	4
27	(382)		
2:01P	395	124	5
02	611	128	5
03	809	138	5
04	1007	145	6
05	1196	145	5
06	1385	157	5
07	1565	150	4
08	1745	98	5
09	1925	70	4
10	2105	47	2
11	2285	44	2
12	2465	70	1
13	2645	269	1
14	2825	282	3
15	3005	285	4
16	3185	292	4
17	3365	300	3
18	3545	306	3
19	3725	317	3
20	3905	328	3
21	4085	317	3
22	4265	315	2
23	4445	304	2
24	4625	296	2
25	4805	295	2
26	4985	282	2
27	5165	248	1
28	5345	190	1
29	5525	196	2
30	5705	194	2
31	5885	193	3
32	6065	190	3
33	6245	200	3
34	6425	213	3
35	6605	214	2
36	6785	214	2
37	6965	210	3
38	7145	206	3
39	7325	198	3

Date and time	Height in meters	Azimuth from; degrees	Velocity m/s
June 27			
2:40P	7505	207	2
41	7685	217	3
42	7865	232	2
43	8045	250	3
44	8225	255	3
45	8405	245	3
46	8585	236	3
47	8765	245	3
48	8945	248	3
49	9125	244	3
50	9305	227	3
51	9485	221	3
52	9665	209	2
53	9845	224	2
54	10025	224	3
55	10205	215	2
56	10385	237	1
28	(383)		
2:41P	395	328	2
42	611	7	3
43	809	36	4
44	1007	42	4
45	1196	35	2
46	1385	184	1
47	1565	164	3
48	1745	157	4
29	(384)		
1:53P	395	124	5
54	611	106	4
55	809	58	4
56	1007	50	5
57	1196	46	4
58	1385	19	3
59	1565	17	5
2:00P	1745	356	5
01	1925	343	3
02	2105	310	1
03	2285	256	1
04	2465	250	2
July 1	(385)		
1:56P	395	166	4
57	611	170	1
58	809	221	1

TABLE II (1928)

Date and time	Height in meters	Azimuth from; degrees	Velocity m/s
July 1	(386)		
2:05P	395	153	4
06	611	137	2
07	809	130	2
08	1007	172	1
09	1196	276	1
2	(387)		
1:59P	395	0	5
2:00P	611	6	6
01	809	14	6
02	1007	19	7
03	1196	14	7
04	1385	8	5
05	1565	11	4
06	1745	19	2
07	1925	281	1
08	2105	280	2
09	2285	271	2
10	2465	179	2
11	2645	142	3
3	(388)		
1:57P	395	116	5
58	611	95	5
59	809	62	4
2:00P	1007	34	5
01	1196	8	5
02	1385	0	7
03	1565	1	7
04	1745	342	6
05	1925	325	5
06	2105	320	6
07	2285	320	6
08	2465	321	6
09	2645	320	6
10	2825	315	5
11	3005	312	5
4	(389)		
1:58P	395	16	4
59	611	11	4
2:00P	809	14	3
01	1007	18	3
02	1196	5	4
03	1385	0	5
5	(390)		
2:03P	395	115	5
04	611	105	9

Date and time	Height in meters	Azimuth from; degrees	Velocity m/s
July 5			
2:05P	809	87	7
06	1007	72	4
07	1196	27	1
08	1385	338	1
09	1565	16	2
10	1745	38	4
11	1925	36	5
12	2105	31	6
13	2285	30	6
14	2465	26	6
15	2645	15	6
16	2825	356	5
17	3005	332	4
18	3185	319	4
19	3365	330	4
20	3545	334	4
6	(391)		
1:58P	395	9	7
59	611	8	7
2:00P	809	10	7
01	1007	19	6
02	1196	18	6
03	1385	18	7
04	1565	16	7
05	1745	16	6
06	1925	30	5
07	2105	47	5
08	2285	66	3
09	2465	75	3
10	2645	60	3
11	2825	44	3
12	3005	38	4
13	3185	42	4
14	3365	46	4
15	3545	49	4
16	3725	44	4
17	3905	39	4
18	4085	33	4
19	4265	43	3
20	4445	49	4
21	4625	49	4
22	4805	52	4
23	4985	53	4
24	5165	59	3
25	5345	64	3
26	5525	52	3
27	5705	50	3

Date and time	Height in meters	Azimuth from; degrees	Velocity m/s
July 6			
2:28P	5885	59	3
29	6065	51	3
30	6245	46	3
31	6425	46	3
32	6605	43	3
33	6785	48	3
34	6965	56	1
35	7145	55	1
36	7325	40	2
37	7505	37	2
38	7685	42	1
39	7865	41	2
40	8045	39	2
7	(392)		
1:57P	395	357	7
58	611	352	6
59	809	354	5
2:00P	1007	346	5
01	1196	334	6
02	1385	335	9
03	1565	331	11
04	1745	331	11
05	1925	324	9
06	2105	323	9
07	2285	336	8
08	2465	349	7
09	2645	358	6
10	2825	7	4
11	3005	12	3
12	3185	5	4
13	3365	359	4
14	3545	357	4
15	3725	353	4
16	3905	353	5
17	4085	350	5
18	4265	345	4
19	4445	345	5
20	4625	342	5
21	4805	343	4
22	4985	342	3
23	5165	325	2
24	5345	295	2
25	5525	285	3
26	5705	280	4
27	5885	262	5
28	6065	245	6
29	6245	227	7

TABLE II (1928)

Date and time	Height in meters	Azimuth from; degrees	Velocity m/s
July 7			
2:30P	6425	222	8
31	6605	222	9
32	6785	220	9
33	6965	215	9
34	7145	213	10
35	7325	211	10
36	7505	205	10
37	7685	205	9
38	7865	208	8
39	8045	203	8
40	8225	210	8
41	8405	228	8
42	8585	232	9
43	8765	226	9
44	8945	220	9
45	9125	217	8
46	9305	217	8
47	9485	217	8
48	9665	223	7
49	9845	232	5
50	10025	264	4
51	10205	290	5
52	10385	305	4
53	10565	308	3
8	(393)		
2:05P	395	303	5
06	611	287	6
07	809	250	1
08	1007	260	1
09	1196	224	1
10	1385	189	3
11	1565	190	5
12	1745	203	7
13	1925	216	8
14	2105	220	8
9	(394)		
2:06P	395	274	5
07	611	267	5
08	809	269	5
10	(395)		
1:50P	395	309	1
51	611	260	1
52	809	—	—
53	1007	51	1
54	1196	—	—

Date and time	Height in meters	Azimuth from; degrees	Velocity m/s
July 10			
1:55P	1385	39	2
56	1565	20	1
57	1745	273	2
58	1925	251	4
59	2105	250	4
2:00P	2285	255	3
11	(396)		
2:02P	395	121	1
03	611	77	1
04	809	58	2
05	1007	25	3
06	1196	7	4
12	(397)		
2:12P	395	97	4
13	611	41	3
14	809	20	5
15	1007	7	4
16	1196	349	3
17	1385	337	3
18	1565	337	3
19	1745	335	5
20	1925	333	6
21	2105	342	7
22	2285	342	7
23	2465	342	7
24	2645	342	7
13	(398)		
2:14P	395	192	4
15	611	218	3
16	809	256	3
17	1007	280	3
18	1196	279	3
19	1385	273	2
20	1565	262	2
21	1745	238	1
22	1925	220	2
23	2105	214	2
24	2285	217	1
25	2465	211	1
26	2645	265	1
27	2825	283	1
28	3005	257	1
29	3185	245	1
14	(399)		
2:19P	395	120	5

Date and time	Height in meters	Azimuth from; degrees	Velocity m/s
July 14			
2:20P	611	96	2
21	809	90	6
22	1007	79	7
23	1196	55	5
24	1385	42	5
25	1565	34	5
26	1745	32	6
27	1925	32	7
28	2105	32	7
29	2285	27	7
30	2465	16	5
31	2645	3	3
32	2825	8	3
33	3005	8	1
34	3185	39	3
35	3365	40	3
16	(400)		
2:14P	395	72	5
15	611	86	8
16	809	109	8
17	1007	139	7
18	1196	160	5
19	1385	175	5
20	1565	191	6
21	1745	204	6
22	1925	207	6
23	2105	213	7
24	2285	232	6
25	2465	240	5
26	2645	240	6
27	2825	260	6
28	3005	270	8
29	3185	270	9
30	3365	265	10
31	3545	258	11
32	3725	264	10
33	3905	266	10
34	4085	267	11
35	4265	277	10
36	4445	272	11
37	4625	266	14
38	4805	266	13
39	4985	263	13
40	5165	265	13
41	5345	260	14
42	5525	255	15
43	5705	263	16

TABLE II (1928)

Date and time	Height in meters	Azimuth from; degrees	Velocity m/s
July 16			
2:44P	5885	267	17
45	6065	269	19
46	6245	275	17
47	6425	281	17
48	6605	281	18
49	6785	280	19
18	(401)		
2:11P	395.	126	5
12	611	160	5
13	809	178	9
14	1007	183	11
15	1196	190	11
16	1385	200	11
17	1565	204	12
19	(402)		
2:24P	395	76	6
25	611	87	14
26	809	108	13
27	1007	137	9
28	1196	171	8
29	1385	176	9
30	1565	185	9
31	1745	193	9
20	(403)		
2:12P	395	80	6
13	611	108	5
14	809	147	6
15	1007	174	6
16	1196	181	5
17	1385	182	5
18	1565	186	5
19	1745	186	7
20	1925	186	7
21	2105	180	9
22	2285	176	11
23	2465	183	11
24	2645	184	10
25	2825	178	9
26	3005	177	10
21	(404)		
2:01P	395	78	3
02	611	80	1
03	809	150	1
04	1007	152	2
05	1196	157	4

Date and time	Height in meters	Azimuth from; degrees	Velocity m/s
July 21			
2:06P	1385	168	5
07	1565	177	5
08	1745	189	6
09	1925	195	6
10	2105	197	6
11	2285	197	6
12	2465	188	7
13	2645	195	8
14	2825	195	9
15	3005	181	9
16	3185	181	9
17	3365	189	11
18	3545	189	11
19	3725	190	10
20	3905	200	9
21	4085	208	9
22	4265	209	10
23	4445	206	11
24	4625	206	12
22	(405)		
2:10P	395	90	2
11	611	86	3
12	809	76	3
13	1007	62	2
14	1196	43	2
15	1385	45	2
16	1565	81	1
17	1745	145	1
18	1925	170	1
19	2105	165	1
20	2285	165	1
21	2465	166	3
22	2645	147	4
23	2825	150	5
24	3005	176	6
25	3185	183	7
26	3365	178	8
27	3545	176	10
28	3725	172	10
29	3905	165	10
30	4085	159	11
31	4265	157	12
32	4445	157	13
33	4625	158	13
34	4805	158	13
35	4985	158	12
36	5165	158	12

Date and time	Height in meters	Azimuth from; degrees	Velocity m/s
July 22			
2:37P	5345	158	13
38	5525	158	14
39	5705	158	15
40	5885	158	13
41	6065	158	12
42	6245	158	13
43	6425	160	10
44	6605	160	10
45	6785	160	13
46	6965	160	11
47	7145	155	10
48	7325	155	11
49	7505	155	13
50	7685	155	12
51	7865	155	10
52	8045	157	11
53	8225	157	12
54	8405	157	11
55	8585	157	11
56	8765	157	11
57	8945	157	12
58	9125	157	11
59	9305	155	11
3:00P	9485	155	11
01	9665	155	11
02	9845	155	11
03	10025	152	11
04	10205	150	12
05	10385	150	13
06	10565	149	13
07	10745	148	15
08	10925	148	17
09	11105	148	17
23	(406)		
2:11P	395	270	9
12	611	270	12
13	809	241	10
14	1007	227	11
15	1196	225	10
16	1385	219	13
17	1565	215	15
18	1745	214	14
19	1925	216	13
20	2105	223	10
21	2285	223	11
22	2465	219	10
23	2645	219	9

TABLE II (1928)

Date and time	Height in meters	Azimuth from; degrees	Velocity m/s
July			
24	(407)		
2:07P	395	115	2
08	611	178	1
09	809	175	1
10	1007	136	1
11	1196	115	2
12	1385	111	2
13	1565	112	2
25	(408)		
2:18P	395	115	5
19	611	169	7
20	809	192	9
26	(409)		
10:12A	395	86	4
13	611	123	3
14	809	171	4
15	1007	179	6
16	1196	173	5
17	1385	179	4
18	1565	220	4
19	1745	230	6
20	1925	226	8
21	2105	224	10
22	2285	221	10
23	2465	214	9
24	2645	217	9
25	2825	220	9
26	3005	222	9
27	3185	222	9
28	3365	225	10
29	3545	224	9
30	3725	220	9
31	3905	219	11
32	4085	213	11
33	4265	207	11
34	4445	208	12
35	4625	208	12
36	4805	203	13
37	4985	198	14
38	5165	204	15
39	5345	202	15
26	(410)		
1:55P	395	65	5
56	611	90	9
57	809	109	5
58	1007	152	2

Date and time	Height in meters	Azimuth from; degrees	Velocity m/s
July			
26			
1:59P	1196	195	2
2:00P	1385	182	4
01	1565	179	4
02	1745	185	5
03	1925	200	5
04	2105	211	5
05	2285	216	5
06	2465	210	6
07	2645	207	6
08	2825	208	6
09	3005	204	6
10	3185	201	7
11	3365	207	7
12	3545	221	7
13	3725	234	8
14	3905	235	7
15	4085	230	7
16	4265	226	7
17	4445	223	6
18	4625	217	6
19	4805	216	5
20	4985	215	9
21	5165	211	12
22	5345	205	11
23	5525	199	11
24	5705	195	13
25	5885	189	12
26	6065	182	10
27	6245	176	7
28	6425	176	6
29	6605	176	7
30	6785	176	9
31	6965	180	9
32	7145	181	9
27	(411)		
1:59P	395	163	3
2:00P	611	158	3
01	809	175	1
02	1007	321	1
03	1196	345	3
04	1385	355	5
05	1565	31	4
06	1745	66	3
07	1925	84	3
08	2105	96	3
09	2285	107	4
10	2465	105	5

Date and time	Height in meters	Azimuth from; degrees	Velocity m/s
July			
27			
2:11P	2645	105	5
12	2825	115	6
13	3005	122	7
14	3185	123	7
15	3365	129	7
16	3545	136	6
17	3725	150	7
18	3905	157	7
19	4085	165	7
20	4265	172	5
21	4445	177	4
22	4625	162	5
23	4805	180	6
24	4985	164	8
25	5165	170	9
26	5345	170	9
27	5525	173	9
28	5705	179	9
29	5885	189	10
30	6065	189	10
31	6245	195	9
32	6425	195	9
33	6605	190	9
34	6785	186	9
35	6965	178	8
36	7145	175	9
37	7325	175	9
38	7505	169	9
39	7685	165	10
40	7865	165	10
41	8045	165	10
42	8225	167	9
43	8405	165	9
44	8585	163	10
45	8765	163	10
46	8945	163	9
47	9125	170	9
48	9305	169	7
49	9485	172	7
50	9665	162	10
51	9845	155	11
52	10025	155	11
53	10205	157	12
54	10385	142	11
55	10565	152	10
56	10745	152	12
57	10925	152	13

TABLE II (1928)

Date and time	Height in meters	Azimuth from; degrees	Velocity m/s
July 28	(412)		
2:02P	395	119	3
03	611	35	2
04	809	357	3
05	1007	2	3
06	1196	314	2
07	1385	271	3
08	1565	219	4
09	1745	200	6
10	1925	194	5
11	2105	193	5
12	2285	190	5
13	2465	183	5
14	2645	176	4
15	2825	180	5
16	3005	193	8
17	3185	196	12
18	3365	189	12
19	3545	184	12
20	3725	148	13
21	3905	181	11
22	4085	177	10
23	4265	177	10
24	4445	177	12
25	4625	177	13
26	4805	173	13
27	4985	163	11
28	5165	157	11
29	5345	160	11
30	5525	162	13
31	5705	162	14
32	5885	161	15
33	6065	160	15
34	6245	160	15
35	6425	160	17
36	6605	160	16
37	6785	160	17
38	6965	159	18
39	7145	160	16
40	7325	159	14
29	(413)		
2:02P	395	100	5
03	611	85	3
04	809	59	3
05	1007	47	3
06	1196	47	3
07	1385	33	3
08	1565	25	3

Date and time	Height in meters	Azimuth from; degrees	Velocity m/s
July 29			
2:09P	1745	18	3
10	1925	11	4
11	2105	10	4
12	2285	358	3
13	2465	319	2
14	2645	279	2
15	2825	265	2
16	3005	258	2
17	3185	262	1
18	3365	246	1
19	3545	225	1
20	3725	202	2
21	3905	187	3
22	4085	204	3
23	4265	218	5
24	4445	218	7
25	4625	218	7
30	(414)		
1:50P	395	81	6
51	611	62	7
52	809	58	7
53	1007	57	8
54	1196	54	7
55	1385	68	5
56	1565	87	4
57	1745	89	4
58	1925	78	5
59	2105	78	5
2:00P	2285	78	4
31	(415)		
10:04A	395	94	2
05	611	67	1
06	809	39	2
07	1007	28	3
08	1196	13	4
09	1385	27	4
10	1565	27	4
11	1745	27	4
12	1925	34	4
13	2105	45	3
14	2285	75	1
15	2465	97	1
16	2645	137	1
17	2825	184	1
18	3005	220	2
19	3185	235	3

Date and time	Height in meters	Azimuth from; degrees	Velocity m/s
July 31			
10:20A	3365	249	3
21	3545	248	3
22	3725	250	3
23	3905	254	3
24	4085	248	3
25	4265	248	4
26	4445	268	4
27	4625	272	5
28	4805	271	1
29	4985	295	1
30	5165	330	1
31	5345	350	2
32	5525	19	1
33	5705	337	2
34	5885	327	3
35	6065	337	2
36	6245	338	1
37	6425	337	2
Aug. 1	(416)		
1:58P	395	125	3
59	611	73	1
2:00P	809	77	1
01	1007	2	3
02	1196	335	3
03	1385	320	4
04	1565	325	4
05	1745	323	3
06	1925	274	2
07	2105	241	2
08	2285	216	2
09	2465	188	1
10	2645	179	1
11	2825	174	2
2	(417)		
10:06A	395	77	4
07	611	113	2
08	809	126	3
09	1007	135	2
10	1196	174	2
11	1385	189	2
12	1565	173	2
13	1745	138	2
14	1925	138	1
15	2105	46	1
16	2285	33	1

TABLE II (1928)

Date and time	Height in meters	Azimuth from; degrees	Velocity m/s
Aug. 2			
10:17A	2465	22	1
18	2645	316	1
19	2825	140	3
20	3005	159	3
21	3185	163	3
22	3365	155	3
23	3545	155	3
24	3725	146	2
25	3905	146	1
26	4085	96	1
27	4265	96	1
28	4445	57	3
29	4625	56	5
30	4805	56	5
31	4985	56	4
32	5165	53	5
33	5345	53	5
34	5525	53	4
35	5705	58	4
36	5885	58	5
37	6065	58	6
38	6245	50	8
39	6425	44	9
40	6605•	44	9
41	6785	42	8
42	6965	42	8
43	7145	30	8
44	7325	26	10
45	7505	26	12
46	7685	27	10
47	7865	56	7
3	(418)		
2:00P	395	191	4
01	611	229	5
02	809	257	5
03	1007	281	5
04	1196	288	6
05	1385	279	7
06	1565	265	7
07	1745	255	6
08	1925	255	5
09	2105	251	7
10	2285	248	7
11	2465	248	7
12	2645	255	6
13	2825	255	5

Date and time	Height in meters	Azimuth from; degrees	Velocity m/s
Aug. 4	(419)		
1:54P	395	8	5
55	611	359	4
56	809	359	4
5	(420)		
1:53P	395	101	5
54	611	97	4
55	809	64	4
56	1007	53	5
57	1196	49	3
58	1385	33	3
59	1565	32	2
2:00P	1745	63	1
01	1925	117	1
02	2105	135	2
03	2285	157	3
04	2465	155	4
05	2645	150	4
06	2825	137	4
07	3005	131	5
08	3185	132	5
09	3365	131	5
10	3545	121	4
11	3725	121	3
12	3905	133	3
13	4085	130	4
14	4265	127	4
15	4445	130	4
16	4625	143	5
17	4805	152	5
18	4985	152	5
19	5165	149	5
20	5345	140	3
21	5525	116	1
22	5705	92	1
23	5885	83	1
24	6065	87	1
25	6245	93	1
26	6425	104	2
27	6605	109	5
28	6785	111	7
29	6965	112	8
30	7145	112	10
31	7325	112	11
6	(421)		
2:06P	395	185	3
07	611	186	2

Date and time	Height in meters	Azimuth from; degrees	Velocity m/s
Aug. 6			
2:08P	809	288	1
09	1007	314	1
10	1196	323	3
11	1385	337	3
12	1565	352	2
13	1745	327	1
14	1925	354	1
15	2105	219	3
16	2285	194	5
17	2465	196	6
18	2645	197	7
19	2825	194	10
20	3005	188	11
21	3185	183	12
22	3365	180	15
23	3545	178	17
24	3725	179	18
25	3905	183	19
26	4085	183	20
27	4265	182	19
28	4445	180	21
29	4625	185	21
30	4805	187	22
31	4985	187	23
32	5165	186	23
33	5345	188	25
34	5525	187	27
35	5705	187	28
6	(422)		
2:00P	395	28	2
01	611	28	4
02	809	27	4
03	1007	22	4
04	1196	22	5
05	1385	26	7
06	1565	30	7
07	1745	42	6
08	1925	57	5
09	2105	57	5
10	2285	47	5
11	2465	44	7
12	2645	37	7
13	2825	26	6
14	3005	31	4
15	3185	27	3
16	3365	39	2
17	3545	85	1

TABLE II (1928)

Date and time	Height in meters	Azimuth from; degrees	Velocity m/s	Date and time	Height in meters	Azimuth from; degrees	Velocity m/s	Date and time	Height in meters	Azimuth from; degrees	Velocity m/s
Aug. 6				**Aug. 6**				**Aug. 6**			
2:18P	3725	148	1	3:06P	12365	257	5	3:54P	21005	232	7
19	3905	295	1	07	12545	249	4	55	21185	233	9
20	4085	324	3	08	12725	269	4	56	21365	233	9
21	4265	340	2	09	12905	252	5	57	21545	233	9
22	4445	28	6	10	13085	237	4	58	21725	233	8
23	4625	25	9	11	13265	220	3	59	21905	233	9
24	4805	7	6	12	13445	225	3	4:00P	22085	233	10
25	4985	5	6	13	13625	251	4	01	22265	234	8
26	5165	5	5	14	13805	240	3	02	22445	238	8
27	5345	5	8	15	13985	237	3	03	22625	236	8
28	5525	352	7	16	14165	226	3	04	22805	248	8
29	5705	333	5	17	14345	215	5	05	22985	235	7
30	5885	351	6	18	14525	212	7	06	23165	235	9
31	6065	351	6	19	14705	213	4	07	23345	235	9
32	6245	351	5	20	14885	230	3	08	23525	227	9
33	6425	351	5	21	15065	230	3	09	23705	227	9
34	6605	349	4	22	15245	230	3	10	23885	226	8
35	6785	337	4	23	15425	252	3	11	24065	235	9
36	6965	314	4	24	15605	250	3	12	24245	235	9
37	7145	269	4	25	15785	276	6	13	24425	249	11
38	7325	239	7	26	15965	292	9	14	24605	238	11
39	7505	239	9	27	16145	285	6	15	24785	230	10
40	7685	226	11	28	16325	258	5	16	24965	241	10
41	7865	206	14	29	16505	257	5	17	25145	240	12
42	8045	199	13	30	16685	245	5	18	25325	240	13
43	8225	211	15	31	16865	231	5	19	25505	239	11
44	8405	211	18	32	17045	231	5	20	25685	239	9
45	8585	204	17	33	17225	230	3	21	25865	239	10
46	8765	204	20	34	17405	220	3	22	26045	235	12
47	8945	208	17	35	17585	226	3	23	26225	235	11
48	9125	205	15	36	17765	226	4	24	26405	236	12
49	9305	203	17	37	17945	227	5	25	26585	236	11
50	9485	200	18	38	18125	227	5	26	26765	236	11
51	9665	200	18	39	18305	227	3	27	26945	237	12
52	9854	197	17	40	18485	227	5	28	27125	241	12
53	10025	197	18	41	18665	227	5	29	27305	245	13
54	10205	197	17	42	18845	227	4	30	27485	236	13
55	10385	189	18	43	19025	227	3	31	27665	224	13
56	10565	198	19	44	19205	227	3				
57	10745	206	18	45	19385	229	3				
58	10925	203	15	46	19565	229	5	**8**	(423)		
59	11105	209	13	47	19745	229	6	2:00P?	395	105	2
3:00P	11285	231	11	48	19925	229	4	01	611	18	1
01	11465	231	9	49	20105	232	7	02	809	337	1
02	11645	228	6	50	20285	232	5	03	1007	306	3
03	11825	224	5	51	20465	232	8	04	1196	289	3
04	12005	224	5	52	20645	232	10	05	1385	234	3
05	12185	226	5	53	20825	232	8	06	1565	197	5

TABLE II (1928)

Date and time	Height in meters	Azimuth from; degrees	Velocity m/s
Aug. 10	(424)		
2:00P	395	83	3
01	611	121	7
02	809	160	6
03	1007	192	6
04	1196	198	5
05	1385	204	4
06	1565	219	5
07	1745	239	7
08	1925	242	7
09	2105	238	7
10	2285	226	7
11	2465	226	7
12	2645	230	9
13	2825	233	9
11	(425)		
10:02A	395	74	5
03	611	100	6
04	809	142	4
05	1007	169	3
06	1196	171	3
07	1385	171	5
08	1565	171	6
09	1745	178	5
10	1925	181	5
11	2105	170	5
12	2285	174	6
13	2465	178	6
14	2645	181	7
15	2825	185	7
16	3005	188	8
17	3185	186	9
18	3365	185	11
19	3545	184	13
20	3725	185	13
21	3905	189	11
22	4085	202	8
23	4265	210	6
24	4445	208	5
25	4625	201	5
26	4805	205	7
27	4985	206	10
28	5165	206	12
29	5345	204	11
30	5525	201	10
31	5705	199	10
32	5885	199	9
33	6065	194	9

Date and time	Height in meters	Azimuth from; degrees	Velocity m/s
Aug. 11			
10:34A	6245	193	10
35	6425	192	11
36	6605	189	11
37	6785	184	12
38	6965	182	12
39	7145	177	11
40	7325	176	13
41	7505	177	13
42	7685	176	11
43	7865	169	11
44	8045	170	11
13	(426)		
1:59P	395	94	5
2:00P	611	84	3
01	809	41	1
02	1007	120	1
03	1196	237	2
04	1385	239	4
05	1565	257	3
06	1745	254	2
07	1925	252	5
08	2105	248	7
09	2285	248	6
10	2465	248	7
11	2645	242	7
12	2825	237	8
13	3005	237	8
14	3185	242	8
15	3365	253	9
16	3545	257	11
17	3725	257	11
18	3905	257	11
19	4085	258	11
20	4265	264	10
21	4445	267	11
22	4625	266	13
23	4805	264	13
24	4985	263	12
14	(427)		
2:07P	395	172	9
08	611	182	21
09	809	182	31
15	(428)		
2:00P	395	165	7
01	611	177	9
02	809	177	9

Date and time	Height in meters	Azimuth from; degrees	Velocity m/s
Aug. 16	(429)		
1:55P	395	109	2
56	611	149	1
57	809	171	1
58	1007	179	1
59	1196	177	1
18	(430)		
1:52A	395	63	4
53	611	110	5
54	809	138	7
55	1007	150	7
56	1196	151	9
57	1385	158	9
58	1565	167	10
59	1745	173	10
12:00M	1925	173	9
0:01P	2105	170	8
02	2285	164	7
03	2465	155	8
04	2645	149	7
05	2825	135	7
06	3005	123	7
07	3185	123	6
08	3365	123	6
09	3545	122	7
19	(431)*		
4:59P	10	66?	9
5:00P	226	66	10
01	424	79	9
02	622	99	10
03	811	108	11
04	1000	126	6
05	1180	178	5
06	1360	176	7
07	1540	175	5
08	1720	142	4
09	1900	140	5
10	2080	135	5
11	2260	134	7
12	2440	133	6
13	2620	132	6
14	2800	132	6
20	(432)		
2:02P	395	89	9
03	611	104	10
04	809	116	17

* This run was made at the flying field, not at Mt. Evans.

TABLE II (1928)

Date and time	Height in meters	Azimuth from; degrees	Velocity m/s
Aug. 20			
2:05P	1007	119	24
06	1196	117	18
07	1385	114	16
08	1565	116	22
09	1745	128	19
10	1925	145	19
11	2105	147	18
12	2285	153	30
13	2465	175	17
14	2645	202	10
21	(433)		
2:11P	395	85	1
12	611	47	1
13	809	26	1
14	1007	26	1
15	1196	27	1
16	1385	27	1
22	(434)		
1:56P	395	85	7
57	611	87	10
58	809	95	6
59	1007	138	6
2:00P	1196	149	7
01	1385	154	7
02	1565	156	8
03	1745	159	10
04	1925	163	10
05	2105	163	10
06	2285	163	10
23	(435)		
10:15A	395	120	5
16	611	152	5
17	809	165	10
18	1007	165	13
19	1196	174	11
20	1385	181	10
21	1565	182	9
22	1745	182	9
25	(436)		
2:23P	395	88	4
24	611	106	6
25	809	139	4
26	1007	189	4
27	1196	189	6
28	1385	189	8

Date and time	Height in meters	Azimuth from; degrees	Velocity m/s
Aug. 25			
2:29P	1565	189	9
30	1745	189	9
31	1925	185	5
32	2105	170	4
33	2285	169	6
34	2465	166	7
35	2645	173	7
36	2825	189	7
37	3005	205	8
38	3185	212	9
39	3365	207	10
40	3545	207	9
41	3725	207	8
42	3905	207	6
43	4085	202	5
44	4265	202	4
44	4445	205	5
45	4625	216	5
46	4805	237	7
47	4985	252	7
48	5165	252	6
49	5345	252	6
50	5525	252	7
51	5705	252	7
52	5885	251	7
53	6065	251	7
54	6245	251	7
55	6425	247	9
56	6605	250	9
57	6785	253	7
58	6965	251	8
59	7145	247	7
3:00P	7325	247	5
01	7505	248	5
02	7685	245	5
03	7865	230	5
04	8045	230	6
05	8245	235	7
06	8405	250	7
07	8585	254	7
08	8765	258	7
27	(437)		
2:02P	395	101	3
03	611	75	5
04	809	44	4
05	1007	22	3
06	1196	23	4

Date and time	Height in meters	Azimuth from; degrees	Velocity m/s
Aug. 27			
2:07P	1385	23	3
08	1565	17	3
09	1745	17	2
10	1925	19	1
11	2105	19	2
12	2285	30	4
13	2465	23	3
14	2645	2	2
15	2825	337	1
16	3005	354	1
17	3185	354	3
18	3365	354	4
19	3545	354	5
20	3725	346	5
21	3905	346	5
22	4085	358	3
23	4265	8	3
24	4445	23	2
25	4625	23	1
26	4805	173	1
27	4985	173	3
28	(438)		
1:56P	395	92	3
57	611	105	3
58	809	125	2
59	1007	127	1
2:00P	1196	73	1
01	1385	63	1
02	1565	63	3
03	1745	58	3
04	1925	63	3
05	2105	75	3
06	2285	71	4
07	2465	68	3
08	2645	60	2
09	2825	71	1
10	3005	150	1
11	3185	4	1
12	3365	16	2
13	3545	4	2
14	3725	316	2
15	3905	282	2
16	4085	270	4
17	4265	259	5
18	4445	243	6
19	4625	234	7
20	4805	231	9

TABLE II (1928)

Date and time	Height in meters	Azimuth from; degrees	Velocity m/s	Date and time	Height in meters	Azimuth from; degrees	Velocity m/s	Date and time	Height in meters	Azimuth from; degrees	Velocity m/s
Aug.				**Aug.**				**Aug**			
29	(439)			**29**				**31**			
2:00P	395	85	3	2:48P	9125	269	32	10:16A	5705	123	5
01	611	54	3	49	9305	269	30	17	5885	121	7
02	809	51	4	50	9485	260	33	18	6065	121	7
03	1007	63	4	51	9665	263	31	19	6245	121	6
04	1196	82	3	52	9845	265	27	20	6425	121	5
05	1385	120	3	54	10025	261	29	21	6605	121	3
06	1565	144	3	55	10205	261	32	22	6785	135	2
07	1745	145	2					23	6965	136	1
08	1925	146	2	**30**	(440)			24	7145	136	2
09	2105	162	2	2:01P	395	303	5	25	7325	128	1
10	2285	169	2	02	611	311	4	26	7505	111	3
11	2465	169	3	03	809	290	2	27	7685	105	2
12	2645	150	3	04	1007	220	3	28	7865	105	2
13	2825	129	3	05	1196	211	6	29	8045	105	2
14	3005	130	5	06	1385	211	7	30	8225	82	2
15	3185	144	5	07	1565	205	9	31	8405	83	3
16	3365	150	6	08	1745	201	9	32	8585	100	3
17	3545	157	6					33	8765	88	5
18	3725	161	6	**31**	(441)			34	8945	102	7
19	3905	160	6	9:47A	395	99	3	35	9125	118	9
20	4085	164	5	48	611	74	4	36	9305	129	8
21	4265	171	5	49	809	105	3	37	9485	132	9
22	4445	178	6	50	1007	124	5	38	9665	138	10
23	4625	181	6	51	1196	138	6	39	9845	138	10
24	4805	182	6	52	1385	151	7	40	10025	135	10
25	4985	182	6	53	1565	150	7	41	10205	132	10
26	5165	179	5	54	1745	150	8	42	10385	132	13
27	5345	183	5	55	1925	154	8	43	10565	132	14
28	5525	185	4	56	2105	154	9				
29	5705	188	4	57	2285	149	11	**Sept.**			
30	5885	195	4	58	2465	143	13	**1**	(442)		
31	6065	205	5	59	2645	143	13	9:58A	395	107	4
32	6245	204	4	10:00A	2825	143	12	59	611	104	3
33	6425	205	4	01	3005	150	11	10:00A	809	122	4
34	6605	238	5	02	3185	164	10	01	1007	138	5
35	6785	267	7	03	3365	162	10	02	1196	142	6
36	6965	276	11	04	3545	152	9	03	1385	133	6
37	7145	276	12	05	3725	140	8	04	1565	127	6
38	7325	276	15	06	3905	125	8	05	1745	130	6
39	7505	272	19	07	4085	125	9	06	1925	132	5
40	7685	275	21	08	4265	125	8	07	2105	134	6
41	7865	277	23	09	4445	131	8	08	2285	133	8
42	8045	277	25	10	4625	139	9	09	2465	127	11
43	8225	274	27	11	4805	139	7	10	2645	127	12
44	8405	274	27	12	4985	139	7	11	2825	122	12
45	8585	276	29	13	5165	139	6	12	3005	111	12
46	8765	267	27	14	5345	130	6	13	3185	96	11
47	8945	267	31	15	5525	123	6	14	3365	98	10

TABLE II (1928)

Date and time	Height in meters	Azimuth from; degrees	Velocity m/s	Date and time	Height in meters	Azimuth from; degrees	Velocity m/s	Date and time	Height in meters	Azimuth from; degrees	Velocity m/s
Sept. 1				**Sept. 2**				**Sept. 2**			
10:15A	3545	84	9	10:45A	7145	30	4	2:27P	2465	358	4
16	3725	88	10	46	7325	26	4	28	2645	10	3
17	3905	92	11	47	7505	13	4	29	2825	16	3
18	4085	76	9	48	7685	3	3	30	3005	16	2
19	4265	60	9	49	7865	3	3	31	3185	36	3
20	4445	65	11	50	8045	32	2	32	3365	23	2
21	4625	70	13	51	8225	27	1	33	3545	8	4
22	4805	70	14	52	8405	0	1	34	3725	9	4
23	4985	70	14	53	8585	0	2	35	3905	21	4
				54	8765	19	2	36	4085	33	5
2	(443)			55	8945	28	3	37	4265	33	5
10:08A	395	105	2	56	9125	28	3	38	4445	29	6
09	611	53	1	57	9305	28	3	39	4625	29	5
10	809	41	3	58	9485	28	2	40	4805	29	5
11	1007	24	3	59	9665	47	3	41	4985	34	5
12	1196	14	3	11:00A	9845	28	3	42	5165	28	5
13	1385	12	3	01	10025	5	4	43	5345	35	6
14	1565	10	3	02	10205	5	4	44	5525	26	4
15	1745	8	3	03	10385	35	4	45	5705	6	2
16	1925	8	4	04	10565	55	5	46	5885	337	2
17	2105	14	5	05	10745	36	5	47	6065	326	2
18	2285	14	5	06	10925	14	6	48	6245	41	1
19	2465	5	4	07	11105	14	5	49	6425	41	2
20	2645	0	4	08	11285	14	5	50	6605	21	2
21	2825	359	4	09	11465	14	4	51	6785	11	3
22	3005	10	4	10	11645	14	4	52	6965	357	5
23	3185	22	4	11	11825	14	4	53	7145	346	3
24	3365	41	3	12	12005	20	4	54	7325	345	3
25	3545	51	4	13	12185	20	4	55	7505	356	3
26	3725	51	5	14	12365	2	4	56	7685	6	4
27	3905	51	6	15	12545	2	4				
28	4085	57	5	16	12725	26	13	**6**	(445)		
29	4265	57	5	17	12905	26	17	2:11P	395	110	5
30	4445	50	6	18	13085	30	12	12	611	122	6
31	4625	50	7	19	13265	37	11	13	809	131	6
32	4805	50	6					14	1007	152	5
33	4985	42	5	**2**	(444)			15	1196	174	6
34	5165	38	5	2:16P	395	181	3	16	1385	186	7
35	5345	44	4	17	611	236	2	17	1565	196	7
36	5525	68	4	18	809	300	3	18	1745	210	5
37	5705	90	4	19	1007	321	4	19	1925	202	4
38	5885	80	4	20	1196	335	5	20	2105	173	6
39	6065	55	4	21	1385	340	6	21	2285	161	8
40	6245	47	3	22	1565	347	7	22	2465	166	8
41	6425	46	3	23	1745	347	6	23	2645	172	7
42	6605	46	3	24	1925	335	5	24	2825	185	6
43	6785	46	3	25	2105	341	4	25	3005	190	8
44	6965	35	4	26	2285	346	5	26	3185	190	9

TABLE II (1928)

Date and time	Height in meters	Azimuth from; degrees	Velocity m/s
Sept. 7	(446)		
2:03P	395	149	8
04	611	154	6
05	809	157	6
06	1007	159	5
07	1196	164	5
08	1385	163	6
09	1565	163	7
10	1745	166	7
11	1925	166	8
8	(447)		
9:58A	395	104	5
59	611	116	5
10:00A	809	142	5
01	1007	174	5
02	1196	190	5
03	1385	199	6
04	1565	199	6
05	1745	179	6
06	1925	167	7
07	2105	163	7
08	2285	163	8
09	2465	172	9
10	2645	176	10
11	2825	166	9
12	3005	154	9
13	3185	154	9
14	3365	157	10
15	3545	161	11
16	3725	161	13
17	3905	161	14
18	4085	159	15
19	4265	158	15
20	4445	158	16
21	4625	158	16
22	4805	155	14
23	4985	153	14
10	(448)		
2:06P	395	121	3
07	611	160	3
08	809	177	4
09	1007	181	4
10	1196	207	5
11	1385	223	6
12	1565	211	5
13	1745	204	5
14	1925	204	5

Date and time	Height in meters	Azimuth from; degrees	Velocity m/s
Sept. 10			
2:15P	2105	201	6
16	2285	201	6
17	2465	201	7
18	2645	201	7
19	2825	187	7
20	3005	176	7
21	3185	165	8
22	3365	159	8
23	3545	154	7
24	3725	159	7
25	3905	159	8
26	4085	172	9
27	4265	182	7
28	4445	190	6
29	4625	194	6
30	4805	194	6
31	4985	201	8
32	5165	201	9
33	5345	192	9
34	5525	190	9
35	5705	187	9
36	5885	187	9
37	6065	195	11
38	6245	205	11
39	6425	205	12
40	6605	208	13
41	6785	208	12
13	(449)		
1:50P	395	139	2
51	611	98	1
52	809	54	1
53	1007	45	2
54	1196	66	2
55	1385	149	1
56	1565	172	2
57	1745	177	2
58	1925	177	4
59	2105	172	7
2:00P	2285	176	10
01	2465	182	11
02	2645	185	11
03	2825	181	10
15	(450)		
4:00P	395	57	3
01	611	51	4
02	809	53	4

Date and time	Height in meters	Azimuth from; degrees	Velocity m/s
Sept. 15			
4:03P	1007	57	3
04	1196	92	1
05	1385	238	1
06	1565	254	1
07	1745	87	1
08	1925	70	2
09	2105	77	3
10	2285	92	3
11	2465	44	3
12	2645	27	1
13	2825	338	1
14	3005	343	1
15	3185	348	1
16	3365	341	2
17	3545	341	2
18	3725	340	3
19	3905	355	5
20	4085	15	7
21	4265	27	9
22	4445	27	9
23	4625	20	9
24	4805	8	11
25	4985	0	11
26	5165	358	11
27	5345	358	11
28	5525	358	10
29	5705	353	10
30	5885	353	11
31	6065	353	11
32	6245	353	13
33	6425	351	17
34	6605	357	20
35	6785	358	22
36	6965	358	24
37	7145	355	25
38	7325	354	25
39	7505	354	29
40	7685	354	29
16	(451)		
11:46A	395	348	2
47	611	107	2
48	809	168	1
49	1007	247	2
50	1196	271	2
51	1385	275	3
52	1565	279	3
53	1745	279	2

TABLE II (1928)

Date and time	Height in meters	Azimuth from; degrees	Velocity m/s
Sept. 16			
11:54A	1925	251	2
55	2105	204	2
56	2285	243	2
57	2465	288	4
58	2645	324	6
59	2825	343	9
12:00M	3005	343	11
0:01P	3185	335	11
02	3365	332	10
16	(452)		
1:48P	395	78	2
49	611	103	3
50	809	124	2
51	1007	163	2
52	1196	241	3
53	1385	260	5
54	1565	269	4
55	1745	261	4
56	1925	255	4
57	2105	249	4
58	2285	245	4
59	2465	261	4
2:00P	2645	305	5
01	2825	332	6
02	3005	342	7
03	3185	342	9
04	3365	338	10
05	3545	336	9
06	3725	335	9
07	3905	335	9
08	4085	344	9
09	4265	350	11
10	4445	350	12
11	4625	341	12
12	4805	338	14
13	4985	334	17
14	5165	334	17
15	5345	334	18
16	5525	331	19
17	5705	333	21
18	5885	335	21
17	(453)		
0:59P	395	82	4
1:00P	611	89	4
01	809	118	3
02	1007	152	2

Date and time	Height in meters	Azimuth from; degrees	Velocity m/s
Sept. 17			
1:03P	1196	145	2
04	1385	157	4
05	1565	166	5
06	1745	172	4
07	1925	172	3
08	2105	160	2
09	2285	143	2
10	2465	138	2
11	2645	137	2
12	2825	136	2
13	3005	137	2
14	3185	144	2
15	3365	161	3
16	3545	175	5
17	3725	184	5
18	3905	195	5
19	4085	195	4
20	4265	195	3
21	4445	193	2
22	4625	192	2
23	4805	192	3
24	4985	207	4
25	5165	219	4
26	5345	236	4
27	5525	244	4
28	5705	249	4
29	5885	253	4
30	6065	253	5
31	6245	253	5
32	6425	253	6
33	6605	265	6
34	6785	276	5
35	6965	284	6
36	7145	284	7
37	7325	284	8
38	7505	284	7
39	7685	285	7
40	7865	291	7
41	8045	290	9
42	8225	285	10
43	8405	280	11
44	8585	282	10
45	8765	275	11
46	8945	266	14
47	9125	270	14
48	9305	272	14
49	9485	269	14
50	9665	269	19

Date and time	Height in meters	Azimuth from; degrees	Velocity m/s
Sept. 17			
1:51P	9845	269	22
52	10025	267	24
53	10205	266	24
21	(454)		
1:58P	395	85	5
59	611	119	7
2:00P	809	137	8
01	1007	156	9
02	1196	174	11
03	1385	181	14
04	1565	182	15
05	1745	180	14
06	1925	178	15
07	2105	174	15
08	2285	169	12
09	2465	169	11
10	2645	168	10
11	2825	168	10
12	3005	168	9
13	3185	168	9
14	3365	155	11
15	3545	149	13
16	3725	149	17
22	(455)		
0:46P	395	80	4
47	611	130	6
48	809	166	8
49	1007	172	8
50	1196	188	7
51	1385	193	9
52	1565	193	11
53	1745	195	11
54	1925	194	12
55	2105	190	12
56	2285	187	12
57	2465	182	13
58	2645	178	12
59	2825	178	12
1:00P	3005	177	13
01	3185	178	11
02	3365	168	8
03	3545	159	10
04	3725	159	12
05	3905	167	14
06	4085	177	14
07	4265	183	14

TABLE II (1928)

Date and time	Height in meters	Azimuth from; degrees	Veloc- ity m/s	Date and time	Height in meters	Azimuth from; degrees	Veloc- ity m/s	Date and time	Height in meters	Azimuth from; degrees	Veloc- ity m/s
Sept.				**Sept.**				**Sept.**			
22				**24**	(457)			**26**			
1:08P	4445	192	13	11:20A	395	75	4	2:04P	809	86	1
09	4625	198	12	21	611	89	2	05	1007	17	1
10	4805	192	13	22	809	125	3	06	1196	340	1
11	4985	189	13	23	1007	133	4	07	1385	279	1
12	5165	187	16	24	1196	153	5	08	1565	220	1
13	5345	185	17	25	1385	164	7				
14	5525	185	17	26	1565	172	7	**27**	(460)		
15	5705	184	17	27	1745	178	9	11:22A	395	58	3
16	5885	184	18	28	1925	175	11	23	611	111	3
				29	2105	161	11	24	809	155	5
23	(456)			30	2285	161	11	25	1007	172	5
0:14P	395	65	3	31	2465	176	11	26	1196	190	5
15	611	55	5	32	2645	176	13	27	1385	210	7
16	809	83	2	33	2825	170	13	28	1565	210	10
17	1007	93	3	34	3005	162	13	29	1745	201	10
18	1196	122	4	35	3185	159	14	30	1925	191	9
19	1385	269	4	36	3365	159	14	31	2105	189	10
20	1565	266	5	37	3545	155	13	32	2285	189	11
21	1745	243	6	38	3725	163	13	33	2465	189	12
22	1925	222	7	39	3905	163	13	34	2645	180	11
23	2105	210	7	40	4085	163	14	35	2825	178	13
24	2285	199	8	41	4265	171	15	36	3005	176	15
25	2465	186	8	42	4445	176	14	37	3185	172	15
26	2645	173	10	43	4625	177	14	38	3365	173	17
27	2825	167	12	44	4805	177	13	39	3545	167	17
28	3005	166	14	45	4985	172	12	40	3725	167	17
29	3185	166	15	46	5165	172	12	41	3905	180	17
30	3365	159	14	47	5345	161	13	42	4085	180	15
31	3545	159	15	48	5525	157	13	43	4265	180	15
32	3725	173	15					44	4445	187	17
33	3905	180	15	**25**	(458)			45	4625	191	17
34	4085	180	16	1:55P	395	92	5	46	4805	200	18
35	4265	180	18	56	611	120	7	47	4985	200	17
36	4445	178	18	57	809	147	9	48	5165	200	17
37	4625	178	14	58	1007	158	9				
38	4805	168	10	59	1196	164	8	**29**	(461)		
39	4985	142	8	2:00P	1385	166	7	1:03P	395	183	10
40	5165	123	9	01	1565	172	9	04	611	192	17
41	5345	111	10	02	1745	180	12	05	809	193	11
42	5525	113	11	03	1925	181	12	06	1007	193	11
43	5705	131	11	04	2105	181	10	07	1196	193	17
44	5885	145	12	05	2285	176	10	08	1385	194	26
45	6065	150	14	06	2465	188	10	09	1565	194	33
46	6245	151	15	07	2645	200	9	10	1745	194	37
47	6425	152	19								
48	6605	154	20	**26**	(459)			**30**	(462)		
49	6785	159	20	2:02P	395	150	1	1:52P	395	281	1
50	6965	161	18	03	611	59	2	53	611	107	2

TABLE II (1928)

Date and time	Height in meters	Azimuth from; degrees	Velocity m/s	Date and time	Height in meters	Azimuth from; degrees	Velocity m/s	Date and time	Height in meters	Azimuth from; degrees	Velocity m/s
Sept.				**Oct.**				**Oct.**			
30				**4**	(464)			**6**			
1:54P	809	138	6	1:19P	395	30	11	1:41P	2645	164	4
55	1007	158	7	20	611	11	7	42	2825	173	5
56	1196	176	9	21	809	9	8	43	3005	184	7
57	1385	188	10	22	1007	12	10	44	3185	189	9
58	1565	200	9	23	1196	20	12	45	3365	194	9
59	1745	195	11	24	1385	23	12	46	3545	202	9
2:00P	1925	190	13	25	1565	27	11	47	3725	203	8
				26	1745	30	11	48	3905	192	7
								49	4085	189	6
Oct.				**5**	(465)			50	4265	196	6
1	(463)			1:50P	395	68	5	51	4445	212	5
1:04P	395	120	5	51	611	106	7	52	4625	233	4
05	611	144	6	52	809	146	8	53	4805	258	5
06	809	156	8	53	1007	167	10	54	4985	256	5
07	1007	157	7	54	1196	169	9	55	5165	247	5
08	1196	151	7	55	1385	163	6	56	5345	254	5
09	1385	151	7	56	1565	178	6	57	5525	262	4
10	1565	147	8	57	1745	197	7	58	5705	291	3
11	1745	142	9	58	1925	210	8	59	5885	352	3
12	1925	139	8	59	2105	209	10	2:00P	6065	356	2
13	2105	135	9	2:00P	2285	199	12	01	6245	22	2
14	2285	141	10	01	2465	194	12	02	6425	47	3
15	2465	149	9	02	2645	189	12	03	6605	73	3
16	2645	154	9	03	2825	183	13	04	6785	63	4
17	2825	153	10	04	3005	175	13	05	6965	49	4
18	3005	152	12	05	3185	167	12	06	7145	43	5
19	3185	151	14	06	3365	161	13	07	7325	42	7
20	3365	148	17	07	3545	157	14	08	7505	40	8
21	3545	145	17	08	3725	155	14	09	7685	39	8
22	3725	138	15	09	3905	158	15	10	7865	40	9
23	3905	138	14	10	4085	161	14	11	8045	40	9
24	4085	139	13	11	4265	165	10	12	8225	40	10
25	4265	143	14	12	4445	174	9	13	8405	39	11
26	4445	151	16	13	4625	184	8	14	8585	40	11
27	4625	153	17					15	8765	38	11
28	4805	153	18	**6**	(466)			16	8945	41	12
29	4985	152	21	1:29P	395	74	5	17	9125	42	13
30	5165	150	18	30	611	113	5	18	9305	45	14
31	5345	149	18	31	809	148	5	19	9485	46	15
32	5525	148	19	32	1007	172	5	20	9665	46	16
33	5705	147	19	33	1196	188	5	21	9845	46	17
34	5885	147	21	34	1385	201	6	22	10025	46	17
35	6065	147	23	35	1565	216	4	23	10205	46	17
36	6245	147	28	36	1745	252	3	24	10385	43	16
37	6425	147	26	37	1925	261	4	25	10565	42	17
38	6605	148	19	38	2105	246	4	26	10745	40	17
39	6785	148	17	39	2285	222	4	27	10925	40	19
40	6965	148	18	40	2465	191	4	28	11105	39	20

TABLE II (1928)

Date and time	Height in meters	Azimuth from; degrees	Velocity m/s	Date and time	Height in meters	Azimuth from; degrees	Velocity m/s	Date and time	Height in meters	Azimuth from; degrees	Velocity m/s
Oct.				**Oct.**				**Oct.**			
6				**9**				**11**			
2:29P	11285	37	21	1:08P	3365	169	9	10:01A	6065	205	3
30	11465	35	24	09	3545	174	11	02	6245	223	4
31	11645	30	25	10	3725	174	12	03	6425	235	4
32	11825	27	25	11	3905	182	13	04	6605	235	6
33	12005	25	25	12	4085	182	13	05	6785	226	7
34	12185	24	25	13	4265	182	12	06	6965	226	7
35	12365	24	25	14	4445	186	11	07	7145	226	7
36	12545	24	25	15	4625	191	10	08	7325	246	7
37	12725	23	25	16	4805	191	11	09	7505	246	9
38	12905	23	24	17	4985	198	12	10	7685	246	9
39	13085	24	22	18	5165	198	13	11	7865	264	9
				19	5345	193	15	12	8045	271	9
7	(467)			20	5525	193	16	13	8225	268	9
0:34P	395	46	4	21	5705	197	17	14	8405	268	9
35	611	71	6	22	5885	197	17	15	8585	278	9
36	809	76	6					16	8765	292	8
37	1007	71	5	**11**	(470)			17	8945	300	8
38	1196	70	5	9:30A	395	59	3	18	9125	300	9
				31	611	61	5	19	9305	298	9
8	(468)			32	809	55	5	20	9485	293	9
1:21P	395	264	1	33	1007	52	4	21	9665	284	8
22	611	246	1	34	1196	57	3	22	9845	270	9
23	809	164	5	35	1385	67	3	23	10025	270	10
24	1007	160	10	36	1565	90	2	24	10205	270	11
25	1196	169	11	37	1745	134	2	25	10385	270	11
26	1385	182	11	38	1925	168	4	26	10565	268	12
27	1565	193	13	39	2105	186	5	27	10745	265	13
28	1745	198	16	40	2285	186	5	28	10925	262	13
29	1925	196	14	41	2465	186	5	29	11105	262	11
30	2105	196	11	42	2645	186	5	30	11285	262	11
				43	2825	186	6	31	11465	262	10
9	(469)			44	3005	186	7	32	11645	247	9
0:52P	395	45	4	45	3185	182	7	33	11825	229	9
53	611	93	1	46	3365	177	7	34	12005	229	12
54	809	188	2	47	3545	177	6	35	12185	235	14
55	1007	232	3	48	3725	172	4	36	12365	230	13
56	1196	296	2	49	3905	155	2	37	12545	222	13
57	1385	296	1	50	4085	96	1	38	12725	222	14
58	1565	217	3	51	4265	66	1	39	12905	218	13
59	1745	203	5	52	4445	66	1	40	13085	218	10
1:00P	1925	203	5	53	4625	94	3	41	13265	226	12
01	2105	209	6	54	4805	103	6	42	13445	226	17
02	2285	201	9	55	4985	125	7	43	13625	233	16
03	2465	189	9	56	5165	141	7	44	13805	233	15
04	2645	178	9	57	5345	154	4	45	13985	227	13
05	2825	174	9	58	5525	176	3	46	14165	227	12
06	3005	165	9	59	5705	201	3	47	14345	222	13
07	3185	163	8	10:00A	5885	210	3	48	14525	222	16

TABLE II (1928)

Date and time	Height in meters	Azimuth from; degrees	Velocity m/s	Date and time	Height in meters	Azimuth from; degrees	Velocity m/s	Date and time	Height in meters	Azimuth from; degrees	Velocity m/s
Oct. 11				**Oct. 12**				**Oct. 12**			
10:49A	14705	222	17	11:05A	7145	210	3	11:53A	15785	210	19
50	14885	226	16	06	7325	211	4	54	15965	210	19
51	15065	223	17	07	7505	213	3	55	16145	210	20
52	15245	231	14	08	7685	216	3	56	16325	210	17
53	15425	231	11	09	7865	216	4	57	16505	206	17
54	15605	231	14	10	8045	229	5	58	16685	200	17
55	15785	231	16	11	8225	238	4	59	16865	191	17
56	15965	229	14	12	8405	247	5	12:00M	17045	194	21
57	16145	229	12	13	8585	255	5	0:01P	17225	195	23
				14	8765	255	5	02	17405	197	23
12	(471)			15	8945	266	5	03	17585	197	27
10:28A	395	78	3	16	9125	280	4	04	17765	199	31
29	611	73	2	17	9305	280	3	05	17945	199	33
30	809	213	3	18	9485	262	1	06	18125	199	36
31	1007	258	3	19	9665	218	1	07	18305	199	37
32	1196	289	5	20	9845	118	1				
33	1385	301	6	21	10025	113	1	**13**	(472)		
34	1565	303	5	22	10205	113	2	1:20P	395	107	6
35	1745	340	7	23	10385	113	4	21	611	133	11
36	1925	351	10	24	10565	113	4	22	809	148	12
37	2105	351	8	25	10745	113	5	23	1007	159	11
38	2285	346	7	26	10925	128	6	24	1196	173	11
39	2465	334	6	27	11105	147	7	25	1385	175	11
40	2645	321	5	28	11285	162	8	26	1565	172	14
41	2825	292	3	29	11465	162	7	27	1745	172	15
42	3005	275	3	30	11645	162	6	28	1925	177	13
43	3185	266	4	31	11825	155	6	29	2105	177	15
44	3365	249	4	32	12005	155	6	30	2285	173	18
45	3545	236	5	33	12185	155	6	31	2465	173	19
46	3725	229	4	34	12365	155	6	32	2645	173	19
47	3905	230	4	35	12545	155	5	33	2825	175	12
48	4085	230	4	36	12725	156	5	34	3005	173	12
49	4265	207	3	37	12905	156	4				
50	4445	192	3	38	13085	192	6	**14**	(473)		
51	4625	191	3	39	13265	192	6	2:08P	395	75	3
52	4805	191	4	40	13445	197	5	09	611	101	3
53	4985	182	5	41	13625	217	6	10	809	119	5
54	5165	168	6	42	13805	227	8	11	1007	125	3
55	5345	164	6	43	13985	232	9	12	1196	138	4
56	5525	171	6	44	14165	237	10	13	1385	148	6
57	5705	171	6	45	14345	237	11	14	1565	148	7
58	5885	171	6	46	14525	237	12	15	1745	144	8
59	6065	161	6	47	14705	237	13	16	1925	138	6
11:00A	6245	154	4	48	14885	227	13	17	2105	140	5
01	6425	159	3	49	15065	223	16	18	2285	148	7
02	6605	159	3	50	15245	223	19	19	2465	145	7
03	6785	162	2	51	15425	214	18	20	2645	150	7
04	6965	204	2	52	15605	210	18	21	2825	157	8

TABLE II (1928)

Date and time	Height in meters	Azimuth from; degrees	Velocity m/s
Oct. 14			
2:22P	3005	163	11
23	3185	161	13
24	3365	161	15
25	3545	163	16
15	(474)		
1:50P	395	76	5
51	611	82	9
52	809	103	9
53	1007	131	9
54	1196	143	10
55	1385	144	11
56	1565	142	13
57	1745	139	15
58	1925	136	15
59	2105	133	14
2:00P	2285	131	15
01	2465	131	16
02	2645	130	16
16	(475)		
11:46A	395	250	4
47	611	252	5
48	809	225	2
49	1007	213	3
50	1196	211	6
51	1385	208	6
52	1565	180	4
53	1745	142	3
54	1925	87	3
55	2105	40	3
56	2285	10	2
57	2465	95	1
58	2645	130	4
59	2825	136	4
12:00M	3005	144	3
0:01P	3185	188	3
02	3365	213	3
03	3545	202	3
04	3725	189	4
05	3905	186	4
06	4085	166	4
07	4265	152	5
08	4445	150	4
20	(476)		
1:27P	395	190	2
28	611	230	4

Date and time	Height in meters	Azimuth from; degrees	Velocity m/s
Oct. 20			
1:29P	809	243	5
30	1007	286	8
31	1196	297	11
32	1385	294	10
33	1565	298	8
34	1745	301	6
35	1925	291	4
36	2105	262	4
37	2285	253	5
38	2465	250	4
39	2645	226	2
40	2825	182	1
41	3005	157	2
42	3185	143	2
43	3365	138	3
44	3545	127	3
45	3725	119	4
46	3905	117	4
47	4085	137	5
48	4265	157	6
49	4445	165	6
50	4625	179	5
51	4805	175	4
52	4985	169	6
53	5165	165	7
54	5345	164	7
55	5525	164	9
56	5705	159	10
57	5885	158	11
58	6065	162	13
59	6245	162	13
2:00P	6425	160	13
01	6605	159	14
02	6785	157	14
03	6965	156	14
04	7145	156	14
05	7325	154	14
06	7505	153	13
07	7685	151	13
08	7865	152	13
09	8045	155	13
10	8225	161	11
11	8405	159	10
12	8585	160	10
13	8765	164	9
14	8945	170	7
15	9125	176	7
16	9305	178	7

Date and time	Height in meters	Azimuth from; degrees	Velocity m/s
Oct. 20			
2:17P	9485	172	7
18	9665	170	7
19	9845	171	7
20	10025	172	8
21	(477)		
1:33P	395	80	4
34	611	100	4
35	809	110	3
36	1007	111	4
37	1196	127	4
38	1385	134	5
39	1565	127	7
40	1745	125	8
41	1925	120	7
42	2105	125	5
43	2285	141	4
22	(478)		
1:57P	395	77	3
58	611	95	3
59	809	128	2
2:00P	1007	163	1
01	1196	209	1
02	1385	242	2
03	1565	258	1
04	1745	282	2
05	1925	266	3
06	2105	241	2
07	2285	215	2
08	2465	201	3
09	2645	205	3
10	2825	208	5
11	3005	208	5
12	3185	208	7
13	3365	208	8
14	3545	200	9
15	3725	192	11
16	3905	190	13
17	4085	188	14
18	4265	184	14
19	4445	184	15
23	(479)		
1:15P	395	67	3
16	611	125	3
17	809	166	4
18	1007	186	5

TABLE II (1928)

Date and time	Height in meters	Azimuth from; degrees	Velocity m/s	Date and time	Height in meters	Azimuth from; degrees	Velocity m/s	Date and time	Height in meters	Azimuth from; degrees	Velocity m/s
Oct.				**Oct.**				**Oct.**			
23				**24**	(480)			**24**			
1:19P	1196	188	6	1:05P	395	79	3	1:53P	9125	144	13
20	1385	184	6	06	611	100	3	54	9305	141	16
21	1565	169	4	07	809	117	1	55	9485	144	18
22	1745	163	3	08	1007	134	1	56	9665	144	21
23	1925	175	2	09	1196	6	1	57	9845	144	22
24	2105	194	1	10	1385	343	2				
25	2285	191	1	11	1565	350	3	**25**	(481)		
26	2465	170	1	12	1745	348	4	0:57P	395	63	2
27	2645	170	2	13	1925	348	3	58	611	101	2
28	2825	181	2	14	2105	356	4	59	809	140	3
29	3005	194	3	15	2285	2	4	1:00P	1007	135	4
30	3185	208	4	16	2465	358	5	01	1196	170	3
31	3365	212	4	17	2645	354	6	02	1385	198	2
32	3545	217	4	18	2825	358	6	03	1565	214	2
33	3725	217	4	19	3005	359	6	04	1745	247	2
34	3905	211	3	20	3185	1	7	05	1925	255	3
35	4085	187	3	21	3365	5	8	06	2105	258	4
36	4265	181	4	22	3545	10	8	07	2285	235	4
37	4445	188	5	23	3725	10	8	08	2465	221	6
38	4625	187	5	24	3905	8	7	09	2645	219	7
39	4805	185	6	25	4085	1	6	10	2825	210	9
40	4985	185	7	26	4265	0	5	11	3005	206	9
41	5165	180	7	27	4445	11	3	12	3185	202	9
42	5345	182	8	28	4625	24	3	13	3365	199	11
43	5525	178	10	29	4805	58	4	14	3545	195	12
44	5705	173	12	30	4985	118	7	15	3725	192	13
45	5885	171	13	31	5165	149	13	16	3905	198	13
46	6065	164	14	32	5345	149	18	17	4085	203	11
47	6245	161	14	33	5525	144	19	18	4265	203	9
48	6425	159	15	34	5705	148	21	19	4445	202	11
49	6605	159	17	35	5885	148	22	20	4625	207	14
50	6785	161	16	36	6065	148	21	21	4805	210	16
51	6965	162	16	37	6245	148	21	22	4985	207	16
52	7145	161	17	38	6425	148	20	23	5165	203	16
53	7325	159	18	39	6605	148	20	24	5345	203	14
54	7505	159	17	40	6785	148	21	25	5525	203	13
55	7685	158	19	41	6965	148	21	26	5705	203	14
56	7865	164	22	42	7145	142	17				
57	8045	164	19	43	7325	142	14	**26**	(482)		
58	8225	168	19	44	7505	142	10	1:25P	395	329	1
59	8405	167	18	45	7685	142	8	26	611	355	1
2:00P	8585	164	19	46	7865	144	7	27	809	33	1
01	8765	159	17	47	8045	144	7	28	1007	146	1
02	8945	157	14	48	8225	150	8				
03	9125	157	15	49	8405	158	6	**28**	(483)		
04	9305	159	15	50	8585	158	4	1:09P	395	70	2
05	9485	159	14	51	8765	158	5	10	611	72	2
				52	8945	144	8	11	809	50	2

TABLE II (1928)

Date and time	Height in meters	Azimuth from; degrees	Velocity m/s	Date and time	Height in meters	Azimuth from; degrees	Velocity m/s	Date and time	Height in meters	Azimuth from; degrees	Velocity m/s
Oct.				**Oct.**				**Oct.**			
28				**29**				**30**			
1:12P	1007	37	2	1:30P	1007	189	2	1:31P	1196	121	9
13	1196	25	3	31	1196	192	3	32	1385	115	12
14	1385	27	3	32	1385	199	3	33	1565	112	11
15	1565	44	4	33	1565	211	3	34	1745	113	11
16	1745	43	5	34	1745	214	3	35	1925	113	11
17	1925	48	5	35	1925	218	4	36	2105	116	13
18	2105	67	5	36	2105	213	5	37	2285	117	15
19	2285	79	5	37	2285	212	5	38	2465	128	13
20	2465	83	4	38	2465	222	5	39	2645	146	11
21	2645	102	3	39	2645	231	6	40	2825	156	10
22	2825	151	4	40	2825	237	6				
23	3005	185	6	41	3005	235	6	**Nov.**			
24	3185	191	7	42	3185	233	7	**1**	(486)		
25	3365	194	5	43	3365	232	8	2:00P	395	112	4
26	3545	179	3	44	3545	228	9	01	611	50	8
27	3725	162	5	45	3725	224	9	02	809	23	16
28	3905	180	5	46	3905	224	9				
29	4085	204	5	47	4085	224	9	**2**	(487)		
30	4265	211	4	48	4265	222	9	11:47A	395	61	5
31	4445	201	4	49	4445	221	10	48	611	78	8
32	4625	205	4	50	4625	221	11	49	809	88	8
33	4805	222	4	51	4805	220	11	50	1007	78	6
34	4985	232	5	52	4985	220	11	51	1196	66	5
35	5165	231	6	53	5165	219	11	52	1385	82	5
36	5345	226	6	54	5345	218	11	53	1565	93	6
37	5525	225	6	55	5525	215	11	54	1745	99	8
38	5705	221	7	56	5705	216	11	55	1925	106	8
39	5885	213	7	57	5885	212	11	56	2105	111	8
40	6065	210	7	58	6065	210	12	57	2285	116	7
41	6245	211	6	59	6245	205	13	58	2465	116	7
42	6425	211	7	2:00P	6425	201	15	59	2645	118	7
43	6605	217	7	01	6605	197	15	12:00M	2825	118	7
44	6785	245	6	02	6785	201	14	0:01P	3005	116	8
45	6965	264	7	03	6965	202	13	02	3185	115	7
46	7145	259	7	04	7145	201	14	03	3365	117	6
47	7325	253	7	05	7325	202	15	04	3545	115	5
48	7505	261	8	06	7505	205	16	05	3725	130	8
49	7685	267	8	07	7685	209	17	06	3905	140	9
50	7865	265	8	08	7865	210	19	07	4085	147	9
51	8045	262	9	09	8045	211	18	08	4265	158	7
52	8225	261	11	10	8225	212	21	09	4445	157	7
53	8405	261	11	11	8405	212	23	10	4625	156	8
54	8585	261	10					11	4805	141	8
				30	(485)			12	4985	140	8
29	(484)			1:27P	395	94	7	13	5165	150	8
1:27P	395	49	2	28	611	94	10	14	5345	147	7
28	611	106	1	29	809	106	9	15	5525	149	8
29	809	168	2	30	1007	119	7	16	5705	151	8

TABLE II (1928)

Date and time	Height in meters	Azimuth from; degrees	Velocity m/s
Nov. 2			
0:17P	5885	155	8
18	6065	165	10
19	6245	172	11
20	6425	171	11
21	6605	174	11
22	6785	176	11
23	6965	174	14
24	7145	172	13
25	7325	175	13
26	7505	174	13
27	7685	164	13
28	7865	156	13
29	8045	152	9
30	8225	150	11
31	8405	149	11
32	8585	155	11
3	(488)		
1:18P	395	78	5
19	611	111	14
20	809	127	11
21	1007	156	9
22	1196	177	11
23	1385	181	14
24	1565	184	14
4	(489)		
1:15P	395	155	20
16	611	171	29
5	(490)		
1:08P	395	348	1
09	611	216	1
10	809	143	6
11	1007	134	10
12	1196	141	10
13	1385	156	13
14	1565	165	15
15	1745	162	19
16	1925	162	21
17	2105	165	24
18	2285	166	27
19	2465	166	26
6	(491)		
10:41A	395	129	7
42	611	133	11
43	809	136	11

Date and time	Height in meters	Azimuth from; degrees	Velocity m/s
Nov. 6			
10:44A	1007	141	13
45	1196	146	15
46	1385	147	17
47	1565	149	18
48	1745	153	19
49	1925	153	23
7	(492)		
0:10P	395	117	7
11	611	139	10
12	809	150	13
13	1007	157	15
14	1196	165	17
15	1385	174	18
16	1465	178	16
17	1645	177	14
8	(493)		
1:18P	395	84	4
19	611	128	9
20	809	148	10
21	1007	171	12
22	1196	177	14
23	1385	170	15
24	1565	170	17
25	1745	170	18
26	1925	177	17
27	2105	187	15
28	2285	193	13
29	2465	194	14
30	2645	193	12
31	2825	196	8
32	3005	186	9
33	3185	182	11
9	(494)		
10:55A	395	67	7
56	611	84	11
57	809	116	8
58	1007	146	9
59	1196	156	12
11:00A	1385	172	17
01	1565	176	20
10	(495)		
0:22P	395	73	3
23	611	71	4
24	809	78	1

Date and time	Height in meters	Azimuth from; degrees	Velocity m/s
Nov. 10			
0:25P	1007	145	2
26	1196	197	3
27	1385	208	5
28	1565	189	6
29	1745	173	6
30	1925	164	4
31	2105	138	3
32	2285	111	3
33	2465	84	3
34	2645	61	2
35	2825	64	2
36	3005	30	1
37	3185	0	2
38	3365	37	2
39	3545	80	3
40	3725	97	3
41	3905	101	2
42	4085	257	1
43	4265	227	1
44	4445	142	1
45	4625	133	3
46	4805	122	4
47	4985	29	1
48	5165	108	5
49	5345	114	11
50	5525	110	9
11	(496)		
11:30A	395	95	1
31	611	75	3
32	809	45	3
33	1007	25	4
34	1196	28	4
35	1385	313	2
36	1565	290	5
37	1745	309	6
38	1925	321	8
39	2105	324	10
40	2285	318	9
41	2465	313	8
42	2645	313	7
43	2825	313	7
44	3005	320	7
45	3185	322	7
46	3365	319	7
47	3545	312	6
48	3725	301	6

TABLE II (1928)

Date and time	Height in meters	Azimuth from; degrees	Velocity m/s	Date and time	Height in meters	Azimuth from; degrees	Velocity m/s	Date and time	Height in meters	Azimuth from; degrees	Velocity m/s
Nov. 11				**Nov. 13**				**Nov. 14**			
11:49A	3905	287	6	1:22P	5345	320	7	0:50P	3185	65	2
50	4085	271	6	23	5525	316	6	51	3365	80	1
51	4265	265	5	24	5705	301	6	52	3545	144	1
52	4445	293	6	25	5885	299	7	53	3725	295	2
53	4625	314	5	26	6065	289	7	54	3905	284	2
54	4805	312	4	27	6245	283	7	55	4085	300	2
55	4985	292	3	28	6425	280	8	56	4265	191	3
56	5165	305	4	29	6605	278	10	57	4445	287	4
57	5345	303	4	30	6785	276	10	58	4625	279	5
58	5525	291	5	31	6965	276	9	59	4805	272	5
59	5705	273	4	32	7145	276	9	1:00P	4985	269	6
12:00M	5885	247	5	33	7325	276	9	01	5165	264	7
0:01P	6065	244	5	34	7505	278	9	02	5345	263	8
02	6245	239	6	35	7685	281	9	03	5525	262	8
03	6425	229	7	36	7865	280	9	04	5705	261	9
04	6605	222	7	37	8045	281	9	05	5885	260	9
05	6785	222	8	38	8225	283	10	06	6065	260	8
06	6965	222	9	39	8405	282	10	07	6245	260	9
07	7145	222	9	40	8585	281	10	08	6425	246	11
				41	8765	274	10	09	6605	244	12
13	(497)			42	8945	263	11	10	6785	248	17
0:55P	395	60	3	43	9125	259	11	11	6965	249	17
56	611	50	3	44	9305	263	11	12	7145	250	17
57	809	50	3	45	9485	262	13	13	7325	251	18
58	1007	42	4	46	9665	288	17	14	7505	251	21
59	1196	32	7	47	9845	290	21	15	7685	252	21
1:00P	1385	28	10	48	10025	292	22	16	7865	253	23
01	1565	26	11	49	10205	296	20	17	8045	251	25
02	1745	25	12	50	10385	296	20	18	8225	250	28
03	1925	23	13	51	10565	292	22	19	8405	249	28
04	2105	20	12	52	10745	293	26	20	8585	248	30
05	2285	20	12					21	8765	247	30
06	2465	23	10	**14**	(498)						
07	2645	20	7	0:35P	395	65	3	**15**	(499)		
08	2825	20	6	36	611	31	3	1:00P	395	60	2
09	3005	20	5	37	809	9	3	01	611	26	1
10	3185	10	4	38	1007	1	4	02	809	347	5
11	3365	357	5	39	1196	1	7	03	1007	350	7
12	3545	338	5	40	1385	359	7	04	1196	358	10
13	3725	319	5	41	1565	0	7	05	1385	6	9
14	3905	312	5	42	1745	2	6	06	1565	6	6
15	4085	314	5	43	1925	4	4	07	1745	3	6
16	4265	320	6	44	2105	8	4	08	1925	354	4
17	4445	329	7	45	2285	15	4	09	2105	319	3
18	4625	325	6	46	2465	25	3	10	2285	311	4
19	4805	324	6	47	2645	38	2	11	2465	305	4
20	4985	330	6	48	2825	63	1	12	2645	294	5
21	5165	322	6	49	3005	68	2	13	2825	291	5

TABLE II (1928)

Date and time	Height in meters	Azimuth from; degrees	Velocity m/s
Nov. 15			
1:14P	3005	269	4
15	3185	252	5
16	3365	239	6
17	3545	229	7
18	3725	234	8
19	3905	239	8
20	4085	240	8
21	4265	244	8
22	4445	245	8
23	4625	238	8
24	4805	230	9
25	4985	230	10
26	5165	237	11
27	5345	236	12
28	5525	234	13
29	5705	231	13
30	5885	229	13
31	6065	230	14
32	6245	230	14
33	6425	229	14
34	6605	228	14
35	6785	227	14
36	6965	226	14
37	7145	224	13
38	7325	226	11
39	7505	228	11
40	7685	227	11
41	7865	229	10
42	8045	234	8
43	8225	232	8
44	8405	230	8
45	8585	238	7
46	8765	244	6
47	8945	248	6
48	9125	253	5
49	9305	254	5
50	9485	251	8
51	9665	248	15
52	9845	247	19
53	10025	247	17
54	10205	248	17
55	10385	247	20
56	10565	247	23
16	(500)		
3:06P	395	55	2
07	611	46	4

Date and time	Height in meters	Azimuth from; degrees	Velocity m/s
Nov. 16			
3:08P	809	37	4
09	1007	37	4
17	(501)		
1:14P	395	63	1
15	611	125	2
16	809	198	3
17	1007	234	4
18	1196	232	4
19	1385	221	4
20	1565	216	4
21	1745	200	5
22	1925	200	5
23	2105	204	5
24	2285	215	3
25	2465	237	2
26	2645	246	2
27	2825	296	2
28	3005	325	3
29	3185	323	3
30	3365	310	3
31	3545	310	4
32	3725	310	5
33	3905	304	5
34	4085	289	3
35	4265	280	3
36	4445	290	4
37	4625	296	4
38	4805	305	4
39	4985	300	3
40	5165	304	4
41	5345	315	5
42	5525	311	6
43	5705	311	5
44	5885	315	6
45	6065	308	8
46	6245	299	11
47	6425	300	10
48	6605	297	13
49	6785	294	16
50	6965	296	15
51	7145	294	15
52	7325	290	16
53	7505	289	17
54	7685	290	20
55	7865	289	22
56	8045	287	23
57	8225	285	25

Date and time	Height in meters	Azimuth from; degrees	Velocity m/s
Nov. 17			
1:58P	8405	285	28
59	8585	287	31
2:00P	8765	290	35
01	8945	293	37
02	9125	294	40
03	9305	294	44
18	(502)		
1:09P	395	45	3
10	611	96	3
11	809	145	3
12	1007	146	3
13	1196	145	2
14	1385	154	1
15	1565	168	2
16	1745	196	2
17	1925	206	3
18	2105	205	3
19	2285	208	3
20	2465	219	3
21	2645	233	3
22	2825	237	3
23	3005	239	3
24	3185	241	3
25	3365	249	3
26	3545	260	3
27	3725	243	3
28	3905	224	3
19	(503)		
0:32P	395	116	5
33	611	141	8
34	809	154	7
35	1007	166	5
36	1196	191	6
37	1385	195	11
38	1565	200	14
39	1745	205	14
40	1925	212	12
41	2105	224	10
42	2285	235	10
20	(504)		
1:20P	395	83	1
21	611	176	3
22	809	203	3
23	1007	238	3
24	1196	250	4

TABLE II (1928)

Date and time	Height in meters	Azimuth from; degrees	Velocity m/s
Nov. 20			
1:25P	1385	257	5
26	1565	260	6
27	1745	253	4
28	1925	218	4
29	2105	205	5
30	2285	203	4
31	2465	188	4
32	2645	173	3
33	2825	145	2
34	3005	121	3
35	3185	102	3
36	3365	92	4
21	(505)		
0:20P	395	355	3
21	611	28	7
22	809	25	9
23	1007	25	12
24	1196	25	15
25	1385	27	14
26	1565	26	14
27	1745	26	13
28	1925	26	11
29	2105	26	9
24	(506)		
0:25P	395	70	3
26	611	65	5
27	809	67	3
28	1007	78	3
29	1196	92	4
30	1385	97	3
31	1565	95	3
32	1745	94	2
33	1925	91	3
34	2105	83	3
35	2285	58	1
36	2465	15	1
37	2645	15	1
38	2825	8	1
39	3005	322	1
40	3185	322	1
41	3365	302	1
42	3545	300	2
43	3725	296	2
44	3905	250	1
45	4085	220	2
46	4265	215	2

Date and time	Height in meters	Azimuth from; degrees	Velocity m/s
Nov. 24			
0:47P	4445	194	2
48	4625	194	3
49	4805	196	4
50	4985	200	4
51	5165	207	5
52	5345	211	6
53	5525	211	6
54	5705	211	6
55	5885	211	6
56	6065	212	7
57	6245	215	8
58	6425	217	9
59	6605	217	9
1:00P	6785	213	10
01	6965	208	10
02	7145	208	11
03	7325	205	11
04	7505	202	11
05	7685	200	11
06	7865	196	12
07	8045	194	11
08	8225	190	12
09	8405	190	11
10	8585	192	11
11	8765	193	10
12	8945	193	9
13	9125	193	9
14	9305	193	10
15	9485	192	11
16	9665	197	11
17	9845	200	11
18	10025	197	10
19	10205	197	11
20	10385	198	12
21	10565	198	12
22	10745	198	11
25	(507)		
0:46P	395	65	3
47	611	114	6
48	809	156	7
49	1007	170	7
50	1196	179	6
51	1385	191	5
26	(508)		
0:37P	395	101	4
38	611	119	9

Date and time	Height in meters	Azimuth from; degrees	Velocity m/s
Nov. 26			
0:39P	809	154	10
40	1007	180	10
41	1196	202	9
42	1385	208	9
43	1565	205	8
44	1745	193	7
45	1925	185	6
46	2105	188	7
47	2285	202	7
48	2465	203	8
49	2645	200	8
50	2825	202	8
27	(509)		
0:25P	395	208	4
26	611	210	3
27	809	238	1
28	1007	200	4
29	1196	190	4
30	1385	209	3
31	1565	224	4
32	1745	222	4
33	1925	223	4
34	2105	224	5
28	(510)		
0:27P	395	206	2
28	611	274	2
29	809	326	3
30	1007	338	3
31	1196	359	1
32	1385	338	1
33	1565	323	1
34	1745	312	1
28	(511)		
0:46P	395	111	5
47	611	139	15
48	809	149	16
49	1007	159	15
50	1196	168	15
51	1385	182	13
52	1565	194	13
53	1745	198	14
54	1925	204	15
55	2105	205	15
56	2285	199	16
57	2465	192	19

TABLE II (1928)

Date and time	Height in meters	Azimuth from; degrees	Velocity m/s	Date and time	Height in meters	Azimuth from; degrees	Velocity m/s	Date and time	Height in meters	Azimuth from; degrees	Velocity m/s
Nov.				**Dec.**				**Dec.**			
28				**3**				**5**			
0:58P	2645	186	21	0:12P	8045	352	22	1:14P	809	234	6
59	2825	183	19	13	8225	352	24	15	1007	238	6
1:00P	3005	181	18	14	8405	357	25	16	1196	245	5
				15	8585	355	27	17	1385	256	4
Dec.				16	8765	355	31	18	1565	288	3
3	(512)			17	8945	354	32	19	1745	305	5
11:30A	395	66	3	18	9125	352	32	20	1925	302	7
31	611	77	3					21	2105	293	7
32	809	110	2	**4**	(513)			22	2285	287	6
33	1007	125	3	0:06P	395	156	1	23	2465	286	6
34	1196	121	3	07	611	170	4	24	2645	286	7
35	1385	107	4	08	809	212	6				
36	1565	105	5	09	1007	253	7	**6**	(515)		
37	1745	98	5	10	1196	266	7	0:05P	395	76	5
38	1925	90	6	11	1385	266	5	06	611	112	9
39	2105	75	5	12	1565	260	5	07	809	129	10
40	2285	41	4	13	1745	261	5	08	1007	128	10
41	2465	41	5	14	1925	260	3	09	1196	126	11
42	2645	36	6	15	2105	245	2	10	1385	126	11
43	2825	21	6	16	2285	223	1	11	1565	122	13
44	3005	13	6	17	2465	238	1	12	1745	124	18
45	3185	3	6	18	2645	300	1	13	1925	125	19
46	3365	356	7	19	2825	269	1	14	2105	123	14
47	3545	1	7	20	3005	264	2	15	2285	115	8
48	3725	3	7	21	3185	253	2	16	2465	108	5
49	3905	352	7	22	3365	262	2	17	2645	105	3
50	4085	356	7	23	3545	259	2	18	2825	112	3
51	4265	345	7	24	3725	262	3	19	3005	115	4
52	4445	329	5	25	3905	269	3	20	3185	120	2
53	4625	336	6	26	4085	271	3	21	3365	156	1
54	4805	340	7	27	4265	279	4	22	3545	201	1
55	4985	340	6	28	4445	272	4	23	3725	214	1
56	5165	334	6	29	4625	270	4	24	3905	214	1
57	5345	333	7	30	4805	267	5				
58	5525	334	7	31	4985	270	5	**7**	(516)		
59	5705	334	7	32	5165	269	5	0:02P	395	166	13
12:00M	5885	334	7	33	5345	260	5	03	611	165	17
0:01P	6065	336	6	34	5525	260	5	04	809	167	17
02	6245	340	6	35	5705	260	5	05	1007	169	17
03	6425	340	6	36	5885	258	5	06	1196	174	17
04	6605	346	5	37	6065	253	5	07	1385	181	15
05	6785	351	9	38	6245	255	5	08	1565	183	15
06	6965	356	12	39	6425	247	5	09	1745	185	15
07	7145	358	15	40	6605	242	5	10	1925	189	15
08	7325	358	18					11	2105	187	15
09	7505	354	18	**5**	(514)						
10	7685	354	20	1:12P	395	44	2	**8**	(517)		
11	7865	354	20	13	611	200	4	1:03P	395	259	5

TABLE II (1928)

Date and time	Height in meters	Azimuth from; degrees	Velocity m/s	Date and time	Height in meters	Azimuth from; degrees	Velocity m/s	Date and time	Height in meters	Azimuth from; degrees	Velocity m/s
Dec.				**Dec.**				**Dec.**			
8				**10**				**11**			
1:04P	611	254	11	0:49P	3005	200	5	0:39P	5345	136	5
05	809	227	6	50	3185	198	5	40	5525	136	5
06	1007	209	5	51	3365	194	5	41	5705	135	4
07	1196	223	5	52	3545	191	5	42	5885	132	4
08	1385	234	6	53	3725	194	5	43	6065	132	3
09	1565	244	8	54	3905	200	6	44	6245	136	3
10	1745	244	12	55	4085	207	9	45	6425	141	3
11	1925	244	13	56	4265	208	13	46	6605	143	2
12	2105	235	17	57	4445	207	16	47	6785	126	2
13	2285	226	18	58	4625	206	17	48	6965	123	3
14	2465	225	18	59	4805	204	17				
15	2645	221	19	1:00P	4985	200	18	**13**	(521)		
16	2825	220	19	01	5165	198	21	0:24P	395	52	2
				02	5345	200	23	25	611	74	5
9	(518)			03	5525	200	30	26	809	76	4
0:15P	395	100	2	04	5705	201	35	27	1007	99	3
16	611	170	4	05	5885	200	31	28	1196	146	4
17	809	201	6	06	6065	201	35	29	1385	193	4
18	1007	209	8	07	6245	201	36	30	1565	210	5
19	1196	213	8					31	1745	205	5
20	1385	201	6	**11**	(520)			32	1925	207	6
21	1565	199	6	0:12P	395	54	3	33	2105	208	5
22	1745	186	8	13	611	75	5	34	2285	229	4
23	1925	179	9	14	809	73	3	35	2465	265	2
24	2105	170	7	15	1007	75	3	36	2645	293	1
25	2285	165	6	16	1196	75	5	37	2825	326	2
26	2465	154	7	17	1385	75	7	38	3005	339	2
27	2645	141	8	18	1565	71	6	39	3185	324	2
28	2825	137	9	19	1745	67	5	40	3365	290	2
29	3005	141	9	20	1925	49	4	41	3545	260	1
30	3185	150	7	21	2105	36	3	42	3725	201	1
31	3365	160	7	22	2285	29	3	43	3905	226	2
				23	2465	27	5	44	4085	245	4
10	(519)			24	2645	22	4	45	4265	263	6
0:35P	395	61	4	25	2825	353	2	46	4445	264	8
36	611	110	5	26	3005	350	1	47	4625	256	8
37	809	124	3	27	3185	106	2	48	4805	258	8
38	1007	184	2	28	3365	137	3	49	4985	259	8
39	1196	248	2	29	3545	171	3	50	5165	255	9
40	1385	269	3	30	3725	179	1	51	5345	260	9
41	1565	253	5	31	3905	144	1	52	5525	268	10
42	1745	265	6	32	4085	115	2	53	5705	270	11
43	1925	274	5	33	4265	126	3	54	5885	271	12
44	2105	244	4	34	4445	125	4	55	6065	274	13
45	2285	235	5	35	4625	135	4				
46	2465	236	4	36	4805	148	4	**14**	(522)		
47	2645	226	3	37	4985	144	5	0:29P	395	65	2
48	2825	202	4	38	5165	139	5	30	611	110	3

TABLE II (1928)

Date and time	Height in meters	Azimuth from; degrees	Velocity m/s
Dec. 14			
0:31P	809	154	3
32	1007	154	2
33	1196	144	1
34	1385	174	2
35	1565	172	1
36	1745	168	2
37	1925	168	4
38	2105	167	5
39	2285	170	5
40	2465	172	5
15	(523)		
0:16P	395	341	4
17	611	54	2
18	809	119	3
19	1007	154	4
20	1196	174	4
21	1385	193	4
22	1565	193	4
23	1745	192	5
24	1925	195	5
15	(524)		
0:32P	395	312	3
33	611	10	2
34	809	89	3
35	1007	112	4
36	1196	148	4
37	1385	173	5
38	1565	198	4
39	1745	200	5
40	1925	197	6
16	(525)		
0:24P	395	50	1
25	611	155	1
26	809	168	4
27	1007	171	5
28	1196	176	6
29	1385	182	6
30	1565	185	7
31	1745	180	9
32	1925	177	10
33	2105	173	11
34	2285	170	12
35	2465	170	12
36	2645	171	11
37	2825	167	10

Date and time	Height in meters	Azimuth from; degrees	Velocity m/s
Dec. 16			
0:38P	3005	161	10
39	3185	157	11
40	3365	157	12
41	3545	158	15
42	3725	158	16
43	3905	157	15
44	4085	157	15
45	4265	156	15
46	4445	160	15
47	4625	164	17
48	4805	163	15
49	4985	163	15
17	(526)		
0:05P	395	9	2
06	611	105	4
07	809	150	7
08	1007	166	8
09	1196	178	7
10	1385	189	9
11	1565	188	10
12	1745	186	11
13	1925	183	13
14	2105	170	13
15	2285	152	13
16	2465	137	14
17	2645	132	14
18	2825	138	14
19	3005	143	14
18	(527)		
11:50A	395	239	4
51	611	232	3
52	809	200	7
53	1007	197	8
54	1196	194	8
55	1385	191	7
56	1565	195	7
57	1745	195	8
58	1925	195	8
59	2105	193	7
12:00M	2285	195	7
0:01P	2465	204	7
02	2645	207	7
03	2825	199	8
04	3005	199	9

Date and time	Height in meters	Azimuth from; degrees	Velocity m/s
Dec. 19	(528)		
12:00M	395	44	3
0:01P	611	84	8
02	809	107	9
03	1007	129	8
04	1196	127	8
05	1385	116	10
06	1565	117	11
07	1745	117	13
08	1925	128	11
09	2105	131	10
20	(529)		
0:14P	395	80	5
15	611	152	7
16	809	165	15
17	1007	168	14
18	1196	178	9
19	1385	189	9
22	(530)		
0:30P	395	55	1
31	611	50	3
32	809	22	2
33	1007	20	3
34	1196	34	5
35	1385	46	7
36	1565	43	7
37	1745	37	7
23	(531)		
0:06P	395	4	1
07	611	247	1
08	809	278	5
09	1007	300	6
10	1196	312	5
11	1385	309	4
12	1565	320	4
13	1745	330	5
14	1925	334	5
23	(532)		
8:50P	395	90	1
51	611	142	5
52	809	161	6
53	1007	157	6
54	1196	147	5
55	1385	146	5
56	1565	151	4

TABLE II (1928)

Date and time	Height in meters	Azimuth from; degrees	Velocity m/s
Dec. 23			
8:57P	1745	159	3
58	1925	151	2
59	2105	162	1
9:00P	2285	196	2
01	2465	207	3
02	2645	221	3
03	2825	227	3
04	3005	227	3
24	(533)		
0:26P	395	59	4
27	611	109	4
28	809	169	4
29	1007	177	5
30	1196	183	5
31	1385	190	5
32	1565	178	5
33	1745	173	5
34	1925	167	4
35	2105	159	4
36	2285	156	4
37	2465	165	3
38	2645	184	3
39	2825	187	4
40	3005	195	5
41	3185	196	7
42	3365	201	6
43	3545	210	5
44	3725	205	4
45	3905	197	4
46	4085	190	4
47	4265	198	5
48	4445	204	5
49	4625	192	4
50	4805	184	4
51	4985	189	3
52	5165	179	3
53	5345	161	2
54	5525	156	2
55	5705	183	2
56	5885	197	2
57	6065	192	2
58	6245	168	3
59	6425	174	5
1:00P	6605	192	5
01	6785	204	4

Date and time	Height in meters	Azimuth from; degrees	Velocity m/s
Dec. 25	(534)		
0:48P	395	85	1
49	611	111	8
50	809	136	7
51	1007	152	5
52	1196	147	5
53	1385	150	7
54	1565	155	7
55	1745	166	6
56	1925	183	5
57	2105	187	5
58	2285	186	6
59	2465	186	6
1:00P	2645	192	5
01	2825	198	5
02	3005	200	4
03	3185	202	4
04	3365	207	4
05	3545	212	5
06	3725	216	5
07	3905	227	6
08	4085	229	6
09	4265	236	5
10	4445	237	4
11	4625	255	5
12	4805	265	6
13	4985	264	7
14	5165	267	8
15	5345	268	9
16	5525	267	9
17	5705	271	7
18	5885	285	7
19	6065	291	7
20	6245	278	7
21	6425	274	8
22	6605	272	8
23	6785	271	9
24	6965	274	8
25	7145	278	7
26	7325	274	7
27	7505	275	8
28	7685	276	9
29	7865	278	9
30	8045	278	10
31	8225	279	10
32	8405	279	11
33	8585	279	11

Date and time	Height in meters	Azimuth from; degrees	Velocity m/s
Dec. 26	(535)		
0:32P	395	285	2
33	611	296	1
34	809	246	2
35	1007	184	2
36	1196	142	3
37	1385	112	4
38	1565	115	5
39	1745	138	5
40	1925	152	6
41	2105	156	6
42	2285	157	7
43	2465	148	8
44	2645	141	8
45	2825	144	9
46	3005	152	10
47	3185	158	11
48	3365	158	12
49	3545	152	13
50	3725	149	13
51	3905	148	13
27	(536)		
0:52P	395	295	2
53	611	262	3
54	809	238	3
55	1007	237	1
56	1196	78	3
57	1385	88	4
58	1565	88	4
29	(537)		
7:10P	395	38	2
11	611	52	5
12	809	47	5
13	1007	61	4
14	1196	63	5
15	1385	54	7
16	1565	46	7
17	1745	37	6
18	1925	26	4
19	2105	35	3
20	2285	46	5
21	2465	46	5
22	2645	28	3
23	2825	319	3
24	3005	287	3
25	3185	272	5
26	3365	262	6

TABLE II (1928–1929)

Date and time	Height in meters	Azimuth from; degrees	Velocity m/s
Dec. 29			
7:27P	3545	262	7
28	3725	262	7
29	3905	262	8
30	4085	262	9
31	4265	257	11
32	4445	257	12
30	(538)		
0:53P	395	76	3
54	611	60	7
55	809	42	9
56	1007	38	11
30	(539)		
1:06P	395	73	3
07	611	60	6
08	809	40	9
09	1007	34	13
10	1196	34	13
11	1385	44	9
12	1565	60	7
13	1745	63	5
14	1925	71	5
15	2105	120	2
16	2285	229	2
17	2465	245	2
18	2645	245	3
19	2825	245	6
20	3005	240	7
21	3185	233	7
22	3365	220	9
23	3545	215	11
24	3725	222	10
25	3905	228	8
26	4085	229	9
27	4265	232	9
31	(540)		
0:30P	395	354	11
31	611	23	10
32	809	48	6
1929 Jan. 2	(541)		
0:16P	395	0	1
17	611	18	3
18	809	350	2

Date and time	Height in meters	Azimuth from; degrees	Velocity m/s
Jan. 2			
0:19P	1007	5	1
20	1196	38	2
21	1385	40	3
22	1565	46	4
23	1745	53	4
24	1925	72	3
25	2105	87	2
26	2285	124	2
27	2465	155	2
28	2645	167	3
29	2825	165	3
30	3005	170	4
31	3185	173	5
32	3365	170	6
33	3545	172	7
34	3725	172	6
35	3905	173	6
3	(542)		
11:56A	395	333	5
57	611	20	8
58	809	45	5
59	1007	97	3
12:00M	1196	147	4
0:01P	1385	169	3
02	1565	191	5
03	1745	198	7
6	(543)		
0:30P	395	233	1
31	611	337	4
32	809	350	7
33	1007	357	8
34	1196	350	9
35	1385	353	9
36	1565	9	8
37	1745	16	7
38	1925	7	6
39	2105	359	6
7	(544)		
0:09P	395	73	4
10	611	79	9
11	809	78	9
12	1007	85	7
13	1196	100	6
14	1385	102	5
15	1565	93	5

Date and time	Height in meters	Azimuth from; degrees	Velocity m/s
Jan. 7			
0:16P	1745	106	6
17	1925	121	7
18	2105	124	7
19	2285	134	6
20	2465	150	6
21	2645	156	7
22	2825	161	7
23	3005	163	7
24	3185	166	7
25	3365	186	7
26	3545	201	9
27	3725	212	10
28	3905	214	11
29	4085	221	15
30	4265	229	17
31	4445	228	19
32	4625	228	21
33	4805	229	23
34	4985	230	24
35	5165	230	25
36	5345	232	27
37	5525	234	27
38	5705	238	28
8	(545)		
0:36P	395	179	5
37	611	180	9
38	809	178	9
39	1007	169	9
40	1196	170	12
41	1385	171	15
42	1565	172	19
43	1745	175	23
44	1925	177	25
45	2105	178	26
9	(546)		
0:04P	395	165	11
05	611	154	13
06	809	156	16
07	1007	165	15
08	1196	181	14
09	1385	192	12
10	1565	192	11
11	1745	189	13
12	1925	188	15
13	2105	185	18

TABLE II (1929)

Date and time	Height in meters	Azimuth from; degrees	Velocity m/s
Jan. 9			
0:14P	2285	186	19
15	2465	186	23
16	2645	185	26
17	2825	183	31
18	3005	183	33
19	3185	183	33
10	(547)		
0:28P	395	0	1
29	611	126	1
30	809	169	3
31	1007	161	2
32	1196	181	2
33	1385	182	3
34	1565	176	6
35	1745	170	7
36	1925	160	7
37	2105	147	8
38	2285	139	9
39	2465	140	10
40	2645	140	11
41	2825	141	11
42	3005	156	11
43	3185	166	15
44	3365	163	19
45	3545	158	20
11	(548)		
0:24P	395	52	4
25	611	71	7
26	809	80	8
27	1007	78	7
28	1196	78	8
29	1385	80	8
30	1565	84	8
31	1745	88	8
32	1925	92	8
33	2105	102	9
34	2285	112	10
35	2465	124	9
36	2645	141	10
37	2825	151	10
38	3005	160	9
39	3185	161	9
40	3365	159	9
41	3545	152	9
42	3725	147	7
43	3905	160	6

Date and time	Height in meters	Azimuth from; degrees	Velocity m/s
Jan. 11			
0:44P	4085	178	7
45	4265	186	7
46	4445	194	7
47	4625	199	8
48	4805	199	10
49	4985	200	9
50	5165	208	7
51	5345	213	8
52	5525	217	11
53	5705	221	11
54	5885	215	15
55	6065	215	19
56	6245	219	19
57	6425	219	20
58	6605	219	22
59	6785	220	22
1:00P	6965	221	23
01	7145	220	22
02	7325	220	23
12	(549)		
0:40P	395	60	4
41	611	110	4
42	809	147	3
43	1007	159	3
44	1196	171	3
45	1385	157	5
46	1565	151	9
47	1745	156	12
48	1925	163	14
49	2105	166	15
50	2285	168	16
51	2465	170	18
52	2645	166	20
53	2825	161	21
54	3005	160	21
55	3185	162	20
56	3365	164	20
57	3545	164	21
58	3725	163	25
59	3905	163	27
13	(550)		
0:25P	395	83	5
26	611	117	11
27	809	141	14
28	1007	150	16
29	1196	155	17

Date and time	Height in meters	Azimuth from; degrees	Velocity m/s
Jan. 13			
0:30P	1385	160	20
31	1565	159	21
32	1745	157	18
33	1925	156	16
34	2105	159	20
35	2285	158	21
36	2465	158	20
37	2645	154	18
14	(551)		
11:40A	395	221	5
41	611	212	6
42	809	180	5
43	1007	166	6
44	1196	170	7
45	1385	172	9
46	1565	179	10
47	1745	183	11
48	1925	183	15
49	2105	183	20
50	2285	183	23
51	2465	181	23
52	2645	179	23
15	(552)		
0:25P	395	75	4
26	611	118	6
27	809	153	12
28	1007	152	17
16	(553)		
0:12P	395	67	6
13	611	90	8
14	809	120	11
15	1007	129	14
16	1196	136	17
17	1385	139	21
18	1565	141	26
19	1745	145	30
20	1925	152	25
21	2105	165	21
22	2285	168	22
23	2465	166	29
24	2645	164	41
25	2825	165	43
26	3005	169	33
27	3185	170	28

TABLE II (1929)

Date and time	Height in meters	Azimuth from; degrees	Velocity m/s
Jan.			
17	(554)		
0:15P	395	81	5
16	611	94	15
17	809	109	13
18	1007	133	12
19	1196	145	10
20	1385	151	9
21	1565	154	8
22	1745	146	9
23	1925	131	9
24	2105	118	8
25	2285	113	6
26	2465	120	3
27	2645	127	3
28	2825	127	2
18	(555)		
12:00M	395	95	11
0:01P	611	117	25
02	809	135	21
03	1007	178	11
04	1196	206	9
05	1385	195	9
06	1565	183	10
07	1745	167	14
08	1925	162	19
09	2105	164	23
10	2285	165	22
11	2465	165	10
12	2645	185	2
19	(556)		
0:18P	395	83	5
19	611	109	5
20	809	150	3
21	1007	201	3
22	1196	199	6
23	1385	188	8
24	1565	188	9
25	1745	188	10
26	1925	188	10
27	2105	187	9
28	2285	177	9
29	2465	163	9
30	2645	161	10
31	2825	163	11
32	3005	171	11
33	3185	180	11

Date and time	Height in meters	Azimuth from; degrees	Velocity m/s
Jan.			
19			
0:34P	3365	183	12
35	3545	184	11
36	3725	188	10
37	3905	184	9
38	4085	163	9
39	4265	166	9
40	4445	183	12
41	4625	189	14
42	4805	191	16
20	(557)		
0:18P	395	64	4
19	611	72	3
20	809	81	1
21	1007	73	2
22	1196	74	3
23	1385	82	3
24	1565	88	4
25	1745	88	5
26	1925	93	4
27	2105	126	3
28	2285	149	3
29	2465	127	3
30	2645	110	4
31	2825	110	3
32	3005	124	3
33	3185	130	3
34	3365	116	3
35	3545	93	3
36	3725	101	3
37	3905	122	3
38	4085	129	4
39	4265	132	4
40	4445	137	4
41	4625	139	4
42	4805	135	4
21	(558)		
0:21P	395	118	7
22	611	136	8
23	809	148	10
24	1007	159	13
25	1196	160	13
26	1385	164	16
27	1565	173	21
28	1745	182	22
29	1925	181	23
30	2105	181	23
31	2285	181	22

Date and time	Height in meters	Azimuth from; degrees	Velocity m/s
Jan.			
22	(559)		
11:55A	395	45	2
56	611	121	3
57	809	136	9
58	1007	138	10
59	1196	140	11
12:00M	1385	146	13
0:01P	1565	150	16
02	1745	154	16
03	1925	155	15
04	2105	157	15
05	2285	163	16
06	2465	161	18
07	2645	158	19
08	2825	152	21
09	3005	142	23
10	3185	132	23
11	3365	128	20
12	3545	127	17
13	3725	128	17
23	(560)		
0:21P	395	110	16
22	611	116	19
23	809	124	25
24	1007	127	29
25	1196	138	15
26	1385	192	5
27	1565	160	7
28	1745	134	12
29	1925	119	13
30	2105	96	15
31	2285	87	16
32	4265	84	16
33	2645	85	14
34	2825	85	13
24	(561)		
0:06P	395	92	10
07	611	106	9
08	809	135	11
09	1007	139	13
10	1196	146	13
11	1385	147	17
12	1565	150	19
13	1745	159	33
14	1925	166	45
15	2105	173	31
16	2285	181	14

TABLE II (1929)

Date and time	Height in meters	Azimuth from; degrees	Velocity m/s
Jan. 24			
0:17P	2465	195	6
18	2645	168	9
19	2825	161	17
20	3005	160	26
21	3185	161	29
22	3365	161	30
25	(562)		
11:50A	395	96	5
51	611	125	5
52	809	147	6
53	1007	145	5
54	1196	130	3
55	1385	114	4
56	1565	97	3
57	1745	63	3
58	1925	60	4
59	2105	90	4
12:00M	2285	128	4
0:01P	2465	132	7
02	2645	129	9
03	2825	130	10
04	3005	130	9
05	3185	129	8
06	3365	129	7
07	3545	128	7
08	3725	132	6
09	3905	134	6
10	4085	143	7
11	4265	149	9
12	4445	154	9
13	4625	155	10
14	4805	155	11
15	4985	152	12
16	5165	144	14
17	5345	140	16
18	5525	139	17
19	5705	139	16
20	5885	136	13
21	6065	116	11
22	6245	92	12
23	6425	93	14
24	6605	94	15
25	6785	97	15
26	6965	102	15
27	7145	106	15

Date and time	Height in meters	Azimuth from; degrees	Velocity m/s
Jan. 26	(563)		
11:50A?	395	150	1
51	611	205	6
52	809	235	7
53	1007	255	9
54	1196	261	10
55	1385	265	9
56	1565	262	7
57	1745	249	8
58	1925	259	6
59	2105	286	5
12:00M	2285	297	6
0:01P	2465	289	7
02	2645	271	7
03	2825	257	5
04	3005	263	4
05	3185	274	4
06	3365	283	3
07	3545	294	4
08	3725	321	3
09	3905	337	4
10	4085	331	4
11	4265	331	5
12	4445	332	6
27	(564)		
0:33P	395	58	3
34	611	102	4
35	809	165	5
36	1007	188	5
37	1196	215	5
38	1385	222	7
39	1565	210	7
40	1745	192	7
41	1925	190	7
42	2105	205	6
43	2285	234	4
44	2465	248	4
45	2645	238	5
46	2825	235	7
47	3005	241	7
48	3185	251	7
49	3365	262	7
50	3545	267	6
51	3725	275	7
52	3905	294	8
53	4085	307	9
54	4265	316	10
55	4445	316	10

Date and time	Height in meters	Azimuth from; degrees	Velocity m/s
Jan. 27			
0:56P	4625	307	11
57	4805	303	12
58	4985	301	12
59	5165	299	12
1:00P	5345	294	13
01	5525	291	13
02	5705	294	12
03	5885	294	15
04	6065	292	15
05	6245	292	15
06	6425	289	16
07	6605	285	16
08	6785	282	15
09	6965	278	17
10	7145	279	16
11	7325	282	17
12	7505	282	17
13	7685	282	15
14	7865	282	13
15	8045	283	15
16	8225	284	16
17	8405	284	16
18	8585	282	17
19	8765	285	18
20	8945	290	17
21	9125	291	14
22	9305	291	13
23	9485	291	15
24	9665	291	15
25	9845	292	13
26	10025	293	14
27	10205	292	14
28	10385	292	12
29	10565	292	11
30	10745	294	12
31	10925	298	11
32	11105	298	11
33	11285	300	12
34	11465	300	11
35	11645	300	11
36	11825	300	11
28	(565)		
0:22P	395	48	5
23	611	70	8
24	809	79	6
25	1007	78	3
26	1196	78	4

TABLE II (1929)

Date and time	Height in meters	Azimuth from; degrees	Velocity m/s	Date and time	Height in meters	Azimuth from; degrees	Velocity m/s	Date and time	Height in meters	Azimuth from; degrees	Velocity m/s
Jan.				**Jan.**				**Jan.**			
28				**30**				**31**			
0:27P	1385	73	7	0:08P	3725	31	13	0:24P	6605	348	6
28	1565	70	11	09	3905	31	11	25	6785	348	6
29	1745	70	12	10	4085	31	12	26	6965	350	6
30	1925	75	11	11	4265	33	13	27	7145	350	6
31	2105	77	12	12	4445	33	15	28	7325	354	7
32	2285	80	13	13	4625	35	15	29	7505	352	6
33	2465	82	13	14	4805	36	18	30	7685	352	7
34	2645	78	15	15	4985	37	21	31	7865	352	7
35	2825	72	17	16	5165	38	21	32	8045	357	8
36	3005	70	17	17	5345	38	21	33	8225	358	9
37	3185	67	18	18	5525	36	23	34	8405	359	10
38	3365	65	18	19	5705	31	24	35	8585	5	12
39	3545	61	17					36	8765	2	13
40	3725	53	17	**31**	(568)			37	8945	4	15
41	3905	53	23	11:50A	395	186	1	38	9125	8	15
42	4085	53	23	51	611	253	2	39	9305	3	15
43	4265	53	23	52	809	267	3	40	9485	3	15
				53	1007	201	3	41	9665	358	15
29	(566)			54	1196	192	6	42	9845	356	17
0:15P	395	122	16	55	1385	195	5	43	10025	356	24
16	611	120	14	56	1565	195	4	44	10205	356	33
17	809	134	9	57	1745	195	4	45	10385	356	42
18	1007	158	8	58	1925	178	4	46	10565	357	46
19	1196	156	11	59	2105	182	4				
20	1385	146	20	12:00M	2285	182	5	**Feb.**			
21	1565	143	32	0:01P	2465	185	6	**2**	(569)		
22	1745	142	39	02	2645	179	5	11:50A	395	113	6
23	1925	142	40	03	2825	172	4	51	611	131	10
				04	3005	168	3	52	809	146	11
30	(567)			05	3185	129	1	53	1007	157	13
11:50A	395	1	1	06	3365	23	2	54	1196	170	14
51	611	41	2	07	3545	17	4	55	1385	183	15
52	809	40	2	08	3725	9	5	56	1565	186	19
53	1007	20	4	09	3905	5	7	57	1745	185	20
54	1196	12	8	10	4085	3	6	58	1925	186	19
55	1385	352	11	11	4265	3	7	59	2105	190	19
56	1565	335	11	12	4445	0	9	12:00M	2285	190	19
57	1745	332	10	13	4625	357	10				
58	1925	336	10	14	4805	0	9	**3**	(570)		
59	2105	339	10	15	4985	0	7	11:52A	395	77	4
12:00M	2285	349	9	16	5165	0	6	53	611	97	7
0:01P	2465	4	9	17	5345	7	5	54	809	128	7
02	2645	16	10	18	5525	4	6	55	1007	159	7
03	2825	21	11	19	5705	0	6	56	1196	172	10
04	3005	21	12	20	5885	348	6	57	1385	164	13
05	3185	21	12	21	6065	348	7	58	1565	164	13
06	3365	28	12	22	6245	348	6	59	1745	165	13
07	3545	31	15	23	6425	348	5	12:00M	1925	167	12

TABLE II (1929)

Date and time	Height in meters	Azimuth from; degrees	Velocity m/s	Date and time	Height in meters	Azimuth from; degrees	Velocity m/s	Date and time	Height in meters	Azimuth from; degrees	Velocity m/s
Feb.				**Feb.**				**Feb.**			
3				**5**				**9**			
0:01P	2105	167	11	11:46A	1196	152	12	11:43A	611	52	5
02	2285	167	10	47	1385	155	14	44	809	51	4
03	2465	167	9	48	1565	158	17	45	1007	51	3
04	2645	167	11	49	1745	160	19	46	1196	62	3
05	2825	171	11	50	1925	158	23	47	1385	47	4
06	3005	181	11	51	2105	155	24	48	1565	29	5
07	3185	190	12	52	2285	156	23	49	1745	30	6
08	3365	191	13	53	2465	156	19	50	1925	30	7
09	3545	191	15	54	2645	144	14	51	2105	26	8
10	3725	191	15	55	2825	125	11	52	2285	22	8
11	3905	187	17	56	3005	119	13	53	2465	22	8
12	4085	187	15	57	3185	125	15	54	2645	23	7
13	4265	187	13	58	3365	129	16	55	2825	25	7
								56	3005	28	6
4	(571)			**6**	(573)			57	3185	42	6
0:02P	395	4	4	11:43A	395	208	1	58	3365	51	5
03	611	68	6	44	611	183	2	59	3545	60	5
04	809	89	6	45	809	248	1	12:00M	3725	90	2
05	1007	105	6	46	1007	312	2	0:01P	3905	146	2
06	1196	129	6	47	1196	312	2	02	4085	178	2
07	1385	152	7	48	1385	339	1	03	4265	206	4
08	1565	157	9	49	1565	21	2	04	4445	212	5
09	1745	151	10	50	1745	22	3	05	4625	202	6
10	1925	145	10	51	1925	22	3	06	4805	199	7
11	2105	140	10	52	2105	36	2	07	4985	196	6
12	2285	137	12	53	2285	45	3	08	5165	183	5
13	2465	132	15	54	2465	34	3	09	5345	180	6
14	2645	132	13	55	2645	11	3	10	5525	183	5
15	2825	137	9	56	2825	335	2	11	5705	181	6
16	3005	135	7	57	3005	254	2	12	5885	178	7
17	3185	119	6	58	3185	216	3	13	6065	186	6
18	3365	102	6	59	3365	176	3	14	6245	195	6
19	3545	103	6	12:00M	3545	135	3	15	6425	214	5
20	3725	120	6	0:01P	3725	130	3	16	6605	230	4
21	3905	143	5	02	3905	136	3	17	6785	241	3
22	4085	148	5	03	4085	141	3	18	6965	245	5
23	4265	142	7	04	4265	126	3	19	7145	259	6
24	4445	146	7	05	4445	99	4	20	7325	268	6
25	4625	145	9	06	4625	85	5	21	7505	268	7
26	4805	138	11	07	4805	80	6	22	7685	261	8
27	4985	139	11	08	4985	83	7	23	7865	256	9
28	5165	145	11	09	5165	96	8	24	8945	258	10
				10	5345	108	9	25	9125	263	11
5	(572)			11	5525	108	10	26	9305	261	13
11:42A	395	132	11	12	5705	109	10	27	9485	261	14
43	611	142	15					28	9665	255	14
44	809	145	17	**9**	(574)			29	9845	257	14
45	1007	149	13	11:42A	395	45	2	30	10025	251	15

TABLE II (1929)

Date and time	Height in meters	Azimuth from; degrees	Velocity m/s
Feb. 9			
0:31P	10205	251	17
32	10385	251	17
33	10565	250	17
34	10745	246	17
35	10925	246	18
36	11105	248	18
37	11285	248	18
38	11465	250	18
39	11645	247	18
40	11825	243	19
41	12005	239	22
42	12185	241	21
43	12365	244	17
11	(575)		
11:46A	395	71	2
47	611	56	4
48	809	31	5
49	1007	29	4
50	1196	28	3
51	1385	28	1
52	1565	11	1
53	1745	164	1
54	1925	182	2
55	2105	193	2
56	2285	207	2
57	2465	207	2
58	2645	197	3
59	2825	195	3
12:00M	3005	194	3
0:01P	3185	200	4
02	3365	196	5
03	3545	192	5
04	3725	185	5
05	3905	185	7
06	4085	191	8
07	4265	191	8
08	4445	191	9
12	(576)		
11:50A	395	65	2
51	611	126	4
52	809	170	5
53	1007	196	6
54	1196	219	7
55	1385	233	7
56	1565	236	8
57	1745	228	7

Date and time	Height in meters	Azimuth from; degrees	Velocity m/s
Feb. 12			
11:58A	1925	211	7
59	2105	204	7
12:00M	2285	204	7
0:01P	2465	203	6
02	2645	215	6
03	2825	232	7
04	3005	236	8
05	3185	230	8
06	3365	223	9
07	3545	219	11
08	3725	216	11
09	3905	216	10
10	4085	216	9
11	4265	215	9
12	4445	214	9
13	4625	209	10
14	4805	207	11
15	4985	206	12
16	5165	204	12
17	5345	204	13
18	5525	204	15
19	5705	207	13
20	5885	207	15
21	6065	207	15
22	6245	208	14
23	6425	208	15
24	6605	208	16
25	6785	208	14
26	6965	208	14
13	(577)		
0:40P	395	62	1
41	611	40	3
42	809	27	6
43	1007	26	10
44	1196	21	11
45	1385	21	9
46	1565	18	8
47	1745	8	8
48	1925	8	10
49	2105	8	13
50	2285	4	15
51	2465	4	15
52	2645	4	14
53	2825	4	13
54	3005	0	11
55	3185	357	8
56	3365	8	4

Date and time	Height in meters	Azimuth from; degrees	Velocity m/s
Feb. 13			
0:57P	3545	85	4
58	3725	101	6
59	3905	125	7
1:00P	4085	146	9
01	4265	146	11
02	4445	139	11
03	4625	139	13
04	4805	149	13
05	4985	150	15
06	5165	148	17
07	5345	148	18
14	(578)		
3:00P	395	306	1
01	611	336	2
02	809	349	6
03	1007	344	11
04	1196	345	12
05	1385	350	11
15	(579)		
11:50A	395	216	3
51	611	205	1
52	809	45	1
53	1007	19	4
54	1196	8	8
55	1385	5	9
56	1565	19	6
57	1745	82	3
58	1925	103	5
59	2105	87	6
12:00M	2285	75	6
16	(580)		
11:55A	395	1	1
56	611	32	4
57	809	46	4
58	1007	68	4
59	1196	103	4
12:00M	1385	131	4
0:01P	1565	149	5
02	1745	162	7
03	1925	169	8
04	2105	169	8
05	2285	147	9
06	2465	130	9
07	2645	128	7
08	2825	125	6
09	3005	124	6

TABLE II (1929)

Date and time	Height in meters	Azimuth from; degrees	Velocity m/s	Date and time	Height in meters	Azimuth from; degrees	Velocity m/s	Date and time	Height in meters	Azimuth from; degrees	Velocity m/s
Feb.				**Feb.**				**Feb.**			
17	(581)			**19**				**21**			
0:05P	395	272	3	1:15P	2465	132	9	11:22A	1385	68	5
06	611	161	3	16	2645	130	9	23	1565	79	4
07	809	164	7	17	2825	133	10	24	1745	107	6
08	1007	178	9	18	3005	140	8	25	1925	140	11
09	1196	183	11	19	3185	150	7	26	2105	155	17
10	1385	175	13	20	3365	152	8	27	2285	157	21
11	1565	166	14	21	3545	158	8				
12	1745	164	16	22	3725	157	8	**22**	(586)		
13	1925	165	19	23	3905	150	10	0:17P	395	126	11
14	2105	165	20	24	4085	151	11	18	611	145	19
15	2285	162	23	25	4265	155	13	19	809	155	21
16	2465	161	24	26	4445	154	14	20	1007	162	19
				27	4625	155	16	21	1196	169	13
18	(582)			28	4805	162	16	22	1385	192	9
0:12P	395	228	2	29	4985	163	17	23	1565	202	10
13	611	256	3	30	5165	160	21	24	1745	196	14
14	809	224	2	31	5345	161	23	25	1925	195	18
15	1007	164	5	32	5525	167	27	26	2105	193	21
16	1196	168	9	33	5705	171	27	27	2285	197	17
17	1385	174	9	34	5885	162	27	28	2465	205	13
18	1565	179	9	35	6065	161	28				
19	1745	181	11					**23**	(587)		
20	1925	180	14	**20**	(584)			0:25P	395	62	6
21	2105	173	17	11:45A	395	51	1	26	611	90	7
22	2285	165	20	46	611	177	3	27	809	125	5
23	2465	162	19	47	809	174	5	28	1007	157	7
24	2645	169	18	48	1007	172	8	29	1196	170	9
25	2825	178	18	49	1196	162	8	30	1385	168	8
26	3005	186	21	50	1385	156	9	31	1565	150	7
27	3185	189	24	51	1565	156	11	32	1745	151	6
28	3365	189	27	52	1745	154	13	33	1925	160	7
29	3545	188	28	53	1925	152	16	34	2105	167	7
30	3725	185	29	54	2105	149	18	35	2285	171	7
31	3905	183	31	55	2285	143	19	36	2465	183	5
32	4085	182	33	56	2465	141	21	37	2645	212	4
				57	2645	138	23	38	2825	220	4
19	(583)			58	2825	135	25	39	3005	156	6
1:04P	395	54	5	59	3005	135	28	40	3185	164	10
05	611	76	8	12:00M	3185	136	31	41	3365	172	12
06	809	73	8	0:01P	3365	133	35	42	3545	177	14
07	1007	70	5	02	3545	131	37	43	3725	179	17
08	1196	97	1					44	3905	181	18
09	1385	111	2	**21**	(585)			45	4085	183	17
10	1565	129	4	11:17A	395	311	3	46	4265	186	17
11	1745	158	6	18	611	5	2	47	4445	186	17
12	1925	162	9	19	809	42	3	48	4625	181	19
13	2105	160	10	20	1007	54	4	49	4805	177	20
14	2285	148	9	21	1196	62	5	50	4985	177	20

TABLE II (1929)

Date and time	Height in meters	Azimuth from; degrees	Velocity m/s
Feb. 24	(588)		
1:00P	395	85	5
01	611	138	5
02	809	145	9
03	1007	145	12
04	1196	148	14
05	1385	153	16
06	1565	156	19
07	1745	159	26
08	1925	161	35
09	2105	161	41
10	2285	161	42
11	2465	162	35
12	2645	164	35
13	2825	166	38
25	(589)		
0:45P	395	148	6
46	611	146	16
47	809	150	11
48	1007	163	11
49	1196	166	19
50	1385	173	21
51	1565	177	23
52	1745	177	21
53	1925	178	20
54	2105	180	19
55	2285	180	19
56	2465	180	19
57	2645	174	19
58	2825	174	23
59	3005	174	27
26	(590)		
0:31P	395	330	3
32	611	86	7
33	809	167	3
34	1007	160	5
35	1196	162	7
36	1385	170	8
37	1565	173	9
38	1745	173	8
39	1925	173	8
40	2105	173	8
41	2285	173	8
27	(591)		
4:20P	395	339	6
21	611	27	5

Date and time	Height in meters	Azimuth from; degrees	Velocity m/s
Feb. 27			
4:22P	809	57	3
23	1007	129	3
24	1196	147	5
25	1385	159	6
26	1565	178	7
27	1745	196	7
28	1925	201	10
29	2105	198	13
30	2285	190	14
31	2465	188	16
32	2645	190	18
33	2825	190	17
34	3005	190	16
28	(592)		
0:20P	395	14	3
21	611	105	4
22	809	129	5
23	1007	136	5
24	1196	141	5
25	1385	164	6
26	1565	174	6
27	1745	159	5
28	1925	145	6
29	2105	156	8
30	2285	162	10
31	2465	160	11
32	2645	153	13
33	2825	151	14
34	3005	152	14
35	3185	155	14
36	3365	159	14
37	3545	160	16
38	3725	165	15
39	3905	168	15
40	4085	171	15
41	4265	172	17
42	4445	174	17
43	4625	175	17
44	4805	179	17
45	4985	180	16
46	5165	176	18
47	5345	173	17
48	5525	173	17
49	5705	167	15
50	5885	161	13
51	6065	161	13

Date and time	Height in meters	Azimuth from; degrees	Velocity m/s
March 1	(593)		
0:10P	395	10	1
11	611	338	9
12	809	216	3
13	1007	199	5
14	1196	202	4
15	1385	209	3
16	1565	227	3
17	1745	220	4
18	1925	210	4
19	2105	210	5
20	2285	206	7
21	2465	201	8
22	2645	204	11
23	2825	204	11
24	3005	195	11
25	3185	196	11
26	3365	182	9
27	3545	168	9
28	3725	178	9
29	3905	174	9
30	4085	169	9
31	4265	166	10
32	4445	165	10
33	4625	160	11
34	4805	159	11
35	4985	159	11
36	5165	161	12
37	5345	161	13
38	5525	161	12
39	5705	161	11
40	5885	165	12
41	6065	173	14
42	6245	173	13
43	6425	173	11
44	6605	178	10
45	6785	185	10
46	6965	192	10
47	7145	199	8
48	7325	216	7
49	7505	233	11
50	7685	240	12
2	(594)		
11:24A	395	70	5
25	611	81	9
26	809	99	7
27	1007	108	6
28	1196	105	6

TABLE II (1929)

Date and time	Height in meters	Azimuth from; degrees	Velocity m/s
March 2			
11:29A	1385	97	8
30	1565	103	8
31	1745	118	8
32	1925	126	8
33	2105	140	8
34	2285	152	8
35	2465	152	8
36	2645	161	8
37	2825	162	7
38	3005	160	6
39	3185	166	7
40	3365	175	7
41	3545	180	8
42	3725	180	9
43	3905	180	10
44	4085	181	11
45	4265	183	13
46	4445	181	15
47	4625	181	15
3	(595)		
0:20P	395	358	2
21	611	273	2
22	809	264	3
23	1007	239	4
24	1196	218	7
25	1385	214	7
26	1565	213	7
27	1745	211	8
28	1925	213	8
29	2105	217	10
30	2285	217	9
31	2465	217	9
32	2645	215	9
33	2825	210	10
34	3005	203	9
35	3185	202	10
36	3365	198	11
37	3545	195	11
38	3725	188	12
39	3905	188	13
40	4085	193	13
41	4265	196	14
42	4445	199	15
43	4625	196	14
44	4805	194	16
45	4985	192	17
46	5165	192	19

Date and time	Height in meters	Azimuth from; degrees	Velocity m/s
March 3			
0:47P	5345	200	23
48	5525	202	24
49	5705	203	25
50	5885	200	21
51	6065	196	19
4	(596)		
1:35P	395	98	4
36	611	145	10
37	809	161	12
38	1007	176	13
39	1196	183	13
40	1385	181	15
41	1565	181	19
42	1745	184	20
43	1925	182	23
44	2105	179	22
45	2285	177	19
46	2465	179	18
47	2645	186	15
48	2825	181	11
49	3005	157	9
50	3185	151	12
51	3365	163	15
52	3545	175	19
53	3725	176	25
54	3905	175	26
5	(597)		
1:10P	395	110	5
11	611	110	9
12	809	121	9
13	1007	146	8
14	1196	164	9
15	1385	169	9
16	1565	176	9
17	1745	187	10
18	1925	197	11
19	2105	184	14
20	2285	190	13
21	2465	186	11
22	2645	182	10
6			
0:15P	395	1	4
16	611	31	7
17	809	45	8
18	1007	58	8

Date and time	Height in meters	Azimuth from; degrees	Velocity m/s
March 6			
0:19P	1196	85	7
20	1385	111	7
21	1565	125	6
22	1745	147	6
23	1925	157	8
24	2105	164	10
25	2285	166	11
26	2465	162	11
27	2645	162	11
28	2825	166	13
29	3005	171	13
30	3185	173	12
31	3365	173	13
32	3545	173	14
33	3725	175	14
34	3905	178	14
35	4085	178	15
36	4265	179	15
37	4445	183	15
38	4625	187	16
39	4805	190	16
40	4985	193	17
41	5165	193	18
42	5345	195	20
43	5525	195	21
44	5705	197	21
7	(599)		
0:15P	395	71	5
16	611	113	9
17	809	129	11
18	1007	140	12
19	1196	148	12
20	1385	148	11
21	1565	134	10
22	1745	129	11
23	1925	129	13
24	2105	134	15
25	2285	139	16
26	2465	144	15
27	2645	152	14
28	2825	156	14
29	3005	158	15
30	3185	158	16
31	3365	159	17
32	3545	164	18
33	3725	173	17
34	3905	184	15

TABLE II (1929)

Date and time	Height in meters	Azimuth from; degrees	Velocity m/s
March 7			
0:35P	4085	183	18
36	4265	183	21
37	4445	181	22
8	(600)		
0:10P	395	103	14
11	611	132	16
12	809	143	15
13	1007	170	13
14	1196	176	16
15	1385	167	17
16	1565	165	16
17	1745	163	17
18	1925	163	18
19	2105	166	15
20	2285	166	15
21	2465	167	14
22	2645	167	13
23	2825	168	14
24	3005	166	16
25	3185	162	17
26	3365	162	18
27	3545	162	19
9	(601)		
0:15P	395	150	16
16	611	160	25
17	809	164	26
18	1007	171	21
19	1196	179	20
20	1385	185	21
21	1565	190	23
22	1745	192	24
23	1925	202	20
24	2105	217	17
9	(602)		
5:15P	395	169	6
16	611	165	10
17	809	153	10
18	1007	165	8
19	1196	185	11
20	1385	191	13
21	1565	191	17
22	1745	190	21
23	1925	186	19
24	2105	181	18
25	2285	175	17
26	2465	171	13
27	2645	163	13

Date and time	Height in meters	Azimuth from; degrees	Velocity m/s
March 9			
5:28P	2825	162	13
29	3005	171	14
30	3185	178	15
31	3365	179	18
32	3545	181	23
33	3725	181	25
10	(603)		
2:20P	395	348	3
21	611	14	5
22	809	118	1
23	1007	88	3
24	1196	94	3
25	1385	153	3
26	1565	157	5
27	1745	148	7
28	1925	147	8
29	2105	155	7
30	2285	181	8
31	2465	185	10
32	2645	175	13
33	2825	168	15
34	3005	161	17
35	3185	158	17
36	3365	162	17
37	3545	168	18
38	3725	167	19
39	3905	165	22
40	4085	165	25
41	4265	163	28
11	(604)		
3:15P	395	290	3
16	611	310	5
17	809	337	4
18	1007	272	1
19	1196	221	2
20	1385	39	2
21	1565	8	3
22	1745	348	5
23	1925	326	3
24	2105	246	3
25	2285	233	6
26	2465	222	7
27	2645	209	8
28	2825	207	10
29	3005	207	13
30	3185	202	15

Date and time	Height in meters	Azimuth from; degrees	Velocity m/s
March 11			
3:31P	3365	197	15
32	3545	192	16
33	3725	185	16
34	3905	181	17
35	4085	183	17
36	4265	186	19
37	4445	186	21
38	4625	186	21
39	4805	186	21
40	4985	188	21
41	5165	188	20
12	(605)		
0:30P	395	84	5
31	611	104	7
32	809	138	6
33	1007	158	6
34	1196	167	6
35	1385	175	9
36	1565	183	11
37	1745	185	13
38	1925	183	14
39	2105	181	16
40	2285	182	17
41	2465	183	19
42	2645	183	19
43	2825	186	19
44	3005	189	21
45	3185	194	23
46	3365	199	24
47	3545	202	25
13	(606)		
1:30P	395	150	4
31	611	172	9
32	809	178	11
33	1007	182	11
34	1196	191	17
35	1385	193	25
36	1565	195	25
37	1745	195	18
38	1925	195	11
39	2105	192	6
40	2285	175	5
41	2465	166	5
42	2645	172	5
43	2825	184	8
44	3005	184	9

TABLE II (1929)

Date and time	Height in meters	Azimuth from; degrees	Velocity m/s
March 13			
1:45P	3185	178	7
46	3365	170	7
14	(607)		
0:12P	395	283	4
13	611	277	5
14	809	280	3
15	1007	295	1
16	1196	298	1
17	1385	296	1
18	1565	250	1
19	1745	191	3
20	1925	176	4
21	2105	177	4
22	2285	174	6
23	2465	175	9
24	2645	183	12
25	2825	188	16
26	3005	192	20
27	3185	191	25
28	3365	187	27
29	3545	185	28
30	3725	185	29
31	3905	185	29
32	4085	185	29
33	4265	188	30
34	4445	188	30
16	(608)		
1:25P	395	18	5
26	611	27	9
27	809	40	8
28	1007	52	7
29	1196	54	5
30	1385	61	2
31	1565	110	1
32	1745	160	2
33	1925	213	5
34	2105	210	7
35	2285	211	7
36	2465	221	7
37	2645	220	9
38	2825	216	10
39	3005	216	11
40	3185	213	14
41	3365	210	18
42	3545	207	20
43	3725	205	21
44	3905	205	23

Date and time	Height in meters	Azimuth from; degrees	Velocity m/s
March 17	(609)		
1:25P	395	167	3
26	611	145	4
27	809	164	3
28	1007	20	8
29	1196	17	12
30	1385	17	17
31	1565	26	18
32	1745	46	13
33	1925	62	10
18	(610)		
0:22P	395	55	2
23	611	53	6
24	809	36	5
25	1007	21	3
26	1196	75	1
27	1385	143	3
28	1565	170	5
29	1745	185	8
30	1925	189	10
31	2105	196	10
32	2285	201	12
33	2465	197	12
34	2645	192	15
35	2825	192	18
36	3005	196	15
37	3185	196	19
38	3365	196	19
39	3545	196	20
40	3725	196	21
41	3905	197	21
42	4085	196	23
43	4265	192	25
44	4445	190	24
45	4625	192	23
46	4805	193	23
19	(611)		
0:30P	395	69	2
31	611	124	5
32	809	158	9
33	1007	170	11
34	1196	186	11
35	1385	194	9
36	1565	188	7
37	1745	188	6
38	1925	188	7
39	2105	188	8

Date and time	Height in meters	Azimuth from; degrees	Velocity m/s
March 19			
0:40P	2285	188	9
41	2465	182	10
42	2645	179	10
43	2825	173	9
44	3005	178	8
45	3185	168	9
46	3365	168	9
47	3545	168	7
48	3725	171	5
49	3905	184	4
50	4085	181	7
51	4265	170	11
52	4445	164	13
53	4625	161	16
54	4805	160	17
55	4985	154	15
56	5165	148	14
57	5345	148	12
58	5525	152	9
59	5705	155	7
1:00P	5885	153	6
01	6065	150	5
02	6245	140	5
03	6425	139	5
04	6605	153	7
05	6785	153	8
06	6965	153	9
07	7145	159	8
08	7325	155	7
09	7505	155	8
10	7685	151	8
11	7865	149	9
12	8045	149	11
13	8225	149	10
14	8405	148	11
15	8585	148	13
16	8765	148	14
17	8945	148	13
18	9125	148	12
19	9305	148	11
20	(612)		
0:45P	395	50	3
46	611	110	8
47	809	134	11
48	1007	142	11
49	1196	147	12
50	1385	145	14

TABLE II (1929)

Date and time	Height in meters	Azimuth from; degrees	Velocity m/s
March 20			
0:51P	1565	143	16
52	1745	145	18
53	1925	148	17
54	2105	147	17
55	2285	142	18
56	2465	140	17
57	2645	143	17
58	2825	147	18
59	3005	154	16
1:00P	3185	158	13
01	3365	158	11
02	3545	158	10
21	(613)		
0:50P	395	252	1
51	611	290	1
52	809	335	1
53	1007	32	1
54	1196	45	1
22	(614)		
6:12P	395	74	4
13	611	63	5
14	809	30	7
15	1007	15	7
16	1196	22	7
17	1385	95	4
18	1565	156	9
19	1745	166	11
20	1925	166	11
21	2105	163	11
22	2285	159	11
23	2465	151	11
24	2645	149	11
25	2825	145	11
26	3005	137	12
27	3185	133	11
28	3365	138	10
29	3545	145	11
30	3725	146	11
31	3905	146	11
23	(615)		
0:25P	395	75	1
26	611	16	4
27	809	29	4
28	1007	83	4
29	1196	124	3

Date and time	Height in meters	Azimuth from; degrees	Velocity m/s
March 23			
2:30P	1385	144	5
31	1565	153	6
32	1745	152	7
33	1925	137	6
34	2105	133	6
35	2285	143	7
36	2465	150	10
37	2645	145	13
38	2825	137	14
39	3005	134	13
40	3185	134	11
41	3365	124	7
42	3545	94	7
43	3725	88	8
44	3905	94	7
45	4085	111	5
46	4265	131	5
47	4445	141	8
48	4625	141	13
49	4805	137	17
50	4985	137	19
51	5165	137	18
52	5345	137	19
53	5525	137	20
24	(616)		
2:10P	395	49	1
11	611	90	7
12	809	101	7
13	1007	121	6
14	1196	149	10
15	1385	159	13
16	1565	161	14
17	1745	160	14
18	1925	163	13
19	2105	164	12
20	2285	161	13
21	2465	159	13
22	2645	156	13
23	2825	152	11
24	3005	144	10
25	3185	137	11
26	3365	127	11
27	3545	119	11
28	3725	114	12
29	3905	110	13
30	4085	105	13
31	4265	98	13

Date and time	Height in meters	Azimuth from; degrees	Velocity m/s
March 27	(617)		
0:55P	395	48	1
56	611	155	1
57	809	193	3
58	1007	215	3
59	1196	236	3
1:00P	1385	233	3
01	1565	251	6
02	1745	263	9
03	1925	260	7
04	2105	251	5
05	2285	226	4
06	2465	199	4
07	2645	199	5
08	2825	207	7
09	3005	202	7
10	3185	193	6
11	3365	193	6
12	3545	193	7
13	3725	190	7
14	3905	183	6
15	4085	183	6
16	4265	189	6
17	4445	193	6
18	4625	193	6
19	4805	193	5
20	4985	193	5
21	5165	193	6
22	5345	196	6
23	5525	196	5
24	5705	196	5
25	5885	203	5
26	6065	207	5
27	6245	219	4
28	6425	223	5
29	6605	231	5
30	6785	244	6
31	6965	256	6
32	7145	256	6
33	7325	259	6
34	7505	262	6
35	7685	262	7
36	7865	262	7
37	8045	262	7
38	8225	262	7
39	8405	262	7
40	8585	251	6
41	8765	245	6
42	8945	244	7

TABLE II (1929)

Date and time	Height in meters	Azimuth from; degrees	Velocity m/s
March 27			
1:43P	9125	244	8
44	9305	244	8
45	9485	236	7
46	9665	228	8
47	9845	220	10
48	10025	225	11
49	10205	225	12
50	10385	225	13
51	10565	235	10
52	10745	228	7
53	10925	220	11
54	11105	220	14
55	11285	225	13
56	11465	225	12
57	11645	225	11
58	11825	221	11
59	12005	212	10
2:00P	12185	207	10
01	12365	209	11
02	12545	212	11
03	12725	212	10
04	12905	205	9
05	13085	205	12
06	13265	205	14
07	13445	205	14
08	13625	214	15
09	13805	214	15
10	13985	215	15
11	14165	215	17
12	14345	215	19
13	14525	213	17
14	14705	213	17
15	14885	213	18
16	15065	213	19
17	15245	213	22
28	(618)		
0:25P	395	34	1
26	611	67	1
27	809	170	1
28	1007	178	2
29	1196	146	2
30	1385	90	1
31	1565	55	1
32	1745	18	1
33	1925	346	1
34	2105	201	1
35	2285	196	2

Date and time	Height in meters	Azimuth from; degrees	Velocity m/s
March 28			
0:36P	2465	202	3
37	2645	205	5
38	2825	193	6
39	3005	186	8
40	3185	188	10
41	3365	183	11
42	3545	177	11
43	3725	175	11
44	3905	175	13
45	4085	175	15
46	4265	175	16
47	4445	175	17
48	4625	175	17
49	4805	175	17
50	4985	175	17
51	5165	175	17
52	5345	175	17
53	5525	172	18
54	5705	172	19
55	5885	172	21
56	6065	172	21
57	6245	172	21
58	6425	172	22
59	6605	172	22
1:00P	6785	172	23
01	6965	172	24
02	7145	172	21
03	7325	173	22
04	7505	173	23
05	7685	172	21
06	7865	172	20
29	(619)		
3:20P	395	65	3
21	611	120	4
22	809	163	4
23	1007	163	5
24	1196	163	5
25	1385	163	6
26	1565	163	6
27	1745	163	6
28	1925	178	6
29	2105	198	5
30	2285	218	5
31	2465	223	5
32	2645	223	6
33	2825	223	7
34	3005	223	6

Date and time	Height in meters	Azimuth from; degrees	Velocity m/s
March 29			
3:35P	3185	223	7
36	3365	223	8
37	3545	223	8
38	3725	216	8
39	3905	203	9
40	4085	203	11
41	4265	198	10
42	4445	188	9
43	4625	188	8
44	4805	188	8
45	4985	191	9
46	5165	191	10
47	5345	191	10
48	5525	191	10
49	5705	191	9
50	5885	188	11
51	6065	188	13
52	6245	188	13
53	6425	188	14
54	6605	185	15
55	6785	180	15
56	6965	179	15
57	7145	178	17
58	7325	184	17
59	7505	184	17
4:00P	7685	184	18
01	7865	184	16
02	8045	182	17
03	8225	182	20
04	8405	184	19
05	8585	184	21
06	8765	184	19
07	8945	184	16
30	(620)		
9:55A	395	60	4
56	611	84	7
57	809	105	6
58	1007	130	6
59	1196	130	7
10:00A	1385	142	9
01	1565	152	9
02	1745	152	9
03	1925	147	9
04	2105	141	9
05	2285	141	7
06	2465	151	7
07	2645	158	7

TABLE II (1929)

Date and time	Height in meters	Azimuth from; degrees	Velocity m/s	Date and time	Height in meters	Azimuth from; degrees	Velocity m/s	Date and time	Height in meters	Azimuth from; degrees	Velocity m/s
March				**March**				**April**			
30				**30**				**1**			
10:08A	2825	158	7	3:06P	1565	181	8	0:38P	2825	148	14
09	3005	158	7	07	1745	183	8	39	3005	153	13
10	3185	158	7	08	1925	187	8	40	3185	156	14
11	3365	158	5	09	2105	191	8	41	3365	153	16
12	3545	168	5	10	2285	191	8	42	3545	150	17
13	3725	179	6	11	2465	188	9				
14	3905	186	7	12	2645	191	9	**1**	(624)		
15	4085	186	8	13	2825	191	9	5:15P	395	324	1
16	4265	186	9	14	3005	191	8	16	611	84	3
17	4445	196	9	15	3185	191	8	17	809	123	4
18	4625	200	10	16	3365	191	9	18	1007	129	5
19	4805	200	10	17	3545	191	11	19	1196	138	5
20	4985	200	10	18	3725	191	12	20	1385	148	6
21	5165	200	9	19	3905	196	12	21	1565	148	7
22	5345	200	9					22	1745	141	7
23	5525	198	9	**31**	(622)			23	1925	141	8
24	5705	198	10	1:00P	395	113	5	24	2105	154	8
25	5885	190	10	01	611	123	9	25	2285	167	11
26	6065	192	11	02	809	136	7	26	2465	177	13
27	6245	192	12	03	1007	159	5	27	2645	183	17
28	6425	192	11	04	1196	163	5	28	2825	179	19
29	6605	192	8	05	1385	176	5	29	3005	172	17
30	6785	192	8	06	1565	199	9	30	3185	173	17
31	6965	187	9	07	1745	209	11	31	3365	183	17
32	7145	187	10	08	1925	213	12	32	3545	183	17
33	7325	191	10	09	2105	217	11	33	3725	177	18
34	7505	193	10	10	2285	215	10	34	3905	175	19
35	7685	196	10	11	2465	203	8				
36	7865	192	13	12	2645	191	7	**3**	(625)		
37	8045	180	15	13	2825	184	9	10:10A	395	331	2
38	8225	180	13	14	3005	184	9	11	611	13	3
39	8405	178	12	15	3185	184	9	12	809	25	2
40	8585	177	14					13	1007	51	1
41	8765	177	14	**April**				14	1196	86	1
42	8945	181	13	**1**	(623)			15	1385	93	3
43	9125	181	14	0:25P	395	73	2	16	1565	110	4
44	9305	181	15	26	611	101	6	17	1745	128	5
45	9485	181	15	27	809	100	7	18	1925	160	5
46	9665	184	14	28	1007	103	6	19	2105	183	7
47	9845	184	14	29	1196	116	4	20	2285	181	8
				30	1385	155	3	21	2465	170	9
30	(621)			31	1565	173	5	22	2645	167	11
3:00P	395	70	4	32	1745	166	7	23	2825	172	12
01	611	124	7	33	1925	160	9	24	3005	173	13
02	809	156	8	34	2105	160	10	25	3185	173	15
03	1007	169	7	35	2285	159	11	26	3365	178	17
04	1196	175	7	36	2465	153	12	27	3545	183	18
05	1385	177	8	37	2645	148	13	28	3725	180	19

TABLE II (1929)

Date and time	Height in meters	Azimuth from; degrees	Velocity m/s
April 3			
10:29A	3905	177	19
30	4085	177	19
31	4265	177	20
3	(626)		
2:55P	395	357	2
56	611	351	2
57	809	353	3
58	1007	349	2
59	1196	16	2
3:00	1385	100	3
01	1565	145	4
02	1745	171	5
03	1925	174	7
04	2105	170	9
05	2285	170	10
06	2465	175	10
07	2645	175	11
08	2825	175	15
09	3005	179	19
10	3185	179	20
11	3365	176	22
12	3545	176	25
13	3725	176	27
14	3905	174	30
5	(627)		
9:50A	395	123	4
51	611	152	4
52	809	191	3
53	1007	213	2
54	1196	193	2
55	1385	193	3
56	1565	201	4
57	1745	193	5
58	1925	181	6
59	2105	181	7
10:00A	2285	194	7
01	2465	196	7
02	2645	196	7
03	2825	202	9
04	3005	202	9
05	3185	193	10
06	3365	192	11
07	3545	203	11
08	3725	209	13
09	3905	207	15
10	4085	202	15
April 5			
10:11A	4265	202	15
12	4445	202	15
13	4625	202	15
14	4805	200	15
15	4985	200	15
16	5165	202	16
17	5345	202	16
18	5525	202	16
19	5705	196	15
20	5885	196	17
21	6065	196	17
22	6245	196	16
23	6425	198	16
24	6605	199	17
25	6785	200	19
26	6965	200	17
27	7145	200	17
28	7325	200	19
29	7505	199	21
30	7685	199	21
31	7865	199	22
5	(628)		
2:12P	395	117	1
13	611	216	1
14	809	235	2
15	1007	212	3
16	1196	207	4
17	1385	207	5
18	1565	220	5
19	1745	237	5
20	1925	237	6
21	2105	231	7
22	2285	231	7
23	2465	225	8
24	2645	213	8
25	2825	210	9
26	3005	207	11
27	3185	197	13
28	3365	193	14
29	3545	197	12
30	3725	195	10
31	3905	200	11
32	4085	205	11
33	4265	204	11
6	(629)		
9:15A	395	85	4
April 6			
9:16A	611	114	4
17	809	170	4
18	1007	179	5
19	1196	179	5
20	1385	195	4
21	1565	221	4
22	1745	237	5
23	1925	245	7
24	2105	247	9
25	2285	247	9
26	2465	247	7
27	2645	231	6
28	2825	217	7
29	3005	213	8
30	3185	213	9
31	3365	210	10
32	3545	209	10
33	3725	207	10
34	3905	206	11
35	4085	206	13
36	4265	206	14
37	4445	206	14
38	4625	206	15
39	4805	205	15
40	4985	205	15
41	5165	205	15
42	5345	204	16
43	5525	204	17
44	5705	204	18
45	5885	204	19
46	6065	201	19
47	6245	205	17
48	6425	205	17
49	6605	205	17
50	6785	205	16
6	(630)		
3:15P	395	106	2
16	611	110	2
17	809	112	2
18	1007	135	1
19	1196	158	1
20	1385	126	1
21	1565	107	1
22	1745	78	1
23	1925	309	1
24	2105	262	2
25	2285	218	3

TABLE II (1929)

Date and time	Height in meters	Azimuth from; degrees	Velocity m/s	Date and time	Height in meters	Azimuth from; degrees	Velocity m/s	Date and time	Height in meters	Azimuth from; degrees	Velocity m/s
April				**April**				**April**			
6				**7**				**9**			
3:26P	2465	202	4	9:50A	4085	181	25	1:33P	1007	71	3
27	2645	202	6	51	4265	181	27	34	1196	159	7
28	2825	202	7	52	4445	186	27	35	1385	166	12
29	3005	203	8	53	4625	188	29	36	1565	170	11
30	3185	203	9	54	4805	184	29	**10**	(634)		
31	3365	209	9	55	4985	180	28	4:00P	395	0	5
32	3545	208	10	56	5165	178	31	01	611	10	8
33	3725	204	10	57	5345	178	33	02	809	24	6
34	3905	204	11	58	5525	177	30	03	1007	27	5
35	4085	204	11	59	5705	177	30	04	1196	25	5
36	4265	204	13	10:00A	5885	177	32	05	1385	21	5
37	4445	202	13					06	1565	185	3
38	4625	202	14	**7**	(632)			07	1745	185	13
39	4805	202	15	3:00P	395	348	4	07:30	1835	185	17
40	4985	202	17	01	611	356	5	**12**	(635)		
41	5165	202	19	02	809	18	4	2:40P	395	254	5
42	5345	202	19	03	1007	38	4	41	611	266	9
43	5525	202	19	04	1196	39	3	42	809	252	6
44	5705	202	20	05	1385	25	2	43	1007	218	3
45	5885	203	20	06	1565	25	1	44	1196	193	2
46	6065	210	21	07	1745	216	1	45	1385	183	3
47	6245	210	25	08	1925	187	5	46	1565	183	4
48	6425	210	26	09	2105	177	9	47	1745	192	6
49	6605	210	25	10	2285	177	12	48	1925	192	9
50	6785	210	25	11	2465	178	14	49	2105	184	11
51	6965	210	26	12	2645	181	15	50	2285	179	11
				13	2825	184	16	**14**	(636)		
7	(631)			14	3005	188	17	9:01A	395	87	1
9:30A	395	62	2	15	3185	188	17	02	611	10	2
31	611	60	5	16	3365	185	18	03	809	18	5
32	809	53	5	17	3545	181	20	04	1007	20	6
33	1007	43	5	18	3725	178	23	05	1196	21	6
34	1196	43	7	19	3905	178	27	**15**	(637)		
35	1385	43	6	20	4085	180	27	3:05P	395	17	5
36	1565	26	4	21	4265	180	27	06	611	19	8
37	1745	0	3	22	4445	180	30	07	809	19	7
38	1925	352	2	23	4625	180	31	08	1007	16	7
39	2105	301	1	24	4805	180	32	09	1196	16	7
40	2285	205	3	25	4985	180	33	10	1385	10	5
41	2465	188	7	26	5165	180	36	11	1565	359	5
42	2645	181	9	27	5345	184	38	**15**	(638)		
43	2825	181	10	28	5525	185	38	9:10A	395	95	4
44	3005	181	11	29	5705	185	39	11	611	58	5
45	3185	178	15					12	809	38	8
46	3365	172	21	**9**	(633)			13	1007	38	11
47	3545	172	24	1:30P	395	35	5				
48	3725	172	23	31	611	32	5				
49	3905	172	23	32	809	33	5				

TABLE II (1929)

Date and time	Height in meters	Azimuth from; degrees	Velocity m/s	Date and time	Height in meters	Azimuth from; degrees	Velocity m/s	Date and time	Height in meters	Azimuth from; degrees	Velocity m/s
April 15				**April 15**				**April 16**			
9:14A	1196	40	11	3:47P	4445	335	15	10:03A	8225	303	20
15	1385	45	9	48	4625	335	17	04	8405	300	22
16	1565	43	7	49	4805	335	17	05	8585	300	23
17	1745	32	9					06	8765	300	25
18	1925	26	11	**16**	(640)			07	8945	300	25
19	2105	24	12	9:20A	395	49	2	08	9125	300	24
20	2285	21	13	21	611	73	3				
21	2465	18	13	22	809	95	3	**16**	(641)		
22	2645	15	13	23	1007	76	4	3:30P	395	71	1
23	2825	10	13	24	1196	68	5	31	611	41	3
24	3005	8	13	25	1385	65	4	32	809	32	3
25	3185	6	15	26	1565	65	3	33	1007	37	4
26	3365	6	17	27	1745	65	3	34	1196	37	3
27	3545	3	19	28	1925	51	3	35	1385	17	3
28	3725	1	19	29	2105	43	3	36	1565	8	2
29	3905	359	19	30	2285	30	3	37	1745	5	1
30	4085	355	18	31	2465	9	4	38	1925	342	1
31	4265	352	18	32	2645	350	5	39	2105	342	1
32	4445	350	18	33	2825	329	7	40	2285	327	1
33	4625	350	17	34	3005	318	7	41	2465	314	1
34	4805	350	18	35	3185	318	6	42	2645	298	1
35	4985	355	18	36	3365	318	5	43	2825	283	3
36	5165	355	19	37	3545	319	5	44	3005	290	4
37	5345	356	19	38	3725	319	5	45	3185	290	4
				39	3905	310	6	46	3365	290	5
15	(639)			40	4085	307	9	47	3545	298	5
3:25P	395	85	2	41	4265	307	12	48	3725	298	7
26	611	70	5	42	4445	307	13	49	3905	294	7
27	809	55	5	43	4625	307	13	50	4085	294	9
28	1007	40	8	44	4805	305	13	51	4265	291	9
29	1196	40	9	45	4985	305	14	52	4445	291	11
30	1385	52	8	46	5165	305	15	53	4625	291	12
31	1565	48	9	47	5345	305	15	54	4805	283	12
32	1745	36	10	48	5525	302	15	55	4985	276	13
33	1925	29	10	49	5705	302	17	56	5165	276	13
34	2105	23	9	50	5885	302	17	57	5345	278	14
35	2285	21	9	51	6065	296	17	58	5525	280	14
36	2465	16	10	52	6245	296	18	59	5705	283	15
37	2645	9	11	53	6425	300	18	4:00P	5885	285	14
38	2825	9	11	54	6605	300	18	01	6065	291	14
39	3005	7	12	55	6785	300	18	02	6245	289	14
40	3185	7	12	56	6965	298	15	03	6425	282	13
41	3365	0	11	57	7145	298	15	04	6605	282	14
42	3545	350	11	58	7345	298	17	05	6785	282	15
43	3725	345	11	59	7505	305	18	06	6965	282	15
44	3905	339	11	10:00A	7685	305	18	07	7145	282	15
45	4085	335	12	01	7865	305	21	08	7325	287	16
46	4265	335	15	02	8045	303	21	09	7505	287	17

TABLE II (1929)

Date and time	Height in meters	Azimuth from; degrees	Velocity m/s
April 16			
4:10P	7685	287	17
11	7865	287	17
12	8045	287	18
13	8585	287	18
14	8405	287	18
17	(642)		
9:22A	395	54	2
23	611	77	3
24	809	125	1
25	1007	134	2
26	1196	140	1
27	1385	160	1
28	1565	311	1
29	1745	300	2
30	1925	285	2
31	2105	281	2
32	2285	281	3
33	2465	285	3
34	2645	293	5
35	2825	289	5
36	3005	281	6
37	3185	287	6
38	3365	292	5
39	3545	289	5
40	3725	289	6
41	3905	289	9
42	4085	288	8
43	4265	283	9
44	4445	283	10
45	4625	283	11
46	4805	283	11
47	4985	289	13
48	5165	289	14
49	5345	288	15
50	5525	285	17
51	5705	282	19
52	5885	279	18
53	6065	279	17
54	6245	279	18
55	6425	279	18
56	6605	279	19
57	6785	279	19
58	6965	279	19
59	7145	279	18
10:00A	7325	279	18
01	7505	279	19
02	7685	277	17

Date and time	Height in meters	Azimuth from; degrees	Velocity m/s
April 17			
10:03A	7865	275	16
04	8045	275	17
05	8225	275	19
06	8405	274	19
07	8585	271	20
08	8765	271	23
09	8945	271	20
10	9125	271	17
17	(643)		
3:30P	395	50	1
31	611	117	1
32	809	163	3
33	1007	176	3
34	1196	174	2
35	1385	183	2
36	1565	185	2
37	1745	175	2
38	1925	202	1
39	2105	226	1
40	2285	217	1
41	2465	255	2
42	2645	263	3
43	2825	264	4
44	3005	258	5
45	3185	252	4
46	3365	252	4
47	3545	252	4
48	3725	252	5
49	3905	263	5
50	4085	260	7
51	4265	250	7
52	4445	250	5
53	4625	250	5
54	4805	258	5
55	4985	258	6
56	5165	258	8
57	5345	253	9
58	5525	247	10
59	5705	247	11
4:00P	5885	247	12
01	6065	245	13
02	6245	242	15
03	6425	242	17
04	6605	242	17
05	6785	242	18
06	6965	242	19

Date and time	Height in meters	Azimuth from; degrees	Velocity m/s
April 17			
4:07P	7145	242	18
08	7325	241	18
09	7505	241	17
10	7685	241	17
11	7865	241	16
12	8045	242	16
13	8225	242	17
14	8405	242	18
15	8585	242	20
16	8765	244	22
17	8945	245	22
18	9125	245	23
19	9305	245	25
18	(644)		
9:25A	395	156	6
26	611	166	17
27	809	170	17
28	1007	180	13
29	1196	194	13
30	1385	194	13
31	1565	192	17
32	1745	190	28
33	1925	190	33
34	2105	188	33
35	2285	186	31
36	2465	186	27
18	(645)		
4:07P	395	178	5
08	611	193	9
09	809	182	9
10	1007	182	9
11	1196	182	9
12	1385	189	10
13	1565	193	10
14	1745	193	11
15	1925	200	15
16	2105	207	19
17	2285	211	19
18	2465	209	19
19	2645	206	19
20	2825	203	21
21	3005	203	22
22	3185	207	22
23	3365	211	23
24	3545	214	22
25	3725	214	19

TABLE II (1929)

Date and time	Height in meters	Azimuth from; degrees	Velocity m/s
April 19	(646)		
9:44A	395	155	7
45	611	157	12
46	809	165	13
47	1007	175	13
48	1196	184	13
49	1385	184	14
50	1565	192	15
51	1745	199	15
52	1925	199	15
53	2105	207	14
54	2285	211	14
55	2465	209	14
56	2645	205	15
57	2825	203	16
58	3005	203	15
59	3185	203	16
10:00A	3365	200	17
01	3545	200	16
02	3725	200	16
03	3905	200	17
19	(647)		
4:05P	395	238	3
06	611	209	4
07	809	192	6
08	1007	185	6
09	1196	180	7
10	1385	180	7
11	1565	183	8
12	1745	192	10
13	1925	204	10
14	2105	211	11
20	(648)		
3:05P	395	300	3
06	611	180	5
07	809	180	8
08	1007	184	6
09	1196	193	6
10	1385	194	6
11	1565	196	7
12	1745	202	7
13	1925	226	5
14	2105	230	6
15	2285	221	10
16	2465	212	16
17	2645	206	21
18	2825	203	23
19	3005	203	22
April 21	(649)		
1:35P	395	114	1
36	611	162	2
37	809	127	2
38	1007	57	2
39	1196	75	1
40	1385	116	1
41	1565	117	2
42	1745	113	1
43	1925	0	2
44	2105	8	3
45	2285	44	1
46	2465	182	2
47	2645	176	3
48	2825	170	3
49	3005	196	4
50	3185	236	5
51	3365	228	7
52	3545	199	7
53	3725	187	7
54	3905	187	8
55	4085	180	9
56	4265	176	9
57	4445	181	10
58	4625	187	12
59	4805	186	13
2:00P	4985	184	13
01	5165	182	11
02	5345	182	11
03	5525	182	12
04	5705	181	12
22	(650)		
9:25A	395	332	2
26	611	178	2
27	809	186	6
28	1007	194	10
29	1196	194	11
30	1385	190	11
31	1565	189	13
32	1745	186	14
33	1925	187	13
34	2105	193	9
35	2285	207	6
36	2465	215	9
37	2645	212	11
38	2825	209	12
April 22	(651)		
3:55P	395	310	3
56	611	317	3
57	809	190	1
58	1007	142	2
59	1196	142	4
4:00P	1385	159	4
01	1565	174	5
02	1745	181	7
03	1925	190	9
04	2105	198	10
05	2285	211	11
06	2465	221	13
07	2645	221	14
08	2825	220	15
09	3005	220	16
10	3185	218	16
11	3365	218	16
12	3545	220	17
13	3725	220	18
14	3905	220	20
15	4085	222	19
16	4265	222	19
23	(652)		
0:10P	395	126	1
11	611	11	2
12	809	12	4
13	1007	26	5
14	1196	35	7
15	1385	35	9
16	1565	31	8
17	1745	36	6
18	1925	55	4
19	2105	82	5
20	2285	109	6
21	2465	120	7
24	(653)		
3:26P	395	20	5
27	611	29	5
28	809	17	5
29	1007	22	7
30	1196	32	9
31	1385	32	9
32	1565	39	7
33	1745	85	3
34	1925	167	6
35	2105	168	8

TABLE II (1929)

Date and time	Height in meters	Azimuth from; degrees	Velocity m/s
April 24			
3:36P	2285	171	9
37	2465	183	10
38	2645	190	10
39	2825	184	11
40	3005	177	11
41	3185	177	12
42	3365	177	12
24	(654)		
11:00A	395	93	1
01	611	68	2
02	809	48	2
03	1007	89	2
04	1196	146	3
05	1385	166	6
06	1565	174	7
07	1745	182	7
08	1925	188	7
09	2105	190	9
10	2285	194	9
11	2465	197	8
12	2645	194	9
13	2825	194	9
14	3005	194	11
15	3185	194	12
16	3365	194	11
17	3545	192	11
18	3725	188	12
19	3905	188	13
20	4085	191	15
21	4265	193	15
22	4445	204	16
23	4625	220	15
24	4805	229	15
25	(655)		
9:12A	395	80	3
13	611	87	3
14	809	123	3
15	1007	139	4
16	1196	149	5
17	1385	171	7
18	1565	182	8
19	1745	184	9
20	1925	187	9
21	2105	187	11
22	2285	177	11
23	2465	167	11

Date and time	Height in meters	Azimuth from; degrees	Velocity m/s
April 25			
9:24A	2645	165	11
25	2825	163	11
26	3005	163	12
27	3185	163	11
28	3365	163	11
29	3545	163	10
30	3725	163	9
31	3905	167	10
32	4085	163	9
33	4265	157	8
34	4445	157	9
35	4625	159	11
36	4805	165	12
37	4985	165	12
38	5165	165	13
39	5345	165	12
40	5525	162	13
41	5705	160	15
25	(656)		
3:50P	395	261	11
51	611	261	17
52	809	115	2
53	1007	99	5
54	1196	118	3
55	1385	124	5
56	1565	129	7
57	1745	136	7
58	1925	143	8
59	2105	153	8
4:00P	2285	176	9
01	2465	195	11
02	2645	202	14
03	2825	203	14
04	3005	197	14
05	3185	189	13
06	3365	175	12
07	3545	159	15
08	3725	149	17
09	3905	141	17
10	4085	126	14
11	4265	117	11
12	4445	120	12
13	4625	124	13
14	4805	127	14
15	4985	136	15
16	5165	142	17
17	5345	147	19

Date and time	Height in meters	Azimuth from; degrees	Velocity m/s
April 25			
4:18P	5525	144	17
19	5705	140	18
20	5885	140	18
21	6065	135	15
22	6245	131	15
23	6425	131	17
24	6605	131	18
26	(657)		
8:52A	395	102	5
53	611	120	9
54	809	137	9
55	1007	142	9
56	1196	148	9
57	1385	161	9
58	1565	174	10
59	1745	181	11
9:00A	1925	174	9
01	2105	157	7
02	2285	138	9
03	2465	125	11
04	2645	123	13
05	2825	128	15
06	3005	136	16
07	3185	144	14
08	3365	160	15
09	3545	163	21
10	3725	156	24
11	3905	147	23
12	4085	142	25
13	4265	138	26
14	4445	135	25
15	4625	135	25
26	(658)		
4:00P	395	174	5
01	611	176	7
02	809	163	7
03	1007	162	6
04	1196	156	6
05	1385	156	5
06	1565	156	3
07	1745	156	2
08	1925	156	2
09	2105	166	3
10	2285	173	3
11	2465	173	4
12	2645	173	5

TABLE II (1929)

Date and time	Height in meters	Azimuth from; degrees	Velocity m/s	Date and time	Height in meters	Azimuth from; degrees	Velocity m/s	Date and time	Height in meters	Azimuth from; degrees	Velocity m/s
April				**April**				**April**			
26				**27**				**27**			
4:13P	2825	173	5	9:28A	5525	165	17	4:30P	5885	267	11
14	3005	172	5	29	5705	165	17	31	6065	270	12
15	3185	161	7	30	5885	165	17	32	6245	273	13
16	3365	161	9	31	6065	165	18	33	6425	273	13
17	3545	161	12	32	6245	169	19	34	6605	269	13
18	3725	164	15	33	6425	171	19	35	6785	261	15
19	3905	162	17	34	6605	171	19	36	6965	258	17
20	4085	158	18	35	6785	171	18	37	7145	258	17
21	4265	155	17	36	6965	170	15	38	7325	258	17
22	4445	145	18	37	7145	169	17	39	7505	258	17
23	4625	138	23	38	7325	169	20	40	7685	258	18
24	4805	136	26	39	7505	169	23	41	7865	259	18
25	4985	136	28	40	7685	169	26	42	8045	259	17
26	5165	138	29	41	7865	170	29	43	8225	258	17
27	5345	140	31	42	8045	170	29	44	8405	254	16
28	5525	147	32	43	8225	170	30	45	8585	250	17
29	5705	144	28					46	8765	246	18
30	5885	143	25	**27**	(660)			47	8945	246	18
				4:00P	395	94	7	48	9125	246	19
27	(659)			01	611	85	7	49	9305	244	17
9:00A	395	85	4	02	809	78	7	50	9485	244	16
01	611	113	4	03	1007	95	5	51	9665	247	14
02	809	145	3	04	1196	105	4	52	9845	247	11
03	1007	154	3	05	1385	115	4	53	10025	243	11
04	1196	135	3	06	1565	120	3	54	10205	237	12
05	1385	147	5	07	1745	120	3	55	10385	237	12
06	1565	175	6	08	1925	122	4	56	10565	237	10
07	1745	193	5	09	2105	137	5	57	10745	232	11
08	1925	200	5	10	2285	145	7	58	10925	233	12
09	2105	200	5	11	2465	155	7	59	11105	234	14
10	2285	200	5	12	2645	166	8	5:00P	11285	234	15
11	2465	194	6	13	2825	178	6				
12	2645	185	6	14	3005	178	6	**28**	(661)		
13	2825	180	7	15	3185	181	9	9:06A	395	179	1
14	3005	180	8	16	3365	183	10	07	611	230	6
15	3185	180	9	17	3545	191	9	08	809	252	6
16	3365	174	10	18	3725	204	7	09	1007	262	5
17	3545	171	11	19	3905	214	7	10	1196	262	6
18	3725	169	11	20	4085	218	7	11	1385	238	6
19	3905	173	13	21	4265	237	6	12	1565	206	5
20	4085	180	16	22	4445	253	6	13	1745	202	5
21	4265	184	15	23	4625	255	6	14	1925	211	3
22	4445	177	13	24	4805	265	9	15	2105	210	4
23	4625	170	12	25	4985	260	11	16	2285	200	5
24	4805	169	12	26	5165	249	12	17	2465	193	4
25	4985	167	12	27	5345	249	12	18	2645	174	3
26	5165	165	13	28	5525	258	11	19	2825	162	5
27	5345	165	16	29	5705	267	11	20	3005	170	8

TABLE II (1929)

Date and time	Height in meters	Azimuth from; degrees	Velocity m/s
April 28			
9:21A	3185	182	9
22	3365	189	11
23	3545	205	10
24	3725	226	9
25	3905	232	9
28	(662)		
3:55P	395	13	5
56	611	30	5
57	809	49	4
58	1007	56	5
59	1196	58	6
4:00P	1385	58	6
01	1565	58	4
02	1745	81	3
03	1925	117	4
04	2105	123	4
05	2285	132	5
06	2465	149	6
07	2645	156	6
29	(663)		
5:05P	395	274	6
06	611	284	4
07	809	280	5
08	1007	287	2
09	1196	62	1
10	1385	25	1
11	1565	101	1
12	1745	120	3
13	1925	111	5
14	2105	104	6
30	(664)		
11:10A	395	112	4
11	611	103	2
12	809	74	1
13	1007	57	2
14	1196	57	2
15	1385	63	2
May 1	(665)		
8:45A	395	77	4
46	611	29	6
47	809	26	8
48	1007	33	8
49	1196	51	5
May 1			
8:50A	1385	53	4
51	1565	46	3
52	1745	48	1
53	1925	37	1
54	2105	48	1
55	2285	147	1
56	2465	209	1
57	2645	230	1
58	2825	242	1
59	3005	274	3
9:00A	3185	281	4
01	3365	278	4
02	3545	287	5
03	3725	296	5
04	3905	294	6
05	4085	290	7
06	4265	294	7
07	4445	285	6
08	4625	282	7
09	4805	286	9
10	4985	284	9
11	5165	289	9
12	5345	289	9
13	5525	289	10
14	5705	289	12
15	5885	289	13
16	6065	289	14
17	6245	284	13
18	6425	280	13
19	6605	280	13
20	6785	285	11
21	6965	289	11
22	7145	289	13
23	7325	287	15
24	7505	282	17
25	7685	280	16
26	7865	280	16
27	8045	280	18
28	8225	280	19
29	8405	280	17
30	8585	285	18
31	8765	285	19
32	8945	285	20
1	(666)		
4:42P	395	359	4
43	611	10	5
44	809	14	4
May 1			
4:45P	1007	14	3
46	1196	33	2
47	1385	46	1
48	1565	47	2
49	1745	324	3
50	1925	266	4
51	2105	259	4
52	2285	269	5
53	2465	283	6
54	2645	283	5
55	2825	286	5
56	3005	287	7
57	3185	286	6
58	3365	287	7
59	3545	299	6
5:00P	3725	298	6
01	3905	297	6
02	4085	298	6
03	4265	297	7
04	4445	293	9
05	4625	292	11
06	4805	293	11
07	4985	293	9
08	5165	292	9
09	5345	286	12
10	5525	285	11
11	5705	285	11
12	5885	283	11
13	6065	283	10
14	6245	286	13
15	6425	287	16
16	6605	278	19
17	6785	278	21
18	6965	277	14
19	7145	274	14
20	7325	274	16
21	7505	274	18
22	7685	274	19
23	7865	275	17
24	8045	275	17
25	8225	274	19
2	(667)		
9:02A	395	84	4
03	611	59	4
04	809	40	3
05	1007	39	5
06	1196	39	5

TABLE II (1929)

Date and time	Height in meters	Azimuth from; degrees	Velocity m/s	Date and time	Height in meters	Azimuth from; degrees	Velocity m/s	Date and time	Height in meters	Azimuth from; degrees	Velocity m/s
May				**May**				**May**			
2				**2**				**3**			
9:07A	1385	38	6	4:14P	2105	10	6	9:28A	2825	352	8
08	1565	31	7	15	2285	12	8	29	3005	344	9
09	1745	27	7	16	2465	12	8	30	3185	331	10
10	1925	21	7	17	2645	12	7	31	3365	341	11
11	2105	13	7	18	2825	11	7	32	3545	341	11
12	2285	8	8	19	3005	8	6	33	3725	331	10
13	2465	1	8	20	3185	2	7	34	3905	327	11
14	2645	358	8	21	3365	0	9	35	4085	327	11
15	2825	357	8	22	3545	4	10	36	4265	327	9
16	3005	356	10	23	3725	5	11				
17	3185	356	11	24	3905	2	11	**3**	(670)		
18	3365	355	11	25	4085	358	10	3:58P	395	100	4
19	3545	355	11	26	4265	357	10	59	611	85	2
20	3725	356	11	27	4445	353	11	4:00P	809	39	2
21	3905	356	9	28	4625	349	10	01	1007	37	3
22	4085	355	9	29	4805	348	9	02	1196	7	2
23	4265	346	9	30	4985	355	10	03	1385	20	1
24	4445	347	7	31	5165	355	12	04	1565	38	1
25	4625	348	9	32	5345	355	13	05	1745	336	3
26	4805	347	10	33	5525	355	13	06	1925	311	4
27	4985	343	9	34	5705	355	13	07	2105	306	4
28	5165	342	10	35	5885	354	13	08	2285	308	5
29	5345	344	12	36	6065	350	13	09	2465	313	6
30	5525	346	13	37	6245	348	14	10	2645	320	7
31	5705	344	16	38	6425	348	16	11	2825	320	6
32	5885	345	14	39	6605	348	17	12	3005	322	7
33	6065	329	15	40	6785	354	16	13	3185	319	7
34	6245	332	17	41	6965	354	17	14	3365	314	8
35	6425	354	17	42	7145	354	18	15	3545	310	9
36	6605	353	19	43	7325	354	18	16	3725	308	11
37	6785	349	16	44	7505	354	18	17	3905	303	10
38	6965	347	15	45	7685	352	18	18	4085	303	12
39	7145	346	14	46	7865	352	16	19	4265	307	13
40	7325	346	15					20	4445	307	14
41	7505	345	14	**3**	(669)			21	4625	307	15
42	7685	340	18	9:15A	395	69	3	22	4805	307	15
43	7865	338	22	16	611	46	3	23	4985	302	15
				17	809	70	1	24	5165	295	15
2	(668)			18	1007	85	1	25	5345	296	15
4:05P	395	325	6	19	1196	50	2	26	5525	301	15
06	611	350	6	20	1385	30	3	27	5705	301	16
07	809	2	5	21	1565	19	3	28	5885	300	23
08	1007	22	4	22	1745	357	3	29	6065	294	22
09	1196	28	5	23	1925	356	4	30	6245	290	17
10	1385	16	7	24	2105	356	4	31	6425	290	21
11	1565	19	5	25	2285	356	5	32	6605	298	25
12	1745	22	5	26	2465	345	7	33	6785	296	25
13	1925	14	5	27	2645	345	7	34	6965	296	24

TABLE II (1929)

Date and time	Height in meters	Azimuth from; degrees	Velocity m/s
May			
4	(671)		
9:02A	395	100	2
03	611	147	1
04	809	159	2
05	1007	159	2
06	1196	168	1
07	1385	168	1
08	1565	216	1
09	1745	247	2
10	1925	287	2
11	2105	299	4
12	2285	299	5
13	2465	297	5
14	2645	295	7
15	2825	292	7
16	3005	292	7
17	3185	292	9
18	3365	295	10
19	3545	297	10
20	3725	300	11
21	3905	306	12
22	4085	310	12
23	4265	310	11
24	4445	307	12
25	4625	304	12
26	4805	304	12
27	4985	307	12
4	(672)		
3:59P	395	205	3
4:00P	611	243	4
01	809	261	5
02	1007	269	5
03	1196	279	4
04	1385	296	3
05	1565	291	3
06	1745	294	3
07	1925	276	4
08	2105	256	7
09	2285	248	8
10	2465	252	8
11	2645	261	9
12	2825	268	10
13	3005	276	11
14	3185	277	11
15	3365	283	12
16	3545	285	12
17	3725	289	11
18	3905	290	13

Date and time	Height in meters	Azimuth from; degrees	Velocity m/s
May			
4			
4:19P	4085	292	13
20	4265	294	13
21	4445	294	15
22	4625	292	16
23	4805	288	18
24	4985	283	20
25	5165	280	21
26	5345	280	22
27	5525	280	21
5	(673)		
4:04P	395	295	4
05	611	236	6
06	809	246	4
07	1007	305	2
08	1196	319	3
7	(674)		
9:10A	395	80	4
11	611	60	3
12	809	68	3
13	1007	87	5
14	1196	84	4
15	1385	107	2
16	1565	109	1
17	1745	83	4
18	1925	90	6
19	2105	90	7
20	2285	92	6
21	2465	85	6
22	2645	75	6
23	2825	79	6
24	3005	75	7
25	3185	74	7
26	3365	74	6
27	3545	74	6
7	(675)		
3:45P	395	10	7
46	611	12	18
47	809	2	26
48	1007	357	34
8	(676)		
3:15P	395	158	4
16	611	158	10
17	809	161	9
18	1007	164	6

Date and time	Height in meters	Azimuth from; degrees	Velocity m/s
May			
8			
3:19P	1196	161	5
20	1385	168	4
21	1565	175	5
22	1745	179	7
23	1925	180	9
24	2105	184	10
9	(677)		
9:02A	395	100	2
03	611	77	2
04	809	90	3
05	1007	73	3
06	1196	62	5
07	1385	73	6
08	1565	91	8
09	1745	95	10
10	1925	92	11
10	(678)		
4:10P	395	20	5
11	611	31	5
12	809	41	5
13	1007	41	4
14	1196	46	4
15	1385	40	4
16	1565	38	3
17	1745	38	3
18	1925	30	2
19	2105	318	1
20	2285	221	2
21	2465	211	3
11	(679)		
9:08A	395	103	4
09	611	76	3
10	809	95	3
11	1007	119	3
12	1196	126	3
13	1385	139	3
14	1565	146	3
15	1745	147	4
16	1925	142	5
17	2105	154	6
18	2285	159	7
19	2465	155	8
20	2645	154	8
21	2825	154	8
22	3005	155	8

TABLE II (1929)

Date and time	Height in meters	Azimuth from; degrees	Velocity m/s	Date and time	Height in meters	Azimuth from; degrees	Velocity m/s	Date and time	Height in meters	Azimuth from; degrees	Velocity m/s
May				**May**				**May**			
11				**12**				**12**			
9:23A	3185	149	7	8:45A	1385	65	4	4:47P	1745	36	2
24	3365	145	6	46	1565	85	4	48	1925	59	2
25	3545	146	5	47	1745	93	3	49	2105	63	1
26	3725	145	5	48	1925	101	3	50	2285	36	1
27	3905	145	6	49	2105	117	3	51	2465	222	1
28	4085	150	6	50	2285	145	3	52	2645	184	1
29	4265	160	7	51	2465	151	5	53	2825	82	2
30	4445	162	7	52	2645	136	5	54	3005	84	4
31	4625	161	8	53	2825	130	5	55	3185	89	6
32	4805	163	9	54	3005	134	5	56	3365	96	7
33	4985	162	9	55	3185	136	5	57	3545	98	8
34	5165	163	9	56	3365	142	5	58	3725	100	9
35	5345	166	11	57	3545	147	7	59	3905	104	9
36	5525	166	12	58	3725	155	8	5:00P	4085	109	9
37	5705	159	12	59	3905	156	9	01	4265	114	9
38	5885	154	12	9:00A	4085	151	9	02	4445	114	9
39	6065	159	13	01	4265	154	9	03	4625	117	8
40	6245	158	13	02	4445	156	7	04	4805	117	9
41	6425	158	12	03	4625	152	9	05	4985	116	9
				04	4805	151	9				
11	(680)			05	4985	146	9	**13**	(683)		
6:40P	395	30	1	06	5165	147	10	4:57P	395	68	5
41	611	23	4	07	5345	151	11	58	611	57	4
42	809	32	3	08	5525	152	12	59	809	49	3
43	1007	14	2	09	5705	150	13	5:00P	1007	42	4
44	1196	7	2	10	5885	150	14	01	1196	32	5
45	1385	23	2	11	6065	150	14	02	1385	6	7
46	1565	67	1	12	6245	150	15	03	1565	6	8
47	1745	98	1	13	6425	154	15	04	1745	18	6
48	1925	104	1	14	6605	156	15	05	1925	23	5
49	2105	123	1	15	6785	150	17	06	2105	37	5
50	2285	187	1	16	6965	149	15	07	2285	56	5
51	2465	179	2	17	7145	147	17	08	2465	59	5
52	2645	161	3	18	7325	145	18	09	2645	61	5
53	2825	153	3	19	7505	145	19	10	2825	77	5
54	3005	146	3	20	7685	145	20	11	3005	88	5
55	3185	146	3	21	7865	145	20	12	3185	88	5
56	3365	146	3	22	8045	145	20	13	3365	88	4
57	3545	140	4	23	8225	145	21	14	3545	88	5
58	3725	150	4					15	3725	94	5
59	3905	159	4	**12**	(682)			16	3905	94	4
				4:40P	395	91	3	17	4085	94	4
12	(681)			41	611	51	3	18	4265	94	5
8:40A	395	97	3	42	809	39	3	19	4445	107	6
41	611	69	3	43	1007	15	3	20	4625	114	6
42	809	49	2	44	1196	10	3	21	4805	114	6
43	1007	49	2	45	1385	22	4	22	4985	114	7
44	1196	49	3	46	1565	28	3	23	5165	117	8

TABLE II (1929)

Date and time	Height in meters	Azimuth from; degrees	Velocity m/s
May 13			
5:24P	5345	116	8
25	5525	112	8
26	5705	110	8
27	5885	102	7
14	(684)		
9:02A	395	101	8
03	611	87	5
04	809	85	7
05	1007	89	10
06	1196	97	13
07	1385	100	14
08	1565	97	14
09	1745	95	13
10	1925	95	13
14	(685)		
7:00P	395	255	5
01	611	240	9
02	809	218	8
03	1007	202	7
04	1196	196	5
05	1385	176	5
06	1565	172	5
07	1745	172	6
08	1925	170	7
09	2105	166	7
10	2285	161	7
11	2465	161	7
12	2645	151	6
13	2825	147	6
14	3005	153	9
15	3185	145	10
16	3365	141	13
17	3545	149	17
18	3725	153	20
19	3905	158	26
20	4085	152	36
21	4265	146	39
22	4445	144	36
15	(686)		
9:10A	395	138	2
11	611	121	3
12	809	125	3
13	1007	132	3
14	1196	145	2
15	1385	145	2
16	1565	145	3

Date and time	Height in meters	Azimuth from; degrees	Velocity m/s
May 15			
9:17A	1745	166	4
18	1925	177	5
19	2105	184	6
20	2285	190	7
21	2465	190	8
22	2645	186	9
23	2825	183	9
24	3005	192	9
25	3185	200	9
26	3365	194	8
27	3545	192	8
28	3725	197	9
29	3905	197	11
30	4085	199	13
31	4265	202	16
32	4445	199	17
33	4625	196	17
34	4805	202	18
35	4985	204	18
36	5165	203	23
37	5345	204	31
38	5525	206	33
39	5705	206	32
15	(687)		
4:05P	395	85	4
06	611	73	5
07	809	60	4
08	1007	60	4
09	1196	81	3
10	1385	132	3
11	1565	151	3
12	1745	161	5
13	1925	135	6
14	2105	105	7
15	2285	93	9
16	2465	93	9
17	2645	113	9
18	2825	126	11
19	3005	126	12
20	3185	124	12
21	3365	125	15
22	3545	126	15
23	3725	123	17
24	3905	121	20
16	(688)		
9:06A	395	85	8

Date and time	Height in meters	Azimuth from; degrees	Velocity m/s
May 16			
9:07A	611	108	7
08	809	127	7
09	1007	142	5
10	1196	127	3
11	1385	90	2
12	1565	119	4
13	1745	109	7
14	1925	96	9
15	2105	91	7
16	2285	98	6
17	2465	109	7
18	2645	104	9
19	2825	115	11
20	3005	122	13
21	3185	103	13
22	3365	94	14
23	3545	100	15
16	(689)		
4:08P	395	284	6
09	611	277	8
10	809	273	10
11	1007	271	12
12	1196	273	7
13	1385	273	2
14	1565	270	3
15	1745	193	2
16	1925	142	3
17	2105	140	1
18	2285	144	3
19	2465	167	3
20	2645	210	5
21	2825	223	7
22	3005	211	7
23	3185	191	8
24	3365	179	7
25	3545	166	6
26	3725	156	6
27	3905	145	5
28	4085	136	5
29	4265	123	4
30	4445	115	5
31	4625	124	7
32	4805	124	8
33	4985	128	6
34	5165	137	6
35	5345	143	7
36	5525	151	8

TABLE II (1929)

Date and time	Height in meters	Azimuth from; degrees	Velocity m/s
May 16			
4:37P	5705	157	9
38	5885	157	11
39	6065	153	11
40	6245	153	11
41	6425	155	15
42	6605	148	20
43	6785	139	20
44	6965	133	20
45	7145	131	21
46	7325	131	21
47	7505	131	23
17	(690)		
8:58A	395	115	1
59	611	114	2
9:00A	809	116	4
01	1007	123	5
02	1196	137	6
03	1385	150	7
04	1565	164	8
05	1745	170	10
06	1925	165	10
07	2105	150	9
08	2285	145	12
09	2465	145	13
10	2645	145	14
11	2825	145	15
12	3005	145	15
17	(691)		
4:22P	395	30	2
23	611	105	2
24	809	106	3
25	1007	101	3
26	1196	100	3
27	1385	97	4
28	1565	100	5
29	1745	110	6
30	1925	125	7
31	2105	138	9
32	2285	135	9
33	2465	129	12
34	2645	136	13
35	2825	149	12
36	3005	127	10
37	3185	95	13
38	3365	95	17
39	3545	95	20

Date and time	Height in meters	Azimuth from; degrees	Velocity m/s
May 18	(692)		
8:45A	395	112	8
46	611	115	11
47	809	120	11
48	1007	126	14
49	1196	132	17
50	1385	136	21
51	1565	138	24
18	(693)		
4:26P	395	156	7
27	611	155	12
28	809	156	11
29	1007	159	7
30	1196	156	5
31	1385	152	6
32	1565	156	7
33	1745	168	7
34	1925	164	7
35	2105	158	6
36	2285	158	6
37	2465	163	3
38	2645	193	2
39	2825	187	3
40	3005	168	3
41	3185	115	1
42	3365	111	1
43	3545	155	2
44	3725	152	4
45	3905	152	5
46	4085	152	6
19	(694)		
9:20A	395	135	6
21	611	132	7
22	809	128	6
23	1007	129	7
24	1196	144	10
25	1385	150	12
26	1565	151	11
27	1745	158	12
28	1925	162	14
29	2105	165	18
30	2285	158	25
31	2465	171	30
32	2645	175	29
33	2825	178	35
34	3005	169	43

Date and time	Height in meters	Azimuth from; degrees	Velocity m/s
May 19	(695)		
6:32P	395	70	11
33	611	76	11
34	809	93	9
35	1007	112	8
36	1196	112	8
37	1385	104	9
38	1565	101	8
39	1745	89	7
40	1925	78	6
41	2105	64	3
42	2285	70	4
43	2465	91	6
44	2645	100	6
45	2825	100	7
46	3005	108	9
47	3185	113	11
48	3365	119	13
49	3545	120	13
50	3725	120	13
51	3905	122	13
52	4085	123	13
53	4265	127	13
54	4445	130	13
55	4625	125	11
56	4805	123	12
57	4985	122	13
58	5165	122	15
59	5345	114	16
7:00P	5525	98	13
01	5705	69	9
02	5885	56	9
21	(698)*		
2:02P	395	172	3
03	611	145	1
04	809	326	1
05	1007	326	2
06	1196	326	2
07	1385	350	2
08	1565	334	3
09	1745	338	3
10	1925	352	4
11	2105	331	2
12	2285	269	1
13	2465	234	1
14	2645	280	1
15	2825	308	2
16	3005	315	2

*Balloon runs Nos. 696 and 697 were defective.

TABLE II (1929)

Date and time	Height in meters	Azimuth from; degrees	Velocity m/s
May 21			
2:17P	3185	315	2
18	3365	306	2
19	3545	306	2
20	3725	311	2
21	3905	313	2
22	4085	322	2
23	4265	350	3
24	4445	7	3
25	4625	13	4
26	4805	10	5
27	4985	24	6
28	5165	31	7
29	5345	30	7
30	5525	26	8
31	5705	26	8
32	5885	26	8
33	6065	29	7
34	6245	29	7
35	6425	29	8
36	6605	29	8
37	6785	34	7
38	6965	34	6
39	7145	34	7
40	7325	34	7
41	7505	49	9
42	7685	51	9
43	7865	53	9
44	8045	53	13
45	8225	53	17
46	8405	53	19
47	8585	53	21
48	8765	53	21
49	8945	53	23
50	9125	53	22
51	9305	53	21
52	9485	53	24
53	9665	53	23
54	9845	53	25
55	10025	53	28
22	(699)		
2:20P	395	328	9
21	611	344	17
22	809	3	14
23	1007	52	7
24	1196	120	6
25	1385	146	7
26	1565	158	6

Date and time	Height in meters	Azimuth from; degrees	Velocity m/s
May 22			
2:27P	1745	180	6
28	1925	203	6
29	2105	203	7
30	2285	203	8
23	(700)		
2:26P	395	152	3
27	611	195	1
28	809	262	2
29	1007	286	3
30	1196	291	3
31	1385	260	4
32	1565	248	5
33	1745	248	6
34	1925	256	6
35	2105	258	6
36	2285	250	9
37	2465	238	9
38	2645	228	8
39	2825	228	7
24	(701)		
10:12A	395	60	2
13	611	3	3
14	809	4	3
15	1007	334	2
16	1196	221	1
17	1385	22	1
18	1565	87	1
19	1745	97	3
20	1925	114	3
21	2105	142	4
22	2285	169	5
23	2465	175	7
24	2645	167	9
25	2825	162	9
26	3005	160	8
27	3185	160	9
28	3365	160	9
29	3545	166	11
30	3725	169	11
31	3905	169	12
32	4085	168	13
33	4265	160	13
34	4445	161	13
35	4625	161	13
36	4805	161	13

Date and time	Height in meters	Azimuth from; degrees	Velocity m/s
May 28	(702)		
9:18A	395	110	3
19	611	76	1
20	809	74	2
21	1007	80	3
22	1196	112	3
23	1385	129	4
24	1565	118	5
25	1745	114	5
26	1925	142	5
27	2105	142	5
28	2285	130	7
29	2465	130	8
30	2645	120	9
31	2825	109	10
32	3005	110	13
33	3185	117	13
34	3365	121	13
35	3545	129	13
36	3725	138	12
37	3905	133	12
38	4085	133	12
39	4265	137	13
40	4445	137	14
41	4625	144	15
42	4805	150	15
43	4985	147	15
44	5165	147	14
45	5345	147	14
29	(703)		
7:33P	395	20	11
34	611	34	7
35	809	57	5
36	1007	65	5
37	1196	74	5
38	1385	90	5
39	1565	106	6
40	1745	108	7
41	1925	108	7
42	2105	110	8
43	2285	117	8
44	2465	120	9
45	2645	129	9
46	2825	149	10
47	3005	163	11
48	3185	174	11
49	3365	182	13
50	3545	180	13

TABLE II (1929)

Date and time	Height in meters	Azimuth from; degrees	Velocity m/s
May 29			
7:51P	3725	175	15
52	3905	170	17
53	4085	166	18
54	4265	169	17
55	4445	169	19
56	4625	169	20
57	4805	169	19
58	4985	172	19
59	5165	172	21
8:00P	5345	172	21
01	5525	172	23
30	(704)		
10:10A	395	150	4
11	611	221	4
12	809	239	4
13	1007	254	4
14	1196	249	4
15	1385	237	3
16	1565	214	5
17	1745	208	7
18	1925	205	7
19	2105	199	7
20	2285	191	7
21	2465	179	7
22	2645	179	8
23	2825	189	9
24	3005	196	11
25	3185	199	13
26	3365	203	11
27	3545	203	12
28	3725	199	15
29	3905	191	14
30	4085	193	14
31	4265	195	22
32	4445	195	23
33	4625	196	18
34	4805	197	20
35	4985	197	22
31	(705)		
9:30A	395	130	4
31	611	203	1
32	809	235	2
33	1007	236	3
34	1196	233	3
35	1385	225	3
36	1565	223	4

Date and time	Height in meters	Azimuth from; degrees	Velocity m/s
May 31			
9:37A	1745	237	5
38	1925	237	5
39	2105	231	4
40	2285	214	5
41	2465	208	7
42	2645	202	9
43	2825	202	9
44	3005	209	9
45	3185	205	9
46	3365	205	8
47	3545	203	10
48	3725	201	11
31	(706)		
4:30P	395	334	4
31	611	350	7
32	809	359	7
33	1007	3	6
34	1196	25	4
35	1385	53	4
36	1565	60	3
37	1745	24	2
38	1925	11	2
39	2105	350	1
40	2285	284	1
41	2465	261	1
42	2645	247	1
43	2825	214	1
44	3005	183	5
45	3185	175	9
46	3365	169	12
47	3545	177	12
48	3725	185	11
49	3905	178	11
50	4085	178	11
51	4265	178	12
June 1	(707)		
8:30A	395	102	5
31	611	31	1
32	809	78	1
33	1007	87	1
34	1196	129	1
35	1385	130	1
36	1565	198	2
37	1745	209	3
38	1925	209	5
39	2105	214	7

Date and time	Height in meters	Azimuth from; degrees	Velocity m/s
June 1			
8:40A	2285	211	7
41	2465	202	8
42	2645	193	5
43	2825	186	6
44	3005	182	6
45	3185	176	6
46	3365	179	7
47	3545	179	9
48	3725	179	9
49	3905	179	9
50	4085	179	8
51	4265	182	8
52	4445	177	11
53	4625	170	12
54	4805	173	13
55	4985	178	13
56	5165	178	11
57	5345	178	10
58	5525	178	12
59	5705	178	13
9:00A	5885	178	14
01	6065	181	14
02	6245	181	16
03	6425	181	16
04	6605	181	16
05	6785	181	13
06	6965	180	15
07	7145	178	17
08	7325	178	16
09	7505	179	13
10	7685	179	14
11	7865	179	18
12	8045	179	20
13	8225	179	18
14	8405	179	18
15	8585	177	19
16	8765	177	19
17	8945	176	21
18	9125	176	23
1	(708)		
4:45P	395	332	8
46	611	4	9
47	809	21	5
48	1007	32	7
49	1196	72	6
50	1385	136	4
51	1565	177	6

TABLE II (1929)

Date and time	Height in meters	Azimuth from; degrees	Velocity m/s	Date and time	Height in meters	Azimuth from; degrees	Velocity m/s	Date and time	Height in meters	Azimuth from; degrees	Velocity m/s
June				**June**				**June**			
1				**2**				**3**	(711)		
4:52P	1745	192	7	9:25A	3185	209	5	1:10P	395	82	7
53	1925	192	5	26	3365	198	9	11	611	57	4
54	2105	192	5	27	3545	198	9	12	809	12	4
55	2285	192	9	28	3725	198	9	13	1007	1	3
56	2465	192	10	29	3905	198	8	14	1196	282	2
57	2645	198	9	30	4085	199	8	15	1385	234	2
58	2825	198	9	31	4265	199	8	16	1565	250	1
59	3005	198	9	32	4445	202	8	17	1745	250	3
5:00P	3185	208	8	33	4625	202	8	18	1925	248	4
01	3365	213	10	34	4805	198	9	19	2105	246	5
02	3545	211	11	35	4985	196	9	20	2285	239	7
03	3725	209	11	36	5165	198	10	21	2465	239	7
04	3905	203	11	37	5345	205	10	22	2645	239	7
05	4085	197	11	38	5525	211	10	23	2825	239	9
06	4265	193	9	39	5705	216	11	24	3005	232	11
07	4445	191	10	40	5885	210	12	25	3185	234	13
08	4625	190	11	41	6065	202	13	26	3365	234	13
09	4805	183	13	42	6245	196	14	27	3545	237	12
10	4985	183	15	43	6425	195	13	28	3725	241	13
11	5165	185	15	44	6605	195	13	29	3905	241	13
12	5345	181	15	45	6785	195	13	30	4085	241	14
13	5525	183	16	46	6965	196	13	31	4265	241	16
14	5705	188	17	47	7145	197	13	32	4445	241	12
15	5885	188	17	48	7325	197	14	33	4625	242	10
16	6065	188	17	49	7505	198	14	34	4805	240	13
17	6245	188	15	50	7685	203	15	35	4985	240	15
18	6425	190	16	51	7865	208	15	36	5165	240	16
19	6605	193	17	52	8045	208	15	37	5345	241	17
20	6785	195	19	53	8225	208	15	38	5525	244	18
21	6965	195	18	54	8405	201	13	39	5705	246	18
22	7145	195	18	55	8585	196	15	40	5885	247	19
				56	8765	196	16	41	6065	246	20
2	(709)			57	8945	194	17	42	6245	245	19
9:10A	395	358	5	58	9125	194	16	43	6425	246	19
11	611	26	5	59	9305	194	15	44	6605	247	20
12	809	33	5	10:00A	9485	194	17	45	6785	247	23
13	1007	37	4	01	9665	198	19	46	6965	243	25
14	1196	42	3	02	9845	199	18				
15	1385	54	3	03	10025	199	15	**3**	(712)		
16	1565	80	3	04	10205	201	13	5:20P	395	39	7
17	1745	106	1	05	10385	201	13	21	611	53	8
18	1925	147	1	06	10565	201	14	22	809	45	7
19	2105	156	3	07	10745	201	17	23	1007	35	6
20	2285	161	5	08	10925	201	17	24	1196	8	5
21	2465	168	4	09	11105	201	15	25	1385	208	4
22	2645	172	5	10	11285	202	16	26	1565	274	5
23	2825	205	3	11	11465	202	17	27	1745	261	4
24	3005	235	2	12	11645	203	15	28	1925	247	5

TABLE II (1929)

Date and time	Height in meters	Azimuth from; degrees	Veloc-ity m/s
June 3			
5:29P	2105	245	5
30	2285	234	6
31	2465	230	7
32	2645	233	8
33	2825	239	9
34	3005	243	9
35	3185	246	11
36	3365	246	11
37	3545	246	13
38	3725	246	15
39	3905	251	15
40	4085	254	15
41	4265	254	17
42	4445	252	17
43	4625	251	16
44	4805	250	15
45	4985	248	15
46	5165	249	16
47	5345	251	17
48	5525	252	19
49	5705	253	21
50	5885	253	20
51	6065	253	21
52	6245	253	20
53	6425	253	22
54	6605	253	26
4	(713)		
9:30A	395	42	5
31	611	34	5
32	809	17	3
33	1007	17	3
34	1196	5	5
35	1385	5	5
36	1565	21	4
37	1745	21	3
38	1925	355	2
39	2105	345	3
40	2285	348	4
41	2465	356	3
42	2645	355	2
43	2825	270	1
44	3005	266	2
45	3185	262	3
46	3365	262	5
47	3545	262	6
48	3725	267	7
49	3905	260	8

Date and time	Height in meters	Azimuth from; degrees	Veloc-ity m/s
June 4			
9:50A	4085	268	10
51	4265	267	10
52	4445	252	9
53	4625	259	10
54	4805	262	11
55	4985	257	11
56	5165	257	14
57	5345	252	13
58	5525	248	12
59	5705	248	13
10:00A	5885	248	13
01	6065	250	14
02	6245	265	15
03	6425	265	16
04	6605	265	16
05	6785	261	19
06	6965	259	19
07	7145	258	20
08	7325	257	21
09	7505	256	21
10	7685	256	22
11	7865	256	26
12	8045	256	28
4	(714)		
4:10P	395	11	6
11	611	21	7
12	809	27	5
13	1007	27	3
14	1196	14	3
15	1385	335	4
16	1565	314	5
17	1745	314	5
18	1925	314	3
19	2105	312	2
20	2285	280	3
21	2465	286	4
22	2645	287	5
23	2825	287	6
24	3005	279	6
25	3185	272	6
26	3365	271	7
27	3545	270	8
28	3725	268	9
29	3905	260	10
30	4085	256	11
31	4265	251	12
32	4445	250	14

Date and time	Height in meters	Azimuth from; degrees	Veloc-ity m/s
June 4			
4:33P	4625	251	16
34	4805	251	17
35	4985	251	17
36	5165	255	17
37	5345	263	17
38	5525	263	18
39	5705	261	18
40	5885	261	19
41	6065	259	21
42	6245	257	21
43	6425	252	21
44	6605	252	20
45	6785	252	22
46	6965	252	22
47	7145	252	20
6	(715)		
4:03P	395	79	5
04	611	93	6
05	809	98	6
06	1007	103	5
07	1196	109	4
08	1385	151	3
09	1565	181	3
10	1745	196	3
11	1925	196	4
12	2105	196	5
8	(716)		
3:00P	395	9	2
01	611	353	1
02	809	351	1
03	1007	5	1
04	1196	171	1
05	1385	175	1
06	1565	175	1
9	(717)		
3:20P	395	228	5
21	611	227	3
22	809	207	1
23	1007	205	1
24	1196	193	1
25	1385	188	3
26	1565	188	4
27	1745	188	4
10	(718)		
10:37A	395	109	6

TABLE II (1929)

Date and time	Height in meters	Azimuth from; degrees	Velocity m/s
June 10			
10:38A	611	128	10
39	809	144	8
40	1007	149	7
41	1196	157	6
42	1385	178	4
43	1565	202	3
10	(719)		
4:04P	395	132	4
05	611	139	6
06	809	139	6
07	1007	135	4
08	1196	114	3
09	1385	120	4
10	1565	133	5
11	1745	149	6
12	1925	165	7
13	2105	183	8
14	2285	202	10
15	2465	210	11
16	2645	210	10
17	2825	210	10
11	(720)		
8:20A	395	176	5
21	611	158	3
22	809	152	5
23	1007	154	6
24	1196	157	7
25	1385	166	7
26	1565	174	7
27	1745	183	8
28	1925	188	7
29	2105	193	7
30	2285	190	6
31	2465	184	7
11	(721)		
3:47P	395	125	4
48	611	139	2
49	809	162	3
50	1007	167	3
51	1196	180	3
52	1385	190	4
53	1565	187	6
54	1745	179	8
55	1925	168	8
56	2105	161	8

Date and time	Height in meters	Azimuth from; degrees	Velocity m/s
June 11			
3:57P	2285	159	7
58	2465	161	5
59	2645	181	5
4:00P	2825	200	7
01	3005	203	8
02	3185	191	8
03	3365	184	9
04	3545	184	9
05	3725	184	10
06	3905	179	12
07	4085	173	15
08	4265	172	17
09	4445	170	18
10	4625	168	19
11	4805	166	19
12	(722)		
9:15A	395	111	5
16	611	120	7
17	809	124	8
18	1007	135	10
19	1196	147	14
20	1385	152	16
21	1565	159	16
22	1745	172	11
23	1925	203	6
24	2105	217	6
12	(723)		
4:00P	395	166	11
01	611	175	12
02	809	183	12
03	1007	188	15
04	1196	193	15
05	1385	190	15
06	1565	186	15
07	1745	187	13
08	1925	183	12
09	2105	181	12
12	(724)		
4:45P	395	165	11
46	611	187	10
47	809	189	13
48	1007	189	15
49	1196	188	15
50	1385	189	17
51	1565	190	21

Date and time	Height in meters	Azimuth from; degrees	Velocity m/s
June 12			
4:52P	1745	188	23
53	1925	184	25
54	2105	187	35
55	2285	188	39
56	2465	188	24
57	2645	188	9
58	2825	188	7
12	(725)		
9:59P	395	197	16
10:00P	611	196	20
01	809	193	15
02	1007	188	11
03	1196	187	14
04	1385	185	21
05	1565	182	24
06	1745	180	24
07	1925	179	16
08	2105	155	6
09	2285	124	5
10	2465	156	8
11	2645	169	12
13	(726)		
9:10A	395	176	19
11	611	183	20
12	809	189	19
13	1007	197	13
14	1196	202	11
15	1385	204	12
16	1565	204	15
17	1745	204	15
18	1925	204	16
19	2105	199	21
20	2285	196	36
21	2465	196	45
13	(727)		
4:22P	395	173	19
23	611	176	20
24	809	184	13
25	1007	192	11
26	1196	200	9
27	1385	201	10
28	1565	196	12
14	(728)		
9:28A	395	174	13
29	611	184	19

TABLE II (1929)

Date and time	Height in meters	Azimuth from; degrees	Velocity m/s
June 14			
9:30A	809	180	15
31	1007	174	14
32	1196	174	13
33	1385	183	9
34	1565	192	8
35	1745	192	11
36	1925	189	14
37	2105	187	16
14	(729)		
3:58P	395	173	13
59	611	198	7
4:00P	809	199	9
01	1007	197	10
02	1196	192	12
03	1385	194	13
04	1565	201	11
05	1745	203	9
15	(730)		
9:03A	395	195	5
04	611	198	5
05	809	198	5
06	1007	204	5
07	1196	212	5
08	1385	216	5
09	1565	219	6
10	1745	226	6
11	1925	222	7
12	2105	222	8
13	2285	226	9
14	2465	230	9
15	2645	232	9
16	2825	229	7
17	3005	218	5
18	3185	199	5
19	3365	194	5
20	3545	194	5
21	3725	193	5
16	(731)		
9:04A	395	350	3
05	611	359	3
06	809	19	3
07	1007	28	3
08	1196	18	2
09	1385	19	1
10	1565	38	1

Date and time	Height in meters	Azimuth from; degrees	Velocity m/s
June 16			
9:11A	1745	64	1
12	1925	79	2
13	2105	77	2
14	2285	70	1
15	2465	70	1
16	(732)		
4:05P	395	61	4
06	611	65	5
07	809	67	6
08	1007	71	4
09	1196	86	3
10	1385	108	2
11	1565	329	1
12	1745	356	2
13	1925	1	2
14	2105	346	3
15	2285	339	4
16	2465	339	4
17	2645	335	4
18	2825	324	4
19	3005	320	4
20	3185	320	4
17	(733)		
8:51A	395	68	5
52	611	39	4
53	809	36	5
54	1007	37	6
55	1196	50	6
56	1385	58	5
57	1565	58	4
58	1745	67	5
59	1925	79	5
9:00A	2105	91	5
01	2285	91	5
02	2465	96	5
03	2645	118	4
04	2825	158	3
05	3005	158	2
06	3185	142	3
07	3365	156	4
08	3545	156	4
09	3725	162	4
10	3905	169	3
11	4085	170	3
12	4265	175	2
13	4445	210	1

Date and time	Height in meters	Azimuth from; degrees	Velocity m/s
June 17			
9:14A	4625	224	1
15	4805	234	1
16	4985	242	2
17	5165	243	2
18	5345	283	3
19	5525	296	5
20	5705	318	9
21	5885	322	13
22	6065	322	15
23	6245	325	15
24	6425	324	17
25	6605	320	19
26	6785	318	20
27	6965	329	21
28	7145	322	22
29	7325	323	23
30	7505	330	22
31	7685	331	25
32	7865	329	25
33	8045	326	23
34	8225	324	25
35	8405	323	25
36	8595	322	27
37	8765	321	26
38	8945	320	25
39	9125	319	28
40	9305	319	31
41	9485	319	31
42	9665	319	29
43	9845	319	28
17	(734)		
3:54P	395	128	4
55	611	70	3
56	809	41	4
57	1007	33	5
58	1196	29	5
59	1385	44	4
4:00P	1565	71	3
01	1745	93	2
02	1925	169	3
03	2105	187	5
04	2285	183	7
05	2465	177	7
06	2645	172	8
07	2825	176	8
08	3005	176	6
09	3185	176	5

TABLE II (1929)

Date and time	Height in meters	Azimuth from; degrees	Velocity m/s
June 17			
4:10P	3365	171	5
11	3545	169	6
12	3725	169	5
13	3905	169	4
14	4085	180	2
15	4265	246	2
16	4445	272	2
17	4625	299	3
18	4805	319	4
19	4985	319	6
20	5165	320	8
21	5345	320	9
22	5525	318	11
23	5705	318	11
24	5885	318	11
25	6065	324	10
26	6245	327	11
27	6425	327	11
28	6605	329	11
29	6785	329	11
30	6965	326	11
31	7145	318	12
32	7325	314	12
33	7505	312	16
34	7685	312	20
35	7865	312	21
36	8045	314	23
37	8225	316	25
38	8405	316	25
18	(735)		
8:51A	395	170	2
52	611	167	2
53	809	247	1
54	1007	270	3
55	1196	285	3
56	1385	294	3
57	1565	280	2
58	1745	255	2
59	1925	240	2
9:00A	2105	239	3
01	2285	230	4
02	2465	216	5
03	2645	205	6
04	2825	207	6
05	3005	213	5
06	3185	222	5
07	3365	226	4

Date and time	Height in meters	Azimuth from; degrees	Velocity m/s
June 18			
9:08A	3545	210	4
09	3725	209	4
10	3905	217	4
11	4085	224	5
12	4265	236	4
13	4445	240	3
14	4625	254	4
15	4805	270	3
16	4985	263	4
17	5165	260	5
18	5345	259	5
19	5525	258	5
20	5705	268	5
21	5885	275	5
22	6065	271	7
18	(736)		
4:40P	395	275	7
41	611	284	5
42	809	311	3
43	1007	332	3
44	1196	351	3
45	1385	20	2
46	1565	54	1
47	1745	95	2
48	1925	123	2
49	2105	133	4
50	2285	164	4
51	2465	187	5
52	2645	213	6
53	2825	223	5
54	3005	221	7
55	3185	221	9
56	3365	212	10
57	3545	207	10
58	3725	201	11
59	3905	208	11
5:00P	4085	220	10
01	4265	229	9
02	4445	230	9
03	4625	234	9
04	4805	238	9
05	4985	236	9
06	5165	236	9
19	(737)		
9:17A	395	111	5
18	611	125	4

Date and time	Height in meters	Azimuth from; degrees	Velocity m/s
June 19			
9:19A	809	127	1
20	1007	53	1
21	1196	15	1
22	1385	356	1
23	1565	318	1
19	(738)		
4:20P	395	41	3
21	611	18	2
22	809	27	3
23	1007	353	3
24	1196	311	3
25	1385	296	3
20	(739)		
9:29A	395	343	6
30	611	325	3
30:30	700	302	1
20	(740)		
3:55P	395	340	7
56	611	354	13
57	809	4	14
58	1007	4	13
59	1196	341	5
4:00P	1385	212	6
01	1565	194	7
02	1745	196	4
03	1925	218	3
21	(741)		
8:55A	395	93	5
56	611	81	3
57	809	56	3
58	1007	55	3
59	1196	48	3
9:00A	1385	40	4
01	1565	36	5
02	1745	30	6
03	1925	24	6
04	2105	15	7
05	2285	12	7
06	2465	355	7
07	2645	341	7
08	2825	340	9
09	3005	336	11
10	3185	325	11
11	3365	321	13
12	3545	321	14

TABLE II (1929)

Date and time	Height in meters	Azimuth from; degrees	Velocity m/s
June 21			
9:13A	3725	321	15
14	3905	321	16
15	4085	321	17
16	4265	323	18
17	4445	335	18
18	4625	336	18
19	4805	330	20
20	4985	327	20
21	5165	323	19
22	5345	320	19
23	5525	320	21
24	5705	320	21
25	5885	317	21
26	6065	312	21
21	(742)		
3:45P	395	49	5
46	611	54	9
47	809	56	11
48	1007	59	12
49	1196	61	11
50	1385	60	3
51	1565	322	2
52	1745	28	3
53	1925	43	5
54	2105	43	4
55	2285	53	3
56	2465	54	2
57	2645	54	1
58	2825	342	3
59	3005	334	6
4:00P	3185	329	7
01	3365	320	8
02	3545	308	10
03	3725	302	12
04	3905	307	12
05	4085	307	13
06	4265	307	14
07	4445	310	16
08	4625	310	16
09	4805	301	16
10	4985	294	17
11	5165	295	17
12	5345	300	17
13	5525	300	18
14	5705	300	18
22	(743)		
8:55A	395	19	3

Date and time	Height in meters	Azimuth from; degrees	Velocity m/s
June 22			
8:56A	611	20	5
57	809	21	4
58	1007	21	1
59	1196	162	1
9:00A	1385	95	1
01	1565	71	2
02	1745	60	3
03	1925	62	2
04	2105	77	1
05	2285	57	1
06	2465	317	2
07	2645	303	2
08	2825	275	2
09	3005	248	4
10	3185	238	5
11	3365	225	4
12	3545	228	4
13	3725	236	5
14	3905	225	3
15	4085	222	6
16	4265	215	6
17	4445	213	5
18	4625	199	5
19	4805	190	5
23	(744)		
8:27A	395	38	4
28	611	40	4
29	809	41	5
30	1007	49	3
31	1196	73	1
32	1385	247	2
33	1565	218	3
34	1745	200	3
35	1925	208	5
36	2105	208	5
37	2285	204	5
38	2465	204	6
39	2645	191	6
40	2825	185	7
41	3005	185	7
24	(745)		
9:23A	395	192	2
24	611	136	3
25	809	155	5
26	1007	159	6
27	1196	162	6

Date and time	Height in meters	Azimuth from; degrees	Velocity m/s
June 24			
9:28A	1385	178	7
29	1565	192	9
30	1745	198	10
31	1925	205	12
32	2105	205	12
33	2285	200	11
34	2465	196	11
35	2645	189	13
36	2825	184	13
37	3005	182	11
38	3185	182	11
39	3365	186	9
40	3545	190	9
41	3725	190	10
42	3905	188	11
43	4085	188	13
44	4265	188	16
45	4445	193	18
46	4625	198	17
47	4805	206	13
48	4985	204	15
49	5165	208	18
25	(746)		
9:02A	395	104	1
03	611	74	2
04	809	39	3
05	1007	9	3
06	1196	353	3
07	1385	357	3
08	1565	20	1
09	1745	67	2
10	1925	85	3
11	2105	104	3
12	2285	137	3
13	2465	161	4
14	2645	181	5
15	2825	187	7
16	3005	187	8
17	3185	181	8
18	3365	174	7
19	3545	166	7
20	3725	174	8
21	3905	187	7
22	4085	200	9
23	4265	194	9
24	4445	184	10
25	4625	176	12

TABLE II (1929)

Date and time	Height in meters	Azimuth from; degrees	Velocity m/s	Date and time	Height in meters	Azimuth from; degrees	Velocity m/s	Date and time	Height in meters	Azimuth from; degrees	Velocity m/s
June 25				**June 26**				**June 27**			
9:26A	4805	170	12	9:10A	3725	177	7	9:11A	5165	117	5
27	4985	170	11	11	3905	177	9	12	5345	113	5
28	5165	170	11	12	4085	177	9	13	5525	112	4
29	5345	164	10	13	4265	177	9	14	5705	145	2
30	5525	160	10	14	4445	177	10	15	5885	197	2
31	5705	158	11	15	4625	180	10	16	6065	204	2
32	5885	158	11	16	4805	180	10	17	6245	214	3
33	6065	158	12	17	4985	182	9	18	6425	201	5
34	6245	158	15	18	5165	185	9	19	6605	195	7
35	6425	159	16	19	5345	185	10	20	6785	193	9
36	6605	162	15	20	5525	185	11	21	6965	197	9
37	6785	170	13	21	5705	188	9	22	7145	206	10
38	6965	172	11	22	5885	193	9	23	7325	217	9
39	7145	165	9	23	6065	191	9	24	7505	234	8
40	7325	165	12	24	6245	184	10	25	7685	245	7
41	7505	165	12	25	6425	181	12	26	7865	250	6
42	7685	165	13	26	6605	179	13	27	8045	245	6
43	7865	165	13	27	6785	175	12	28	8225	231	6
44	8045	165	11	28	6965	173	12	29	8405	229	7
45	8225	165	10	29	7145	173	11				
46	8405	166	9					**27**	(749)		
47	8585	166	11	**27**	(748)			3:50P	395	59	6
48	8765	173	13	8:45A	395	90	5	51	611	53	9
49	8945	173	14	46	611	83	3	52	809	53	8
50	9125	170	16	47	809	62	2	53	1007	56	5
51	9305	170	17	48	1007	60	1	54	1196	61	4
52	9485	170	17	49	1196	63	1	55	1385	59	4
53	9665	170	17	50	1385	73	1	56	1565	52	4
				51	1565	88	2	57	1745	52	3
26	(747)			52	1745	114	2	58	1925	51	3
8:52A	395	71	3	53	1925	95	2	59	2105	54	2
53	611	118	1	54	2105	62	2	4:00P	2285	84	2
54	809	92	1	55	2285	62	2	01	2465	100	2
55	1007	91	1	56	2465	66	2	02	2645	95	3
56	1196	34	1	57	2645	66	3	03	2825	105	3
57	1385	343	1	58	2825	81	2	04	3005	104	3
58	1565	5	1	59	3005	125	2	05	3185	93	3
59	1745	96	1	9:00A	3185	131	2	06	3365	126	5
9:00A	1925	111	3	01	3365	155	3	07	3545	140	7
01	2105	119	3	02	3545	185	3	08	3725	140	7
02	2285	126	3	03	3725	170	4	09	3905	144	5
03	2465	147	4	04	3905	172	4	10	4085	144	5
04	2645	158	6	05	4085	175	3	11	4265	141	5
05	2825	158	7	06	4265	163	2	12	4445	138	6
06	3005	159	7	07	4445	175	3	13	4625	131	5
07	3185	166	7	08	4625	162	5	14	4805	125	5
08	3365	166	9	09	4805	141	6	15	4985	119	7
09	3545	166	8	10	4985	121	6	16	5165	117	10

TABLE II (1929)

Date and time	Height in meters	Azimuth from; degrees	Veloc- ity m/s	Date and time	Height in meters	Azimuth from; degrees	Veloc- ity m/s	Date and time	Height in meters	Azimuth from; degrees	Veloc- ity m/s
June				**June**				**June**			
27				**28**				**28**			
4:17P	5345	116	12	9:49A	7505	128	11	4:48P	8405	185	9
18	5525	116	11	50	7685	128	10	49	8585	185	9
19	5705	119	10					50	8765	185	9
20	5885	128	8	**28**	(751)			51	8945	192	9
21	6065	128	8	4:04P	395	70	6	52	9125	195	9
22	6245	128	11	05	611	67	5	53	9305	201	9
				06	809	93	5	**29**	(752)		
				07	1007	136	2	9:03A	395	84	2
28	(750)			08	1196	162	2	04	611	122	2
9:10A	395	71	4	09	1385	226	2	05	809	160	3
11	611	59	4	10	1565	207	1	06	1007	177	5
12	809	82	3	11	1745	159	2	07	1196	188	7
13	1007	79	4	12	1925	159	2	08	1385	193	7
14	1196	91	4	13	2105	154	1	09	1565	202	6
15	1385	119	5	14	2285	134	2	10	1745	215	6
16	1565	132	6	15	2465	132	2	11	1925	222	7
17	1745	143	6	16	2645	193	2	12	2105	225	7
18	1925	146	5	17	2825	233	4	13	2285	225	6
19	2105	149	6	18	3005	236	5	14	2465	225	5
20	2285	150	8	19	3185	223	6	15	2645	206	6
21	2465	148	8	20	3365	217	7	16	2825	192	7
22	2645	149	8	21	3545	208	7	17	3005	180	7
23	2825	141	7	22	3725	204	7	18	3185	170	8
24	3005	139	6	23	3905	204	7	19	3365	172	9
25	3185	136	6	24	4085	201	9	20	3545	178	9
26	3365	130	7	25	4265	194	11	21	3725	181	11
27	3545	128	6	26	4445	194	9	22	3905	178	11
28	3725	130	7	27	4625	196	8	23	4085	178	12
29	3905	126	6	28	4805	196	8	24	4265	178	12
30	4085	117	5	29	4985	198	7	25	4445	178	12
31	4265	144	5	30	5165	199	7	26	4625	173	11
32	4445	164	7	31	5345	200	7	27	4805	173	13
33	4625	159	7	32	5525	197	7				
34	4805	146	8	33	5705	187	7	**29**	(753)		
35	4985	148	8	34	5885	177	6	4:33P	395	283	7
36	5165	148	7	35	6065	169	4	34	611	261	7
37	5345	141	7	36	6245	160	4	35	809	214	6
38	5525	136	9	37	6425	160	6	36	1007	198	7
39	5705	136	9	38	6605	166	7	37	1196	193	6
40	5885	136	10	39	6785	172	7	38	1385	188	8
41	6065	136	11	40	6965	176	6	39	1565	188	9
42	6245	135	11	41	7145	178	5	40	1745	195	9
43	6425	135	11	42	7325	176	7	41	1925	201	9
44	6605	134	11	43	7505	178	7	42	2105	197	9
45	6785	128	11	44	7685	187	7	43	2285	178	8
46	6965	124	11	45	7865	190	8	44	2465	164	8
47	7145	124	11	46	8045	188	8	45	2645	166	9
48	7325	128	12	47	8225	185	9	46	2825	166	9

TABLE II (1929)

Date and time	Height in meters	Azimuth from; degrees	Velocity m/s
June			
30	(754)		
8:55A	395	40	2
56	611	62	2
57	809	110	3
58	1007	122	3
59	1196	133	3
9:00A	1385	139	3
01	1565	141	4
02	1745	142	4
03	1925	145	4
04	2105	152	3
05	2285	175	3
06	2465	203	3
07	2645	208	4
08	2825	204	4
09	3005	200	5
10	3185	197	6
11	3365	200	7
12	3545	210	9
13	3725	216	10
14	3905	212	9
15	4085	206	8
16	4265	204	7
17	4445	203	7
18	4625	199	6
19	4805	196	5
20	4985	190	5
21	5165	183	4
22	5345	159	3
23	5525	137	4
24	5705	145	6
25	5885	159	6
26	6065	160	7
27	6245	161	6
28	6425	165	6
29	6605	165	6
30	6785	168	5
31	6965	181	4
32	7145	189	4
33	7325	195	5
34	7505	211	5
35	7685	215	5
36	7865	213	5
37	8045	216	5
30	(755)		
4:15P	395	78	7
16	611	80	5
17	809	82	4

Date and time	Height in meters	Azimuth from; degrees	Velocity m/s
June			
30			
4:18P	1007	82	2
19	1196	101	1
20	1385	123	1
21	1565	108	2
22	1745	103	4
23	1925	101	4
24	2105	112	3
25	2285	124	3
26	2465	141	2
27	2645	168	3
28	2825	195	3
29	3005	188	4
30	3185	182	5
31	3365	184	6
32	3545	179	9
33	3725	173	10
34	3905	165	10
35	4085	162	10
36	4265	162	8
37	4445	162	7
38	4625	177	7
39	4805	195	6
40	4985	197	7
41	5165	188	7
42	5345	186	7
43	5525	194	7
44	5705	195	7
July			
1	(756)		
8:55A	395	101	2
56	611	45	4
57	809	50	5
58	1007	57	5
59	1196	38	3
9:00A	1385	25	3
01	1565	40	2
02	1745	50	2
03	1925	45	2
04	2105	46	2
05	2285	71	2
06	2465	88	4
07	2645	105	5
08	2825	121	7
09	3005	122	6
10	3185	118	4
11	3365	135	3
12	3545	160	3

Date and time	Height in meters	Azimuth from; degrees	Velocity m/s
July			
1			
9:13A	3725	165	4
14	3905	166	5
15	4085	167	4
16	4265	164	4
17	4445	169	5
18	4625	166	5
19	4805	175	5
20	4985	191	7
21	5165	202	7
22	5345	204	6
23	5525	183	5
24	5705	152	5
25	5885	145	5
26	6065	141	8
27	6245	139	10
28	6425	135	8
29	6605	131	8
30	6785	129	8
31	6965	127	9
32	7145	126	9
33	7325	128	9
34	7505	130	9
35	7685	129	9
36	7865	128	11
37	8045	128	11
1	(757)		
4:06P	395	116	4
07	611	57	2
08	809	343	3
09	1007	334	5
10	1196	337	6
11	1385	341	7
12	1565	338	6
13	1745	339	5
14	1925	343	5
15	2105	16	3
16	2285	79	5
17	2465	56	5
18	2645	17	6
19	2825	13	7
20	3005	13	9
21	3185	18	9
22	3365	33	8
23	3545	32	8
24	3725	24	9
25	3905	22	9
26	4085	23	10

TABLE II (1929)

Date and time	Height in meters	Azimuth from; degrees	Velocity m/s
July 1			
4:27P	4265	23	11
28	4445	25	12
29	4625	33	13
30	4805	39	14
2	(758)		
8:51A	395	125	2
52	611	340	1
53	809	323	1
54	1007	357	1
55	1196	65	1
56	1385	85	1
57	1565	148	1
58	1745	177	2
59	1925	179	3
9:00A	2105	181	4
01	2285	181	5
02	2465	185	6
03	2645	185	6
04	2825	175	6
05	3005	186	4
06	3185	178	3
07	3365	156	4
08	3545	168	4
09	3725	183	5
10	3905	188	6
11	4085	191	6
12	4265	197	6
13	4445	184	5
14	4625	167	6
15	4805	173	6
16	4985	173	5
17	5165	159	4
18	5345	167	4
19	5525	174	5
20	5705	174	6
21	5885	162	6
22	6065	162	7
23	6245	162	7
24	6425	172	5
25	6605	170	3
26	6785	144	3
27	6965	144	5
28	7145	144	7
29	7325	144	9
30	7505	148	10
31	7685	149	11
32	7865	149	11

Date and time	Height in meters	Azimuth from; degrees	Velocity m/s
July 2			
9:33A	8045	149	9
34	8225	149	7
35	8405	153	8
36	8585	152	7
3	(759)		
4:10P	395	92	6
11	611	71	8
12	809	57	8
13	1007	47	8
14	1196	51	6
15	1385	32	4
16	1565	13	3
17	1745	74	1
18	1925	229	1
19	2105	229	1
20	2285	200	2
21	2465	164	2
22	2645	148	3
23	2825	148	4
24	3005	152	6
25	3185	156	7
26	3365	155	7
27	3545	147	7
28	3725	128	7
29	3905	115	5
30	4085	118	5
31	4265	118	5
4	(760)		
8:55A	395	70	3
56	611	102	3
57	809	157	3
58	1007	191	5
59	1196	202	6
9:00A	1385	218	5
01	1565	218	5
02	1745	204	4
03	1925	184	4
04	2105	173	6
05	2285	162	6
06	2465	142	5
07	2645	128	5
08	2825	130	5
09	3005	145	4
10	3185	155	4
11	3365	171	5
12	3545	178	4

Date and time	Height in meters	Azimuth from; degrees	Velocity m/s
July 4			
9:13A	3725	158	3
14	3905	149	4
15	4085	147	4
16	4265	142	5
17	4445	133	5
18	4625	117	6
19	4805	111	6
20	4985	125	7
21	5165	137	8
22	5345	141	8
23	5525	154	8
24	5705	152	7
25	5885	156	7
26	6065	156	8
27	6245	154	7
28	6425	153	7
29	6605	131	6
30	6785	98	6
31	6965	85	6
32	7145	94	5
33	7325	103	5
34	7505	81	5
35	7685	58	3
36	7865	45	4
37	8045	29	5
38	8225	359	5
39	8405	351	4
40	8585	345	5
41	8765	345	5
42	8945	345	5
43	9125	345	4
44	9305	349	2
45	9485	29	1
46	9665	29	1
47	9845	29	1
48	10025	29	2
49	10205	29	2
50	10385	29	2
51	10565	29	1
52	10745	170	2
53	10925	145	4
54	11105	117	4
55	11285	104	5
56	11465	107	6
57	11645	111	5
58	11825	110	5
59	12005	110	5
10:00A	12185	110	5

TABLE II (1929)

Date and time	Height in meters	Azimuth from; degrees	Velocity m/s	Date and time	Height in meters	Azimuth from; degrees	Velocity m/s	Date and time	Height in meters	Azimuth from; degrees	Velocity m/s
July 4				**July 4**				**July 6**			
10:01A	12365	110	5	4:15P	4265	131	8	9:01A	1385	86	5
02	12545	111	6	16	4445	127	6	02	1565	86	4
03	12725	111	6	17	4625	116	5	03	1745	91	5
04	12905	110	7	18	4805	109	6	04	1925	94	5
05	13085	100	7	19	4985	130	7	05	2105	94	5
06	13265	87	8	20	5165	130	7	06	2285	117	5
07	13445	74	8	21	5345	115	7	07	2465	131	6
08	13625	73	9	22	5525	115	8	08	2645	115	6
09	13805	73	11	23	5705	122	8	09	2825	94	6
10	13985	77	12	24	5885	122	9	10	3005	88	6
11	14165	84	14	25	6065	121	9	11	3185	81	7
12	14345	92	20	26	6245	121	10	12	3365	75	7
13	14525	102	24	27	6425	120	10	13	3545	82	7
14	14705	112	27	28	6605	121	8	14	3725	80	8
15	14885	110	29	29	6785	121	8	15	3905	80	9
16	15065	111	34	30	6965	121	7				
17	15245	117	37	31	7145	123	5	**7**	(763)		
18	15425	115	32	32	7325	125	5	3:55P	395	238	5
19	15605	113	32	33	7505	123	6	56	611	260	10
20	15785	113	41	34	7685	112	8	57	809	266	11
21	15965	113	43	35	7865	102	9	58	1007	278	7
22	16145	113	36	36	8045	93	9	59	1196	297	6
23	16325	113	36	37	8225	91	8	4:00P	1385	308	5
24	16505	113	40	38	8405	88	9	01	1565	4	3
				39	8585	79	9	02	1745	38	4
				40	8765	73	8	03	1925	22	5
4	(761)			41	8945	77	8	04	2105	14	3
3:54P	395	85	5	42	9125	89	7	05	2285	14	1
55	611	75	7	43	9305	96	5	06	2465	323	2
56	809	78	3	44	9485	102	4	07	2645	309	3
57	1007	92	3	45	9665	102	3				
58	1196	100	3	46	9845	96	3	**8**	(764)		
59	1385	86	2	47	10025	66	3	4:00P	395	8	7
4:00P	1565	65	2	48	10205	106	2	01	611	5	16
01	1745	43	2	49	10385	154	3	02	809	10	18
02	1925	87	2	50	10565	154	3	03	1007	14	18
03	2105	129	3	51	10745	154	2	04	1196	19	19
04	2285	150	5	52	10925	155	2	05	1385	21	21
05	2465	157	6	53	11105	154	3	06	1565	11	21
06	2645	158	6	54	11285	158	2	07	1745	357	18
07	2825	154	6	55	11465	158	1	08	1925	354	15
08	3005	140	4								
09	3185	117	4	**6**	(762)			**9**	(765)		
10	3365	113	6	8:56A	395	45	3	8:57A	395	75	3
11	3545	126	8	57	611	57	3	58	611	39	3
12	3725	138	8	58	809	83	4	59	809	47	3
13	3905	136	9	59	1007	83	5	9:00A	1007	52	4
14	4085	133	9	9:00A	1196	83	5	01	1196	52	4

TABLE II (1929)

Date and time	Height in meters	Azimuth from; degrees	Velocity m/s
July 9			
9:02A	1385	52	5
03	1565	49	6
04	1745	44	6
05	1925	44	6
06	2105	55	7
07	2285	79	7
08	2465	97	7
09	2645	108	6
10	2825	114	6
9	(766)		
3:50P	395	142	1
51	611	40	1
52	809	332	1
53	1007	340	1
54	1196	2	1
55	1385	1	1
56	1565	182	1
57	1745	184	1
58	1925	155	1
59	2105	168	3
4:00P	2285	149	1
01	2465	348	4
02	2645	345	5
03	2825	349	2
04	3005	5	2
05	3185	38	2
06	3365	43	2
07	3545	41	3
10	(767)		
9:32A	395	193	3
33	611	258	7
34	809	261	2
35	1007	225	4
36	1196	219	5
37	1385	211	6
38	1565	205	6
39	1745	210	5
40	1925	212	4
41	2105	195	3
42	2285	185	4
43	2465	185	4
11	(768)		
8:52A	395	355	5
53	611	1	5
54	809	7	5

Date and time	Height in meters	Azimuth from; degrees	Velocity m/s
July 11			
8:55A	1007	27	5
56	1196	40	6
57	1385	49	5
58	1565	69	3
59	1745	73	1
9:00A	1925	80	2
01	2105	90	3
02	2285	90	3
03	2465	95	4
04	2645	102	5
05	2825	104	5
06	3005	121	4
07	3185	156	4
08	3365	152	3
09	3545	136	4
10	3725	145	5
11	3905	157	5
12	4085	168	7
13	4265	174	7
14	4445	174	7
15	4625	174	8
12	(769)		
3:50P	395	336	2
51	611	15	1
52	809	1	1
53	1007	186	1
54	1196	188	1
55	1385	141	2
56	1565	74	3
57	1745	57	3
58	1925	335	5
59	2105	328	7
4:00P	2285	0	4
13	(770)		
8:52A	395	41	3
53	611	20	2
54	809	21	1
55	1007	48	1
56	1196	70	1
57	1385	50	2
58	1565	24	3
59	1745	9	4
9:00A	1925	357	4
01	2105	355	5

Date and time	Height in meters	Azimuth from; degrees	Velocity m/s
July 14	(771)		
9:03A	395	75	3
04	611	104	1
05	809	163	2
06	1007	180	3
07	1196	190	3
08	1385	190	2
09	1565	184	2
10	1745	226	2
11	1925	244	4
12	2105	237	5
13	2285	240	6
14	2465	240	5
15	2645	236	5
16	2825	231	4
17	3005	239	4
18	3185	251	3
19	3365	266	4
20	3545	267	4
21	3725	267	5
22	3905	267	5
23	4085	267	4
24	4265	267	5
25	4445	272	5
26	4625	279	4
27	4805	279	5
28	4985	279	6
29	5165	282	6
30	5345	282	7
31	5525	275	7
32	5705	273	9
33	5885	273	9
14	(772)		
4:05P	395	78	5
06	611	73	3
07	809	94	1
08	1007	140	2
09	1196	182	3
10	1385	197	3
11	1565	194	5
12	1745	193	7
13	1925	193	9
14	2105	193	8
15	2285	196	7
16	2465	203	7
17	2645	211	9
18	2825	208	9
19	3005	199	7

TABLE II (1929)

Date and time	Height in meters	Azimuth from; degrees	Velocity m/s
July 14			
4:20P	3185	194	5
21	3365	205	7
22	3545	209	7
23	3725	207	7
24	3905	209	7
25	4085	209	7
26	4265	207	8
27	4445	211	9
28	4625	218	9
29	4805	222	8
30	4985	226	7
31	5165	226	7
32	5345	221	9
33	5525	222	9
34	5705	225	9
35	5885	234	8
36	6065	232	9
37	6245	229	11
38	6425	230	12
39	6605	228	13
40	6785	230	15
41	6965	230	17
42	7145	230	17
43	7325	228	16
44	7505	228	15
45	7685	228	15
46	7865	226	16
47	8045	223	17
16	(773)		
8:52A	395	123	4
53	611	162	5
54	809	182	7
55	1007	190	9
56	1196	194	9
57	1385	194	7

Date and time	Height in meters	Azimuth from; degrees	Velocity m/s
July 16			
8:58A	1565	196	7
59	1745	205	7
9:00A	1925	209	9
01	2105	205	11
02	2285	201	13
03	2465	199	15
04	2645	197	16
05	2825	196	16
17	(774)		
9:40A	395	172	11
41	611	172	16
42	809	167	14
43	1007	167	14
44	1196	167	15
45	1385	165	17
46	1565	164	19
47	1745	161	20
48	1925	160	20
49	2105	160	20
18	(775)		
8:58A	395	109	1
59	611	104	1
9:00A	809	163	1
01	1007	182	2
02	1196	185	3
03	1385	188	4
04	1565	198	3
05	1745	198	3
06	1925	194	3
07	2105	210	3
08	2285	221	5
09	2465	212	4
10	2645	212	4

Date and time	Height in meters	Azimuth from; degrees	Velocity m/s
July 18			
9:11A	2825	212	5
12	3005	212	5
13	3185	217	5
14	3365	217	7
15	3545	224	8
16	3725	231	9
17	3905	237	10
18	4085	240	12
19	4265	240	13
20	4445	236	14
21	4625	236	15
22	4805	240	16
23	4985	243	18
24	5165	244	19
25	5345	246	19
26	5525	246	18
27	5705	248	23
28	5885	246	31
29	6065	239	35
30	6245	241	34
31	6425	243	33
19	(776)		
9:10A	395	92	4
11	611	115	5
12	809	145	5
13	1007	170	5
14	1196	184	7
15	1385	191	7
16	1565	186	7
17	1745	181	8
18	1925	185	8
19	2105	198	7
20	2285	211	6
21	2465	218	7
22	2645	218	7

VI. DIRECTION AND VELOCITY OF THE WIND FROM OBSERVATIONS OF PILOT-BALLOONS AND FROM THE ANEMOGRAPH AT MOUNT EVANS, FROM JULY, 1927, TO JULY, 1929

BY

CLARENCE R. KALLQUIST, July 21, 1927, to May, 1928

WILLIAM S. CARLSON, May 28 to July 10, 1928

LEONARD R. SCHNEIDER, July 11, 1928, to July, 1929

(Wind directions shown by arrows oriented with regard to cardinal directions; wind velocities in meters per second. Beneath each run there is given in order from top to bottom the time on 45th meridian, the air pressure at the surface in millibars with indication of falling (F), stationary (S) or rising (R) barometer; and the air temperature in degrees centigrade, with the same abbreviations for falling stationary or rising temperature. Elevations in thousand of meters.)

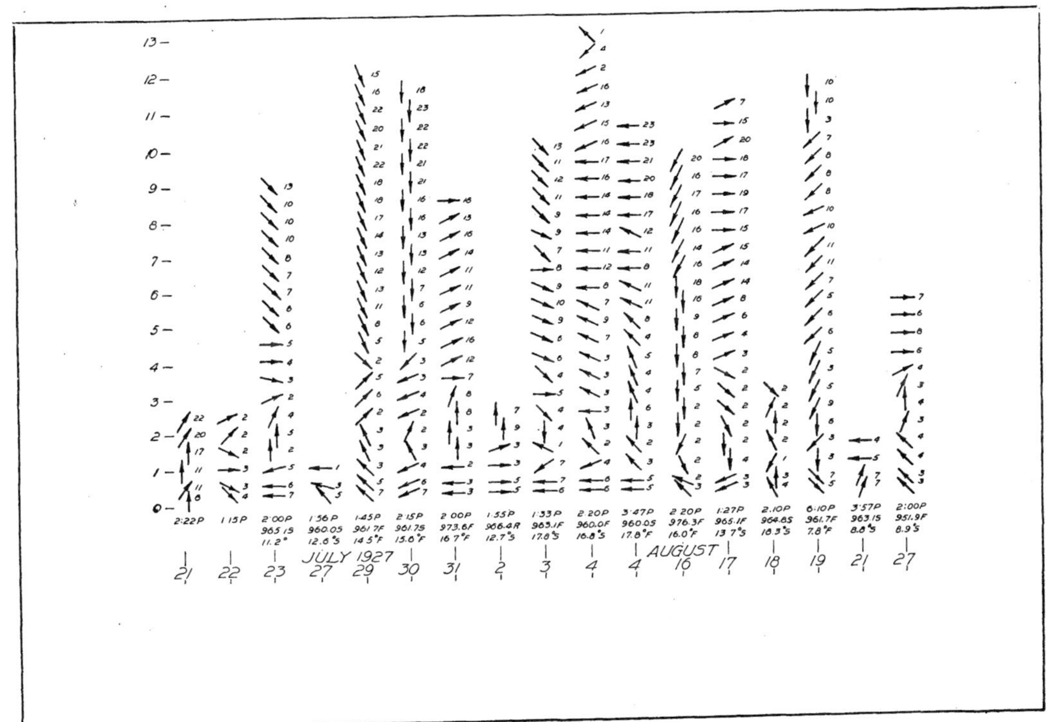

SEPTEMBER 1927

2 2 3 4 5 6 6 7 7 8 8 9 10 11 13 13 14 15 16 16 17

SEPTEMBER

17 18 19 20 21 22 23 24 25 25 26 27 28 29 30 1

OCTOBER

2 3 4 5 8

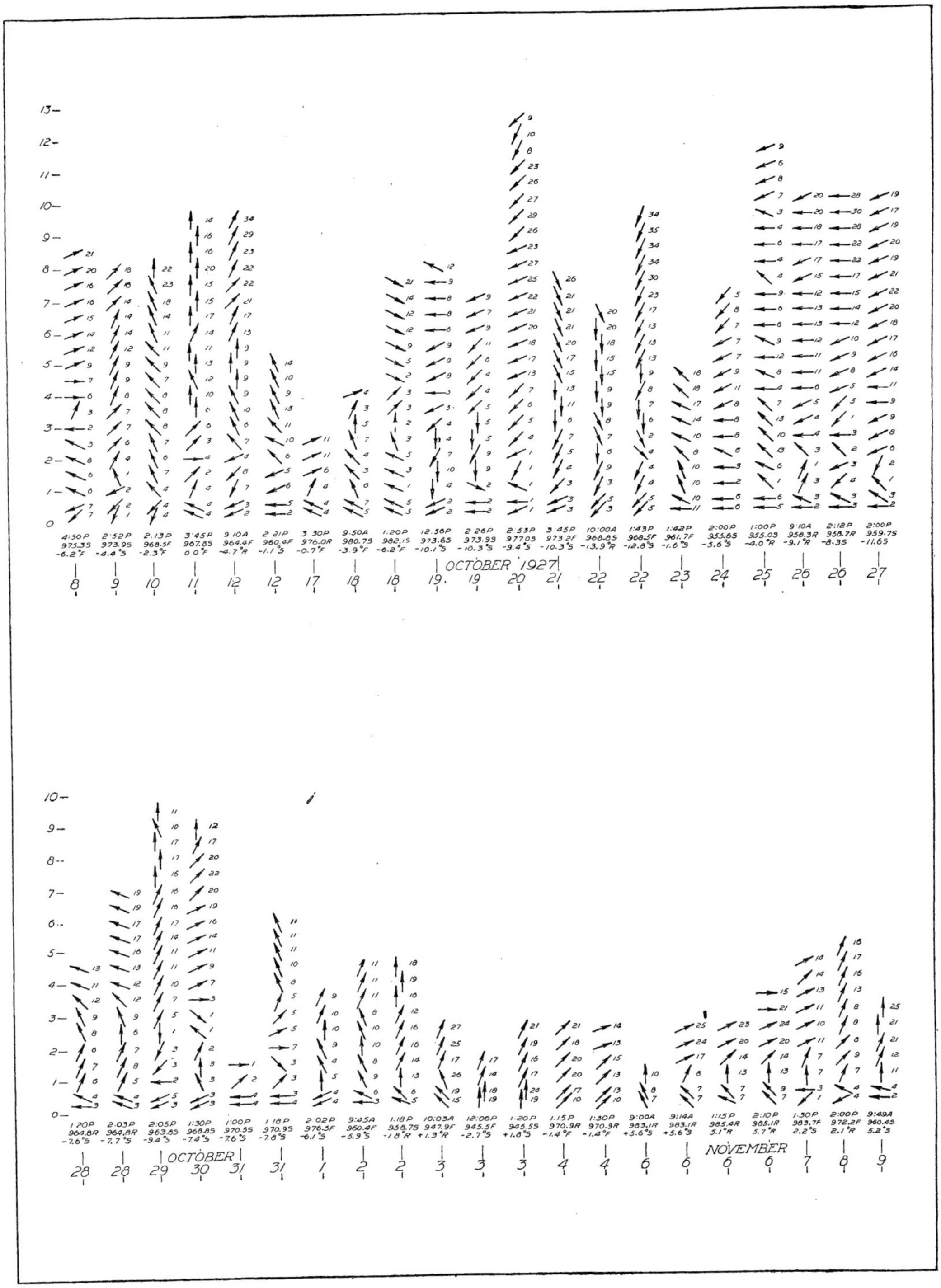

OCTOBER 1927
OCTOBER
NOVEMBER

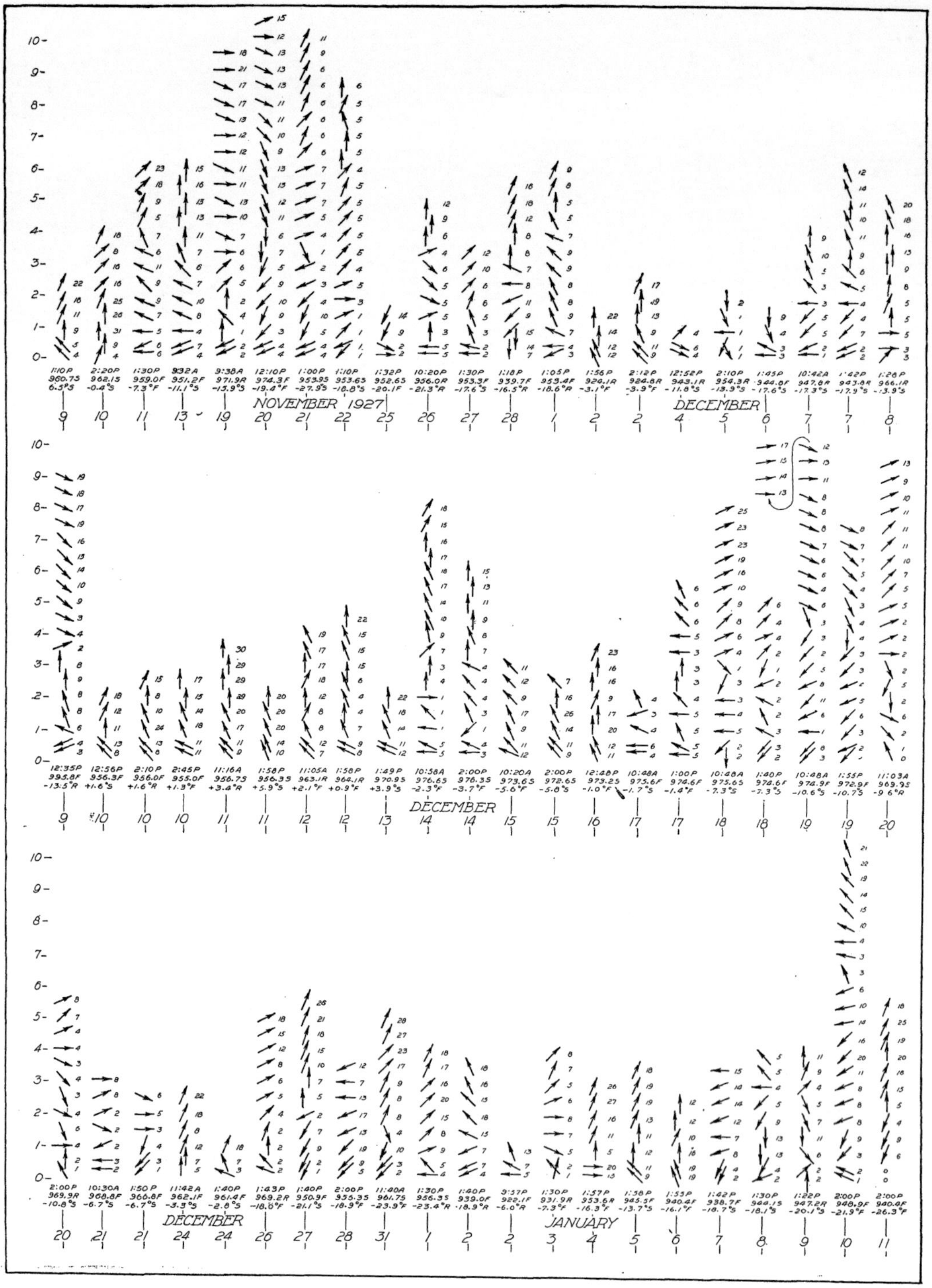

NOVEMBER 1927

DECEMBER

DECEMBER

DECEMBER

JANUARY

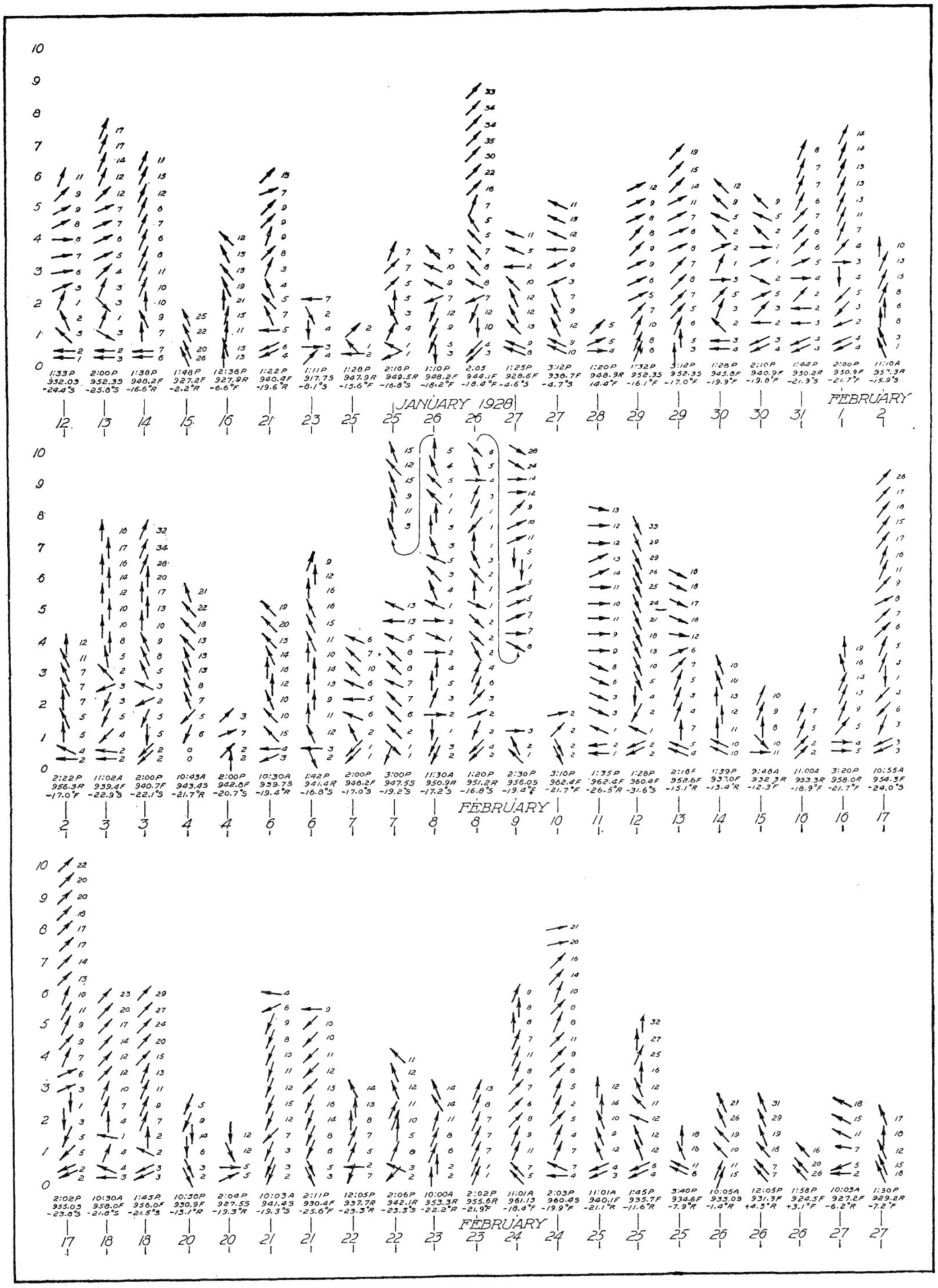

JANUARY 1928

FEBRUARY

FEBRUARY

FEBRUARY

FEBRUARY 1928

MARCH

MARCH

MARCH

APRIL

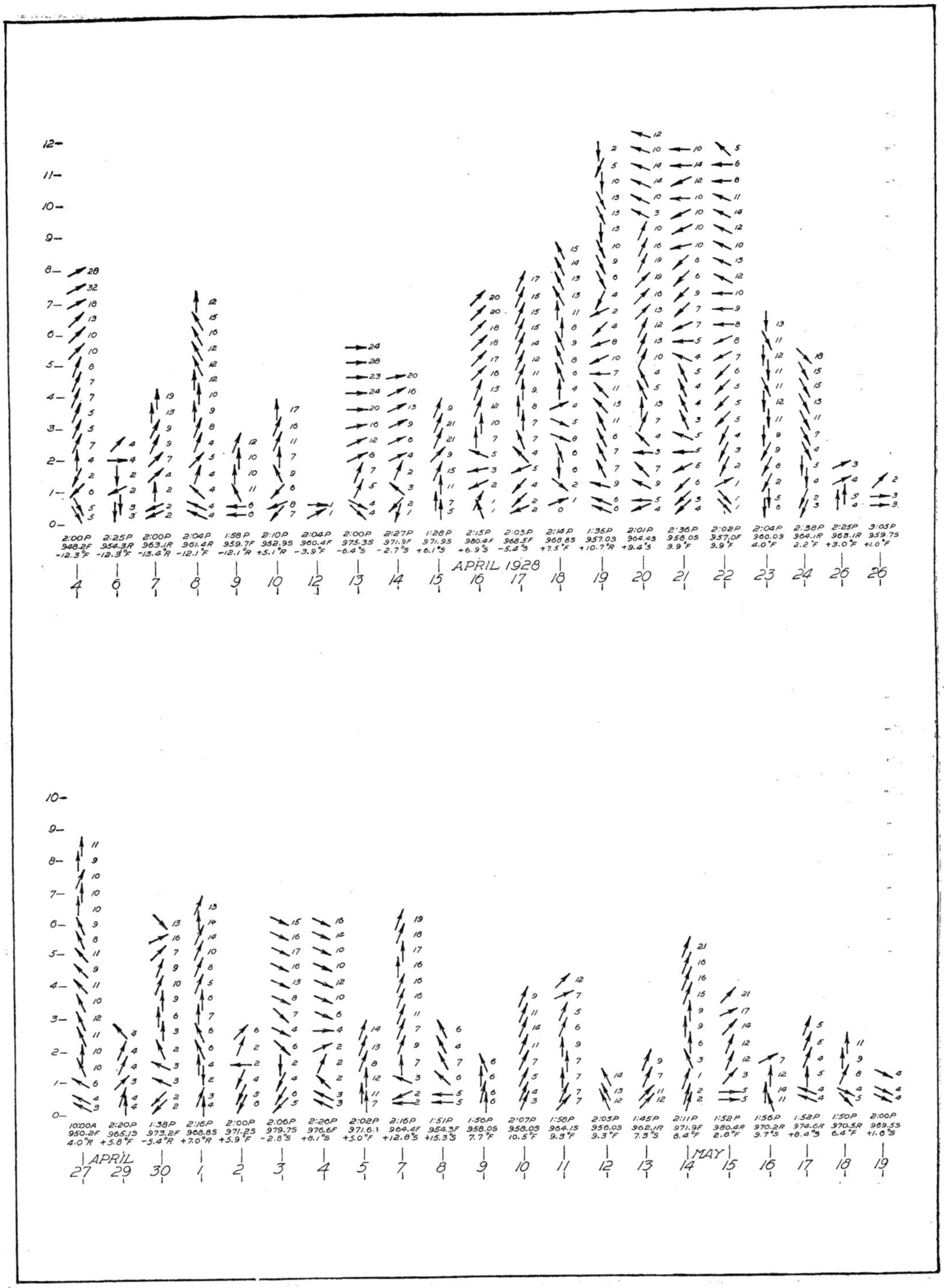

APRIL 1928

APRIL

MAY

JUNE 1928

JULY

JULY

AUGUST 1928

AUGUST SEPTEMBER

SEPTEMBER OCTOBER

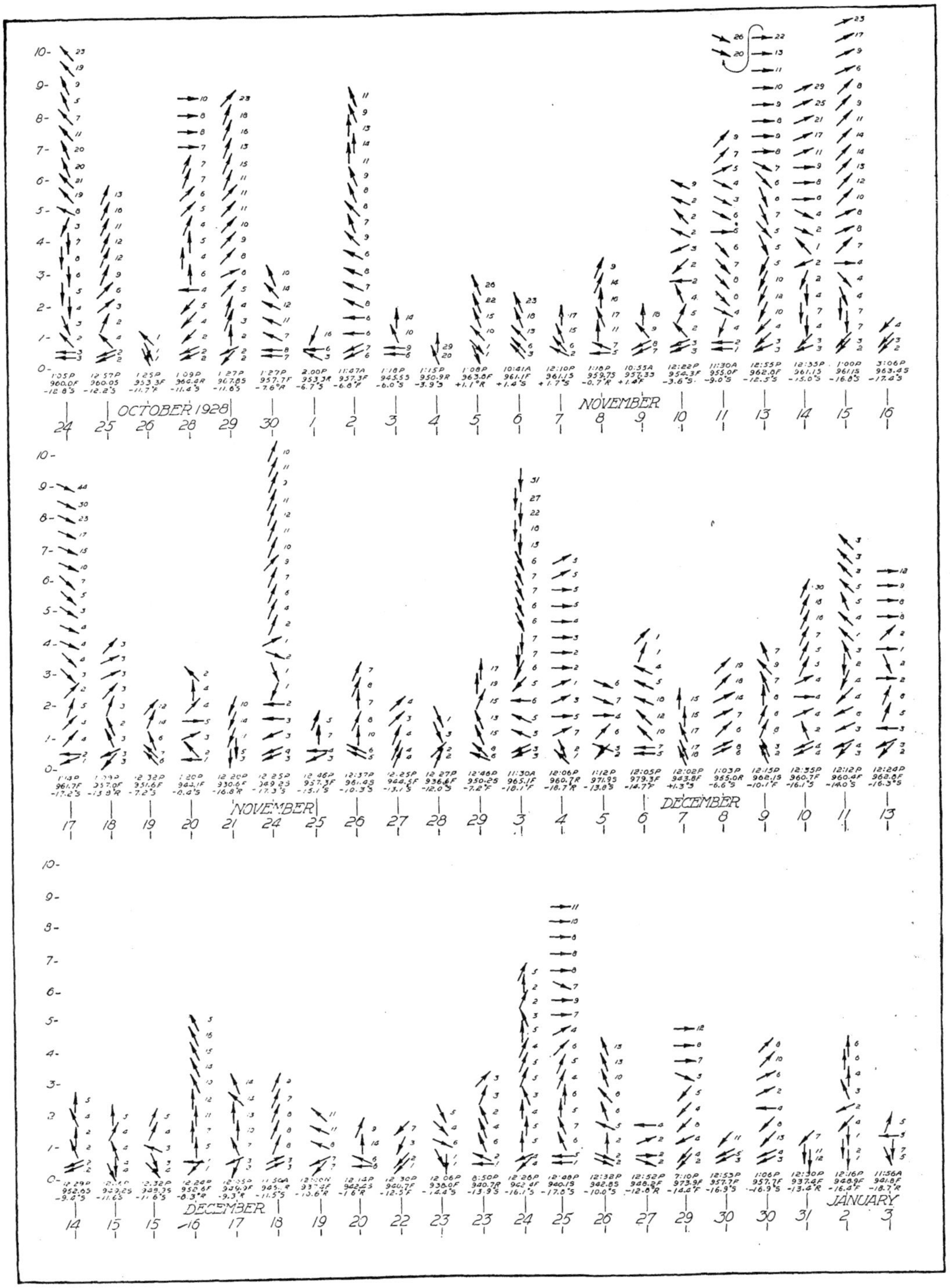

OCTOBER 1928 — NOVEMBER

24 25 26 28 29 30 1 2 3 4 5 6 7 8 9 10 11 13 14 15 16

NOVEMBER — DECEMBER

17 18 19 20 21 24 25 26 27 28 29 3 4 5 6 7 8 9 10 11 13

DECEMBER — JANUARY

14 15 15 16 17 18 19 20 22 23 23 24 25 26 27 29 30 30 31 2 3

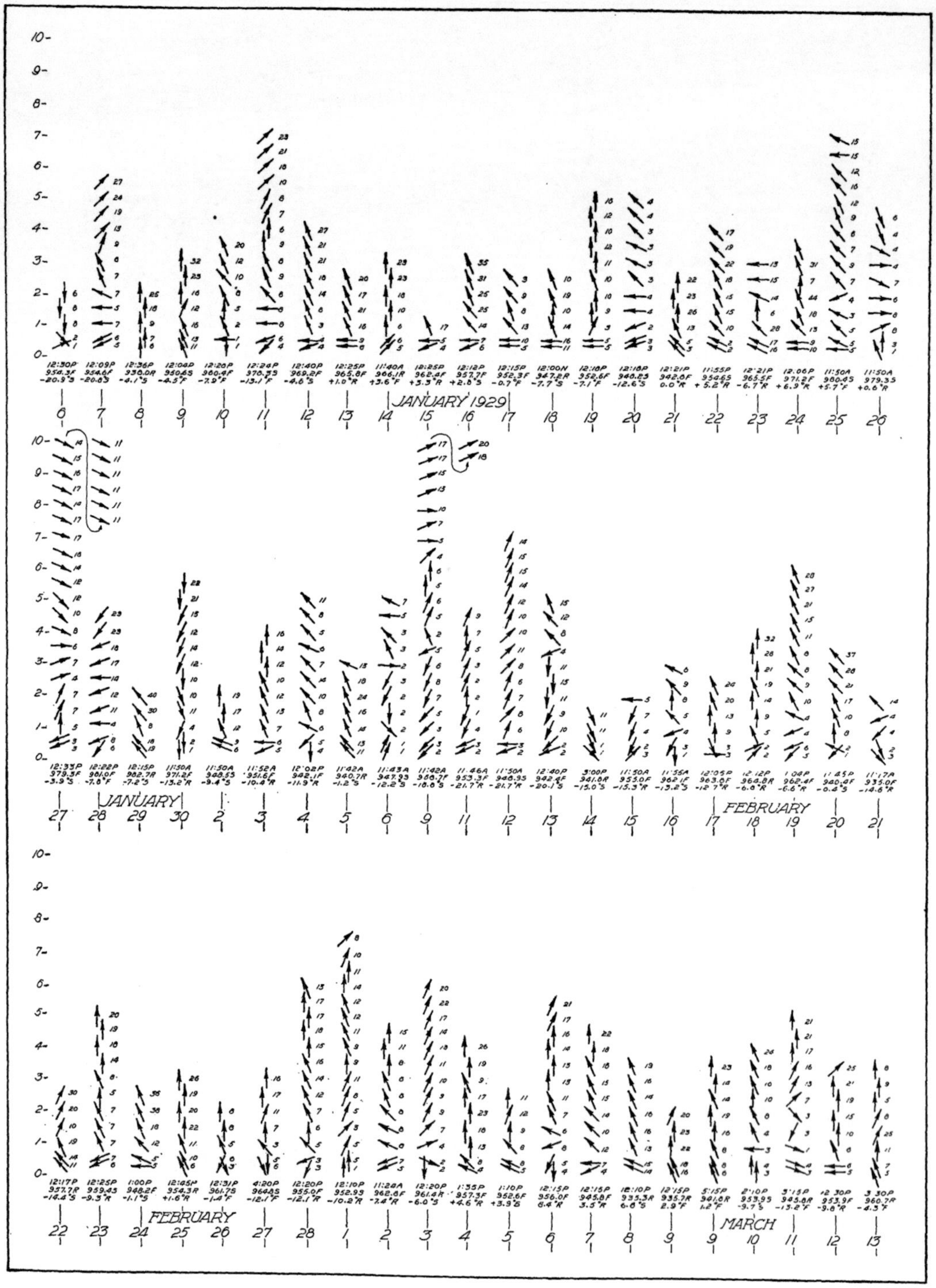

JANUARY 1929

JANUARY

FEBRUARY

FEBRUARY

MARCH

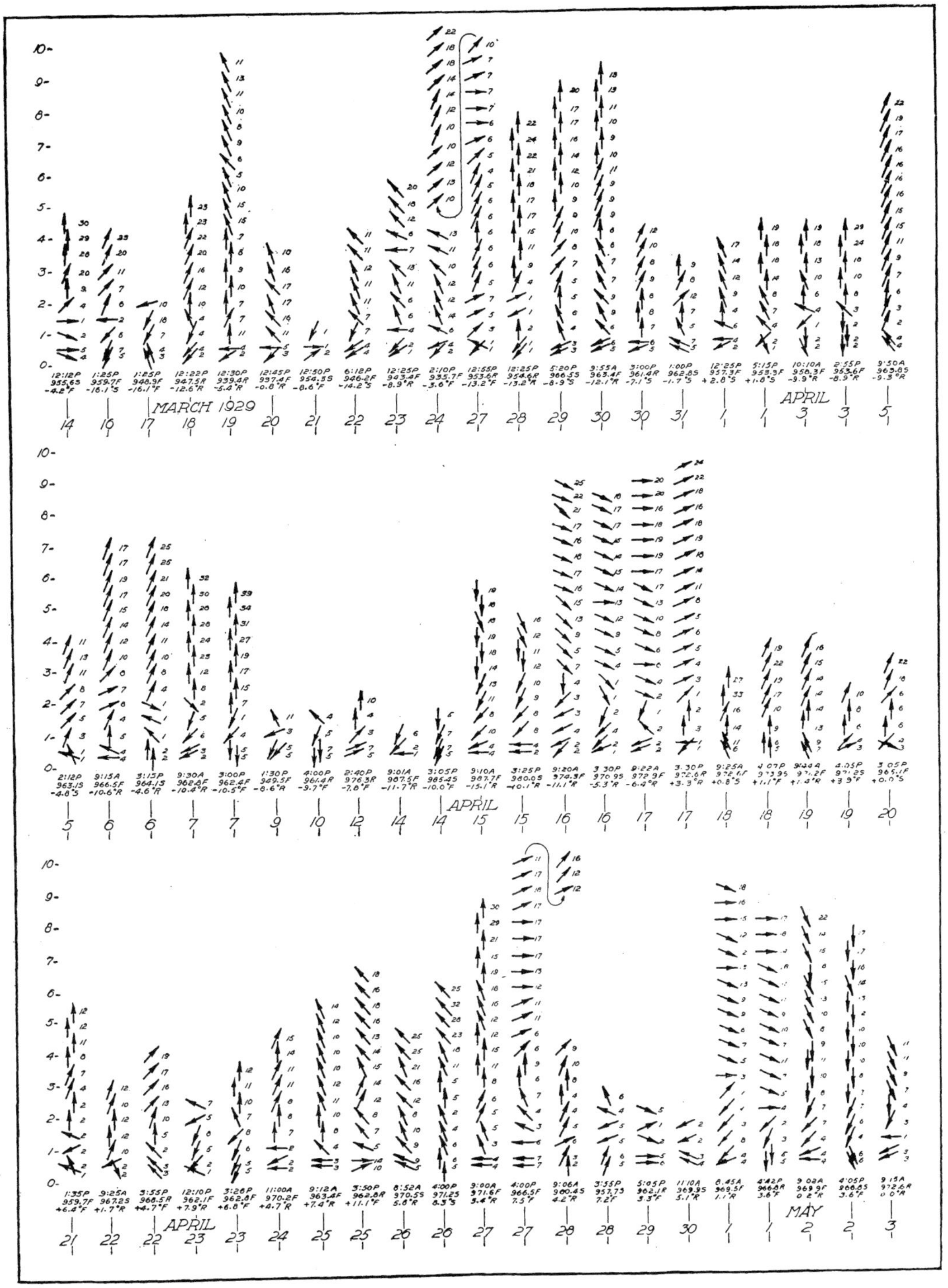
MARCH 1929
APRIL
APRIL
APRIL
MAY

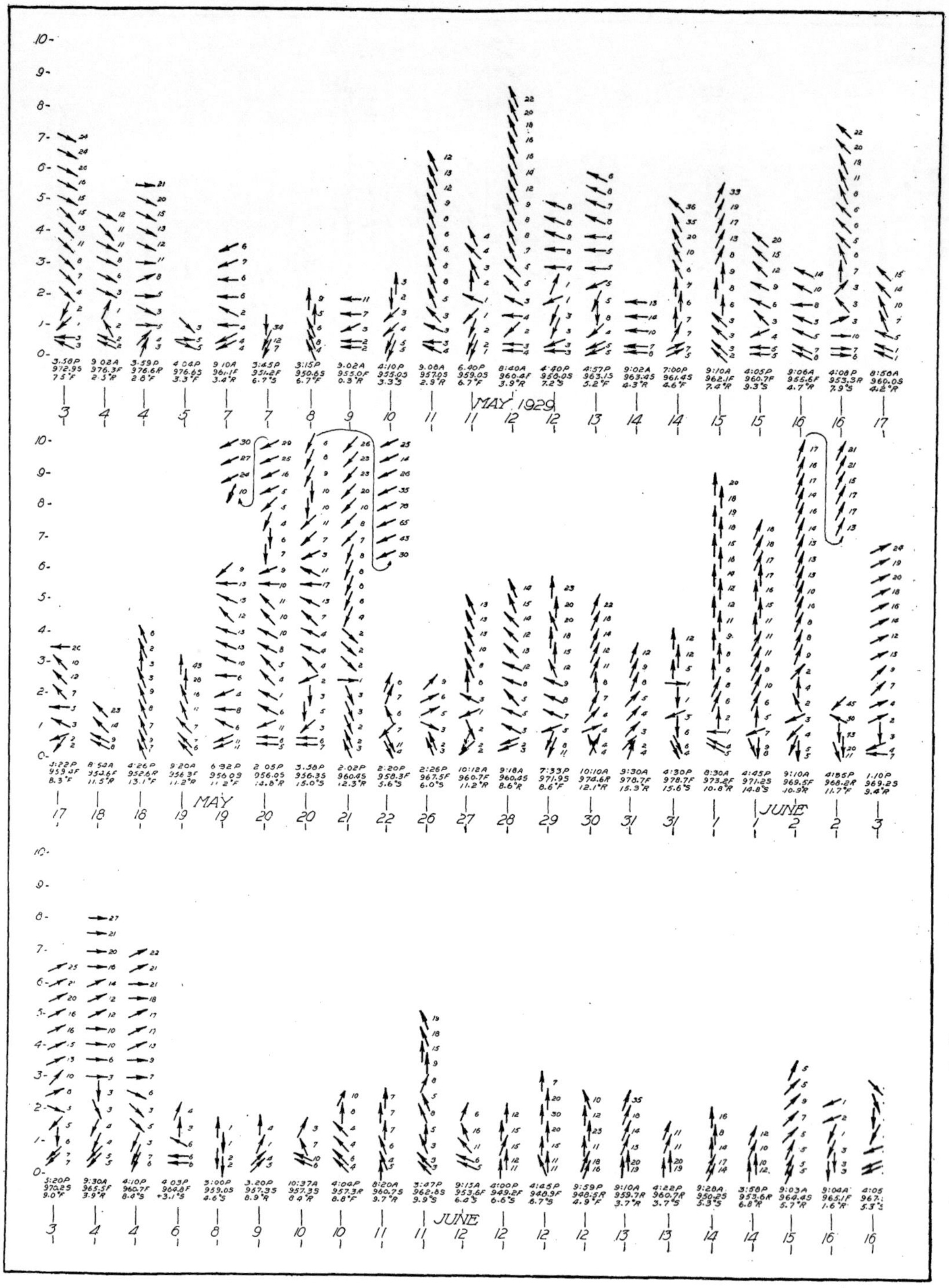

MAY 1929

MAY

JUNE

JUNE

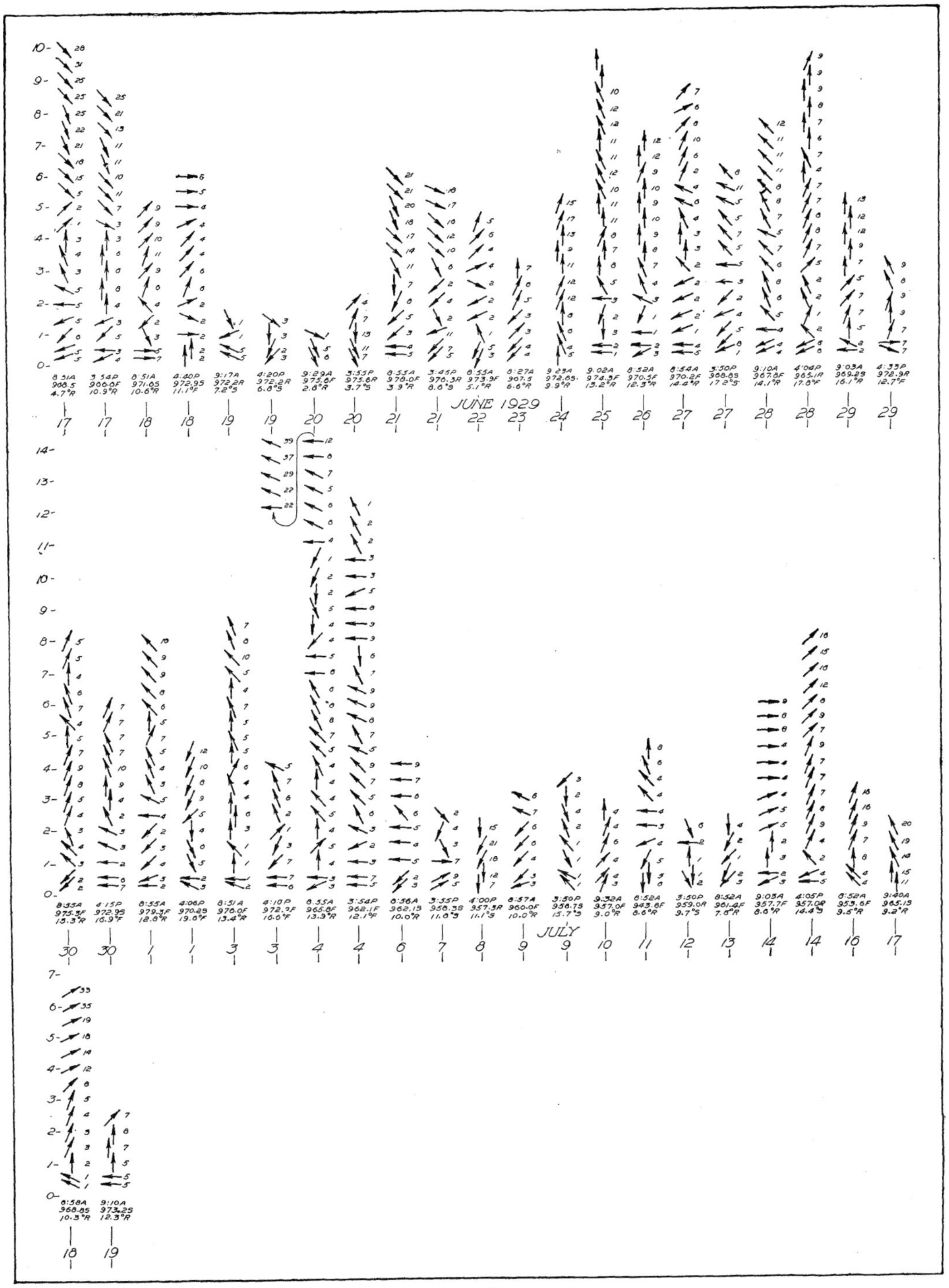

VII. FREE ATMOSPHERE DATA FROM ASCENSIONS OF *BALLONS SONDES* AT CAMP MICHIGAN, 1926

BY

S. P. FERGUSSON

TABLE III (1926)

Date and time, 1926	Height	Pressure	Temperature	Δt	Humidity		Wind	
				100m.	Relative	Vapor pressure	Direction	Velocity
	m.	mb.	°C	°C	per ct.	mb.		m/s
Aug. 4								
4:08P		1002.7	19.0		35	7.69	....	....
4:11	250		16.2		36	6.63	269	2.8
4:12	342	964.1	15.1	1.17	36	6.18	289	2.3
	500		13.4		36	5.54	288	1.6
	750		10.7		37	4.76	277	1.6
4:14	829	909.7	9.9	1.07	37	4.51	270	1.9
	1000		8.4		37	4.08	249	2.0
	1250		6.2		38	3.60	175	1.2
	1500		4.0		39	3.17	120	1.4
4:18	1894	798.3	0.5	0.88	40	2.53	86	1.7
Aug. 13								
1:05P	8	1002.0	12.8		62	9.17	....	....
1:06	138	984.6	9.0	2.92	72	8.27	225	0.3
	250		8.7		69	7.76	118	0.4
	500		8.0		61	6.55	230	1.6
1:10	523	941.6	7.9	0.29	60	6.39	211	1.7
1:11	750		6.6		63	6.14	176	1.2
	1000		5.2		66	5.83	150	1.4
	1250		3.8		69	5.53	75	1.8
1:17	1301	856.1	3.5	0.56	70	5.50	16	3.5
	1500		2.0		64	4.51	58	2.1
	2000		−1.9		48	2.51	64	1.6
1:20	2242	761.1	−3.8	0.78	40	1.78	45	2.5
	2500		−5.3		39	1.53	38	1.5
1:24	2835	705.9	−7.2	0.57	38	1.27	44	2.8
	3000		−7.8		39	1.24	55	3.4
	3500		−9.7		43	1.16	75	3.6
	4000		−11.6		47	1.07	83	2.0
1:33	4334	579.4	−12.9	0.38	50	1.01	79	1.0
	4500		−14.6		52	0.90	153	0.5
1:37	4965	534.8	−20.1	1.18	60	0.62	209	2.0
	5000		−20.1		59	0.61	215	2.3
1:39	5257	514.0	−20.1	0.00	50	0.52	213	2.5

VIII. OBSERVATIONS OF PILOT-BALLOONS NEAR MARGIN OF OTTO NORDENSKJÖLD GLACIER, JULY 5 AND 7, 1926

BY

WILLIAM HERBERT HOBBS

The Ascensions on August 5 were made at Camp Cooley 335 metres above sea-level and one kilometre distant from the ice; those on the 7th were made on the ice-cap about two kilometres from its edge, and from a height of 320 metres.

Date and time	Height in meters	Azimuth from; degrees N=0°	Velocity m/s
Aug. 5, 1926			
8:42A	335		
43	546	108	4
44	742	87	3
45	937	24	2
46	1122	91	3
47	1309	87	3
48	1485	99	5
49	1662	98	4
50	1839	141	3
51	2015	143	5
52	2193	139	5
53	2370	154	5
54	2547	145	5
55	2724	165	2
56	2900	132	5
57	3077	109	7
58	3255	98	5
59	3432	91	5
9:00A	3609	125	5
01	3787	133	4
02	3963	152	4
03	4139	148	3
04	4316	172	4
05	4493	170	5
06	4670	164	4
07	4847	169	5
08	5025	182	5
09	5202	188	4

Date and time	Height in meters	Azimuth from; degrees N=0°	Velocity m/s
Aug. 5			
9:10A	5378	172	5
11	5554	165	5
12	5739	167	8
13	5906	185	5
5			
1:48P	335		
49	1456	75	6
50	1625	69	3
51	1797	99	5
52	1969	111	5
53	2141	114	9
54	2313	118	10
55	2485	134	9
56	2657	137	11
57	2829	121	9
58	3001	113	9
59	3173	118	9
2:00P	3345	137	7
01	3517	140	8
02	3689	143	10
03	3861	150	8
04	4033	141	11
05	4205	122	10
06	4377	155	11
7			
3:40P	320		

Date and time	Height in meters	Azimuth from; degrees N=0°	Velocity m/s
Aug. 7			
3:41P	543	262	2
42	748	258	6
43	1163	124	1
45	1355	257	1
7			
3:59P	320		
4:00P	543	262	7
01	748	255	4
02	952	190	1
03	1148	102	2
04	1343	105	2
07	1900	115	4
08	2087	119	3
09	2378	151	7
7			
4:22P	320		
23	530	264	9
24	722	281	6
25	915	117	1
26	1098	99	2
27	1282	92	2
28	1457	129	5
29	1632	100	7
30	1807	33	2
31	1982	122	3
32	2155	71	1
33	2322	168	1

IX. PREVAILING KIND AND TOTAL AMOUNT OF CLOUD AT CAMP MICHIGAN, JULY TO SEPTEMBER, 1926

BY
S. P. FERGUSSON

TABLE IV

Prevailing Kind and Total Amount of Cloud at Camp Michigan

Date, 1926	A.M. 7	8	9	10	11	12	P.M. 1	2	3	4	5	6	7	8	9	10
July 21	AK 7	AK 9	AK 9	AK 9	AK 9	AK 9	AK 9	AK 9	AK 9	AK 9	AK 9	AK 8	AK 7	AK 7	AK 5	AK 2
22	SK ∞	SK ∞	SK ∞	SK ∞	SK 3	SK 5	SK 6	SK 6	SK 4	SK 6	AK 6	AK 7	AK 7	AK 8	AK 8	AK 7
23	AK 6	AK 5	AK 6	AK 6	AK 7	AK 6	AK 7	AK 8	AK 8	AK 9	AK 10	SKN 10	SKN 10	SKN 10	SN 10	SN 10
24	SKN 10	SKN 10	SN 10	fS 10	fS 10	fS 9	SK 9	SK 8	SK 8	SKN 10	SN 10	SN 10	SN 10	SN 10	......	......
25	......	......	S 4	fS 4	fS 4	fK–K 6	fK–K 6	KN 8	KN 10	SK 6	AK 6	......	......	......	......	......
26	C 3	C 3	C 2	CS 3	CS 4	CS 4	CS 2	CS 1	C ∞	C ∞	C ∞	C ∞	AK ∞	AK ∞	AK ∞	AK ∞
27	......	AK 5	AK 5	AK 5	AK 5	AK 5	AK 5	AK 4	AK 4	AK 4	CS 5	CS 5	CS 6	CS 6	CS 7	CS 6
28	AK 2	AK 3	AK 3	AK 4	AK 4	AK 4	C 2	C 1	C 1	C 2	C 5	C 5	......	......	AK 5	......
29	K ∞	K ∞	K 1	SK 3	SK 5	SK 6	SK 7	SKN 7	SKN 8	SKN 8	SKN 8	SKN 7	SKN 8	AK–SK 8	AK–SK 9	SKN 10
30	0	0	0	......	......	......	......	K ∞	......	......	......	AK 3	......	......	C 1	......
31	S 10	S 10	S 10	C 3	C 4	C 3	C 3	C 3	C 4	C 6	CS 7	CS 9	......	CS 9	......	......

TABLE IV (*Continued*)

PREVAILING KIND AND TOTAL AMOUNT OF CLOUD AT CAMP MICHIGAN

Date, 1926	A.M.						P.M.									
	7	8	9	10	11	12	1	2	3	4	5	6	7	8	9	10
August 1	....	ASN 10	ASN 10	AS 10	AS 10	CS 9	CS 9	CS 9	CS 9	C–CS 8	C–CS 8	C–CS 6	C–CS 6	C–CS 3	C–CS 2	C–CS 00
2	....	0	0	C 00	C 00	C 00	C 00	CK 00	0	0	0	CK 00	CK 00	0	0	0
3	....	AK 3	AK 3	AK 3	AK 4	AK 3	AK 1	AK 1	AK 00	AK 00	CK 00	CK 00	CK 00	CK 00	CK 00	....
4	....	0	0	0	C 00	C 1	C 00	C 1	C 1	C 1	C 1	C 00	C 00	C 00	C 1	C 1
5	C 2	CS 3	CS 3	CS 3	CS 2	C 1	C 00	C 1	C 1	C 00	C 1	CS 1	C 1	C 3	C 3	C 4
6	AK 5	AK 4	AK 5	AK 4	AK 5	CS 3	CS 5	CS 5	CS 4	CS 6	CS–AS 6	CS–AS 6	CS–AS 6	CS–AS 00	....	....
7	AK 4	AK 8	AK 9	AK 9	AK 9	AK 9	AK 6	AK 7	SK 9	SK 10	SK 10	SK 10	SK 10	SK 10	SKN 10°	SKN 10
8	SN 10	SN 10	SN 10	SKN 10	SKN 10	SKN 10	SKN 10	SK 10	SK 10	SK 10	SK 10	SK 10	SK 10	SK 10	SK 10	SK 10
9	AK–SK 10	AK–SK 10	AK–SK 9	AK 4	AK 3	AK 5	SK 7	SK 9	SKN 8	SKN 10	SKN 10	SKN 10	SKN 10	SKN 10	SKN 10	....
10	SK 9	SK–AK 7	AK–SK 7	AK 8	AK 7	AK 6	AK–SK 6	AK–SK 5	AK–SK 5	AK–SK 8	AK–SK 7	AK–SK 6	AK–SK 5	AK 2	AK 3	AK 3
11	S* 10	S 10	S 10	S 5	S 2	fS 00	fK 00	K 00	O	O	O	AK 00	AK 00	AK 00	AK 00	O 0

TABLE IV (*Continued*)

PREVAILING KIND AND TOTAL AMOUNT OF CLOUD AT CAMP MICHIGAN

Date, 1926	A.M.						P.M.									
	7	8	9	10	11	12	1	2	3	4	5	6	7	8	9	10
August 12	AK	AK	AK	AK	AK	AK	AK	AK	AK	AK	AK	AK				
	00	00	00	00	00	00	3	4	3	2	2	1	......	......	......	......
13	S	S	fS	fS	C	C	C	C	C	C	C	AK	AK	AK	C	C
	10	10	9	00	00	1	1	1	3	3	1	2	3	2	1	1
14	S	S	AK	AK	AK	AK	AK	AK	AK	AK	AK	C	C	C	C	C
	10	10	7	7	7	7	6	6	6	6	6	5	5	6	5	5
15	S	AK	AK	AK	AK	AK	AK	AK	AK	SK	SK	SK	SKN	SKN	SKN	SK
	10	10	10	10	10	10	9	10	10	10	10	10	10	10°	9°	7
16	0	SK	SK	SK	SK	K	K	K	K	K	K	SK	0	0	0	0
		00	00	00	00	00	00	1	1	1	00	00				
17																
	0	0	0	0	0	0	0	0	0	0	0	0	0	0	0	0
18										SK	SK	SK	SK			
	0	0	0	0	0	0	0	0	0	00	1	1	00	0	0	0
19					C	C							fS	fS	fS	fS
	0	0	0	0	00	00	0	0	0	0	0	0	00	1	1	1
20	AK	AK	AK	AK	AK	SKN	SKN	SKN	SKN	SKN	SKN	SKN	SKN	SKN	SKN	SKN
	6	6	6	8	10	10	10	10	10	10	10	10	10	10	10	10
21	SKN	SKN	SKN	SKN	SKN	SKN	SKN	SKN	SKN	SKN	SKN	SKN	SKN	SKN	SKN	SKN
	10	9	10	10	10	10	10	10	10	10	10	10	10	10	10	10
22	......	SK	SK	SK	SK	SK	SK	SK	SK	SKN	SKN	SKN	SKN	SKN	SKN	......
		5	7	6	6	7	7	8	9	9	9	9	9	10	10	......

TABLE IV (*Concluded*)

PREVAILING KIND AND TOTAL AMOUNT OF CLOUD AT CAMP MICHIGAN

Date, 1926	A.M.						P.M.									
	7	8	9	10	11	12	1	2	3	4	5	6	7	8	9	10
August 23	....	SKN 9	SKN 10	SKN 8	SKN 8	SKN 7	SKN 7	SKN 7	SKN 9	SKN 7	SKN 6	SKN 6	SKN 6	SKN 6	SKN 6	SKN ....
24	....	C 2	C 5	C 9	CS 10	CS 10	CS–CK 10	CS–CK 10	vCS 10	vAS 10	AS 10	ASN 10	ASN 10	ASN 10	ASN 10	 ....
25	....	SN 10	SN° 10	SN 10	SN 10	SN 10	SN 10	SN 10	SN 10	SN 10	SN 10	SN 10	SN 10	SN 10	SN 10	....
26	SN 10	SN 10	SN° 10	SN 10	SN 10	SN 10	SN 10	SN 10	ASKN 10	ASKN 10	ASN 10	ASN 10	ASN 10	ASK	ASK	....
27	K 5	K 5	K 6	K 6	K 5	K 4	K 4	K 5	K 4	K 3	K 4	K 6	AK 7	AK 8	AK 10	....
28	SK–fK oo	SK–fK oo	fK oo	SK–fK oo	SK–fK oo	SK–fK oo	SK–fK oo	SK–FK oo	C oo	C oo	C oo	C oo	C oo	AK 3	AK 9	....
29	AK 4	CK–AK 2	CK–AK 3	CK–AK 3	CK 1	CK oo	CS? oo	CS oo	CS 1	C 1	C 2	C 4	C 3	C 3	C 5	C 6
30	C 5	C 6	C 7	C 7	C 5	C 4	C 2	C 2	C 2	C 1	C oo	SK oo	o	o	o	o
31	o	o	o	C oo	C 2	C 4	C 4	C 4	C 4	C 4	C 3	C 4	C 4	C 3	C 1	....

X. OBSERVATIONS OF CLOUDS, IN DETAIL, AT CAMP MICHIGAN, JULY TO SEPTEMBER, 1926

BY

S. P. FERGUSSON

TABLE V

Observations of Clouds, in Detail, at Camp Michigan

Date and time, 1926	Level 1–5	Kind	Density 0–4	Amount 0–10	Azimuth from	Velocity: Relative, mm/m	Velocity: Absolute, m/s	Position	Height in meters
July 21									
11 A.	3	AK, SK	3–4	9	330	22		N^{d}	
11	5	S, fS	0	00	0?	20?		NW	900
6:30 P.	3	AK	4	8	292	20		W^{d}	
6:30	5	fS	1–2	00				SW^{d}	
9	3	AK	2–4	8	288	10		N	
9	4	SK		00				W^{d}	
* **22**									
8 A.	4	SK	0	00				SW^{h}	
11	4	K	0	00				NW	
1 P.	3?	AK	2–4	00	250	4	4	W	1100
1	4	SK	1	5	289	7		SW^{d}	
3	3?	AK, SK	4–1	6	245	7		SW^{d}	
3	5	fS	1	00	224	60		S^{d}	
6	2?	CS, CK	1	00	247	26		N^{d}	
6	?	AK–SK–fAK	0–4	6	245	19	4	SW^{d}	1900
23									
10 A.	3	AK, SK	3–1	7	288	8			
1 P.	3	AK, SK	3–1	5	310	5			
1	5	fS	3	00	192	60		S^{d}	
2	4	K	3	2				N	800
5	4	SKN	1–2	9?	WNW	slow		S^{d}	
5	5	S, fS		2	S				300
8	4	SKN	4	?					
8	5	S, fS		9	S				250
* **24**									
8 A.	5	S, fS	4	10					
12 M.	4	SK	0–2	?					
12	5	S, fS		7	S				200
5 P.	4	SK	0–2						
5	5	fS	3	7	S			N^{d}	300
8	4	SK	?	?				S	
8	5	fS, SK		9	192	30		N^{d}	
* **25**									
9 A.	3?	AK, SK		00				SW^{h}	
9	5	S, fS	0–2	4	60	16		E^{d}	
9	"	"	"	"	78	45		"	
9	5	K, fK		00	?			NW	
12 M.	?	C, CK	2	00	328	34		W	
12	3	AK	00	00	4	28		N	
12	4?	fK,K		6	240				

TABLE V (*Continued*)

Date and time, 1926	Level 1–5	Kind	Density 0–4	Amount 0–10	Azimuth from	Velocity		Position	Height in meters
						Relative, mm/m	Absolute, m/s		
July									
* **26**									
7:30 A.	1	C, CS	0–3	3	328	8		NE^{d}	
11	1	CS, C	3–2	3	315	8		NE^{d}	
2 P.	1	C						E^{h}	
2	3	AK	1	00				SW^{h}	
2	4	K	1	00				NW	
5	3	AK	1	00				W^{h}	
8	3	AK		00				W^{h}	
27									
8 A.	1	C	1–2	00	248	10?		SE	
8	3	AK		6	234	12		SW^{d}	
11	1	C	1–3	1	259	14		S^{d}	
11	3?	AK	0–2	6	250	15		S^{d}	
3 P.	1	C	1–3	2	274	13		E^{d}	
3	2	AK	0–3	4	260	14		SW^{d}	
5	1	C	0–2	3	269	10		SW^{d}	
5	2	CK–AK		3	260	12			
9	1	C, CK	3–4	2	264	11		N^{d}	
9	1	CS	0–3	5	256	10			
9	3	AK	1–2	00	150	20		W^{d}	
* **28**									
8 A.	3	AK	0–2	3	295	10			
8	5	S, fS	0	00	300	60		S	300
11	2	CK, AK	0–2	4	300	10		S^{d}	
11	3	AKt	0–1	2	321	5		N^{d}	
11	5	S, fS	0–2	00	193	45			
11	"	"	"	"	230	110		S^{d}	300
6 P.	1	C, CS	0–2	5	292	12			
6	2	AK	0–1	00	10	3		N	
9:35	3	AK	0–1	4	78	9		N^{d}	2600
9:35	3	AK	0–2	2	29	7		NE^{d}	
* **29**									
9 A.	5	K, fK	1–3	1				NW^{d}	
9	5	fK	0–2		324	18		NW^{d}	
4 P.	5?	SKN–fSK	0–2	8	320	10		N^{d}	
4	"	"	"	"	304	20		"	
9	3?	AK, SK	0–2	7	5	12		NE^{d}	
9	"	"	"	"	65?	5		"	
30									
9 A.		O							
2 P.	4	K	0–1	00		0		N^{d}	1600
6	1	C	0–2	2	357	30?		W^{d}	
6	3?	AK–SK	0–2	2	282	9		N^{d}	
6	"	"	"	"	40	6?		"	
9	1	C	0–3	1	353	24		W	

TABLE V (*Continued*)

Date and time, 1926	Level 1–5	Kind	Density 0–4	Amount 0–10	Azimuth from	Velocity Relative, mm/m	Velocity Absolute, m/s	Position	Height in meters.
July									
* **31**									
9 A.	5	S		10					
11	1	C–CS, CK	0–2	3	320	21		SW^d	
11	5	fSK	0	00				W^h	
4 P.	1	C, CS, CK	0–3	6	311	21		W^d	
4	4	K	0–1	00				NW^h	970
6	1	C, CS	0–4	9	310	18		W^d	
6	4	fSK	0–1	00	140	19		S^d	
8:20	1	C–CK–CS	0–4	9	299	19		SW^d	
8:20	4	fSK	0–3	3	152	20		NW^d	
August									
* **1**									
9 A.	3	ASN	0–1	10					
11	1	C, CS	0–1	8					
11	3	ASN	0–1	6					
2 P.	1	C, CS, CK	0–2–3	9	280	?		W^d	
2	4	K	0–1	00				W^h	
6	1	C, CK	0–3	5	274	16		W^d	
6	2	AK	0–1	3	80	6		S^d	
6	4	fK	0–1	00				NW^h	
9:30	1	C–CS	0–3	3	275	15		NW^d	
9:30	3	CK, AK	0–1	00	75	6		NW	
2									
9 A.	1	C	0–2	00	W?			W^h	
12 M.	1	C	0–3	00				SW^h	
5 P.	1	C–CS		00				SW^h	
9		0		0					
* **3**									
9 A.	3?	CK, AK	0–2	3	SW?			S^d	
12 M.	3?	CK–AK	0–2	4	208?	5		S^d	
5 P.		0		0					
8	2?	CK–AK	0–1	00	240	2?		E^h	
4									
9 A.		0		0					
11	1	C	0–1	00				W^h	
2 P.	3	AK		00				W^h	
6	1	C	0–2	1	192	4		W^d	
6	3	AK	0	00				W^h	
9:30	1	C–CS	0–2	1	195	3		NW	
5									
9 A.	1	C–CS	0–2	3	246	28		SW^d	
2 P.	1	C–CS	0–3	3					
9	1	C–CS	0–3	3	198	30		SW^d	
9	2	CK–AK	0–1	00	172	15			

TABLE V (*Continued*)

Date and time, 1926	Level 1–5	Kind	Density 0–4	Amount 0–10	Azimuth from	Velocity		Position	Height in metres
						Relative, mm/m	Absolute, m/s		
August									
* **6**									
8 A.	3	AK–AS	0–2	4					
11	3	AK		5	174	30			
11:30A.	2?	CS–AK	0–1	1	155	29		Sd	
11:30	4	K	0–1					NWh	1400
3:30P.	1	C, CS	2–4	3?	124	28		Sd	
2:30	2?	CS,CK	2–4	2	130	27			
2:30	4	K	0–1	00				NWh	
7	1?	C–CS–CK	0–4	6	125	22		Wd	
7	2?	AK, CK	0–1	00	132	19			
7	3?	vCS–AS		00				Wh	
7	4?	SK	2	00				SWh	
7									
11 A.	1	C	0–2	00					
11	3	AK	0–4	9	142				3400
11	4	SK	1	00				SWh	
11	5	fS	0	00	170	130			
6 P.	4	SK	1–2	10	135	30			
6	5	S, fS		4	175	90	2	Ed	250
* **8**									
9 A.	4?	SK		10	....	0			
11	5	S, fS		7	170?			SEd	270
8:30P.	4	SK?	0–4	10	?	0?		Ed	
8:30	5	S, fS	0–1	4	355	4	1	Nd	350?
9									
9 A.	1	C, CS	0–2	00				SWh	
9	3?	AK–SK	3–0	9	175	5			
9	5	fS	2	00	70	18			
3 P.	3	AK, SK	4–1	9	178	6			
3	4	K,fK	0–3	2	230?	8			970?
3	5	fS	0	00					
5	4	SK, SKN	0–4	10	182	22		SWd	
5	5?	SK-fK	0–1	3	175	25			
9	4	SK	2–4	9	S	?			
9	5	S, fS	1	4	22	36			
10									
11 A.	1	C	2–3	1	215	39		W	
11	3	AK, fAK	1–3	8	210	45	9	Ed	2238
11	5	K	0–3	00	SW				690

TABLE V (*Continued*)

Date and time, 1926	Level 1–5	Kind	Density 0–10	Amount 0–4	Azimuth from	Velocity		Position	Height in meters
						Relative, mm/m	Absolute, m/s		
August 10									
6 P.	4	SK–fK	0–2	5	202	24			
6	5	fS	0–1	00					300
11									
9 A.	5	S	0–1	10					60
11	5	S	0–1	2					300
2 P.	4	SK–fK	1	00				NW	
6	3?	AK	0	00				SW	
9	3?	AK	0	00					
12									
9 A.		0		0					
12 M.	3?	CK–AK		3	225	15		SW^d	
6 P.	3	AK	0–1	1	212	12		E	
6	5	fS	0–1	00	220	100	3	S^d	300
* **13**									
9 A.	5	fS	0–1	9					
2 P.	1	C, CS	0–3	00	315	8			
7	3	AK–SK	2–4	2	100	12			
7	5	fS	0	00	172	150			
8:30	1	C	0–2	1	231	3		W^h	
8:30	3	AK		00	122	9		E^d	
8:30	5	fS	0						200
* **14**									
9 A.	3	AK, SK	3–0	7	198	10			
9	"	"			185	9		S^d	
9	5	fS	0	2					300
11	1	C	0–1	00				SW^h	
11	3	AK, SK	3–0	7	195	12	3		2500
11	4	S, SK, K	0–2	00				NW^h	
4 P.	1	C	0–2	3	165	13		SW^d	
4	3	AK	0–4	6	147	27			
4	4?	K-fK	0–1	00				W^h	
* **15**									
8:20 A.	3	AK	3	9	123	9			
8:20	5	S, fSK		2					500
2 P.	3	AK–SK	2–1	9	38	18		S^d	
2	4	fK, SK, KN		1	355	8		NW^d	
6	3	AK–SK		10	40	14	4	N^d	2700
6	4	K, fK		00				NE^d	

TABLE V (*Continued*)

Date and time, 1926	Level 1–5	Kind	Density 0–4	Amount 0–10	Azimuth from	Velocity: Relative, mm/m	Velocity: Absolute, m/s	Position	Height in meters
August									
* **16**									
9 A.	5	fS, fK		00				NW^h	
11	4	fK, SK, K	0–1	00	72	20	3	W^d	1450
"	"	"			"	26			
3 P.	4	K–fK		1	48	21		W^d	
"	"	"			"	33			
6	4	K, fK		00				W^h	
17									
9 A.		0		0					
9 P.									
* **18**									
9 A.		0		0					
3 P.									
3:45	4	fK–SK		00	225	25	6	NW	2400
6	4	fK, SK		00				NW^h	
9		0		0					
19									
9 A.		0		0					
11	1?	CS	0	00	268	20		E^h	
3 P.		0		0					
7	5	fS		00				N	
* **20**									
9 A.	3	AK	0–3	6	249	20		W^d	
11	3	AK, SK	3–0	9	245	20		W^d	
11	4	SK	0	00				W^d	
2:30 P.	4	SN	3	10					1700
2:30	5	fSK	1	1	192	46	1		300
6	4	SKN		10					
* **21**									
9 A.	4	SKN	2–3	9	258	60			
9	5	fS		3				E	
6 P.	1	C, CS	1–2	?					
6	4	SKN		9	260			W^d	
6	5	fS		1					
* **22**									
9 A.	4	SK, K		7	28	30			
9	5?	fS		00	60?	20			
3 P.	4	SK	1–4	8	45	16	3		1800
"		"			358	16	2		900
5	1	C	0–3	1	352	20			
5	4	SK		8	O	16			
9	5?	fSK		6	N?	?			

TABLE V (*Continued*)

Date and time, 1926	Level 1–5	Kind	Density 0–10	Amount 0–10	Azimuth from	Velocity: Relative, mm/m	Velocity: Absolute, m/s	Position	Height in meters
August									
* **23**									
9 A.	4	SK, SKN	2–4	9	248	9			
11	2?	CS–AS?	4	?	359	8			
11	4	K, SKN	2–4	6	245	10			
11	5	fS, SKN	0–2	3	N?	?			900
3 P.	2?	CS–AS		?	355	9			
3	4	K, SKN, fK	1–3	9	350	10	1	N^{d}	900
"	"	"			245	20			875
8:30	4?	SK–fK	0–1	6	174	7			
* **24**									
8 A.	1	C–CS	0–2	5	231	53		SW^{d}	
3 P.	1	vCS–AS		10					
3	2	CK–AK	0–2	2	210	42		S^{d}	
6:30	2?	vCS–AS	4–3	10					
6:30	4	fSK		00	101	50			
25									
9 A.	4?	SN	4	10	W	?			
9 P.	5	fS		00	W				300
* **26**									
9 A.	4	SN		10	W			E^{d}	
9	5	fS		1	WNW				400
3 P.	3?	ASN–SKN	4–2	10	218	10		E^{d}	
3	4?	K, fSKN		3	248	28	1		500
6:40	1	CS	0–2	00				W^{h}	
6:40	4	ASKN		10	215	23			
6:40	4	K, fK	2–3	2	264	12			
6:40	5	fS	0	00	....	0			600
7:40	1	C	1–2	?	288	6			
27									
9 A.	3?	AK	0–2	8		12		S^{d}	
9	5	fS		00					
3 P.	1	C	1–2	1	4	20		W^{d}	
3	4?	K, fK	2	4	NW				
"		"			43	15	1		990
9	3	AK	0–2	10	344	23			
9	5?	SK–fK	3	1					
28									
9 A.	4	SK–fK	0	00				SE^{h}	
3 P.	4	SK–fK		00				h	
7	1	C	0–2	00	302	23			
7	4	SK–fK		00	198	17			

TABLE V (*Concluded*)

Date and time, 1926	Level 1–5	Kind	Density 0–4	Amount 0–10	Azimuth from	Velocity		Position	Height in meters
						Relative, mm/m	Absolute, m/s		
August 29									
9 A.	1?	C	0	00				S^h	
9 P.	2	CK–AK	0–2	4	289	15		N^d	
4	1	C–CS	0–2	1	267	18		SW	
7	1	C	0–2	3	265	28		SW^d	
7	2?	CK		00				SW	
30									
9 A.	1	C–CK	0–3	7	239	21		SE^d	
3 P.	1	C–CK		2	234	20		SE^d	
3	4	SK	0	00	....	0		NE	
7		O		0					
31									
9 A.		O		0					
12 M.	1	C	0–1	4	258	20		SW^d	
3 P.	1	C–CK	0–1	4	262	23		SW^d	
7	1	C		4	268	22			

XI. ATLAS OF CLOUD TYPES OBSERVED AT MOUNT EVANS

PHOTOGRAPHS BY DAVID M. POTTER AND LEONARD R. SCHNEIDER
IDENTIFICATIONS BY S. P. FERGUSSON AND LEONARD R. SCHNEIDER

ABBREVIATIONS (INTERNATIONAL) OF NAMES OF CLOUDS

Cirrus	C	Cumulus	K
Cirro-stratus	C S	Strato-cumulus	S K
Cirro-cumulus	C K	Cumulo-nimbus	K N
Alto-cumulus	A K	Nimbus	N
Alto-stratus	A S	Stratus	S

COMPOUNDS OR MIXTURES

Cirro-stratus veil......................^{v}C S (a very thin uniform C S)

Fracto-cumulus, fracto-stratus, fracto-strato-cumulus, etc. ....f K, f S, f S K, etc.

Clouds indicating rain ....A S N, A K N, A S K N, S N (alto-stratus nimbiformis etc.)

C–C S, A K–A S, etc., indicate mixture of these types; C, C S, A K, A S, etc., indicate forms present at the same level, but not mixed.

PLATE VII

A. C S and C K or A K. Camp Lloyd, August 5, A.M., 1928 (photograph by David M. Potter)

B. C S or high A S. Camp Lloyd, August 5, P.M., 1928 (photograph by David M. Potter)

PLATE VIII

A. ^{v}C S, K and f k. Camp Lloyd, August 9, noon, 1928 (photograph by David M. Potter)

B. C K or A K. Camp Lloyd, August 27, A.M., 1928 (photograph by David M. Potter)

PLATE IX

A. C bands and plumes. Eastern sky from Lower Sand Flat, August 23, A.M., 1928 (photograph by David M. Potter)

B. C with plumes rather low. Camp Lloyd, undated (photograph by David M. Potter)

PLATE X

A. C S or high A S. Camp Lloyd, August 10, P.M., 1928 (photograph by David M. Potter)

B. C, C S, possibly C K mixed; K and f K. Camp Lloyd, August 27, A.M., 1928 (photograph by David M. Potter)

PLATE XI

A. High C S veil, low C with plumes. East of Camp Lloyd, undated (photograph by David M. Potter)

B. Sheet of C, C S in parallel bands; K. Camp Lloyd, August 2 or 3, P.M., 1928 (photograph by David M. Potter)

PLATE XII

A. Dense C or C S. Camp Lloyd, July 24, P.M., 1928 (photograph by David M. Potter)

B. C K with C fringes; K (?) below. Camp Lloyd, August 5, A.M., 1928 (photograph by David M. Potter)

PLATE XIII

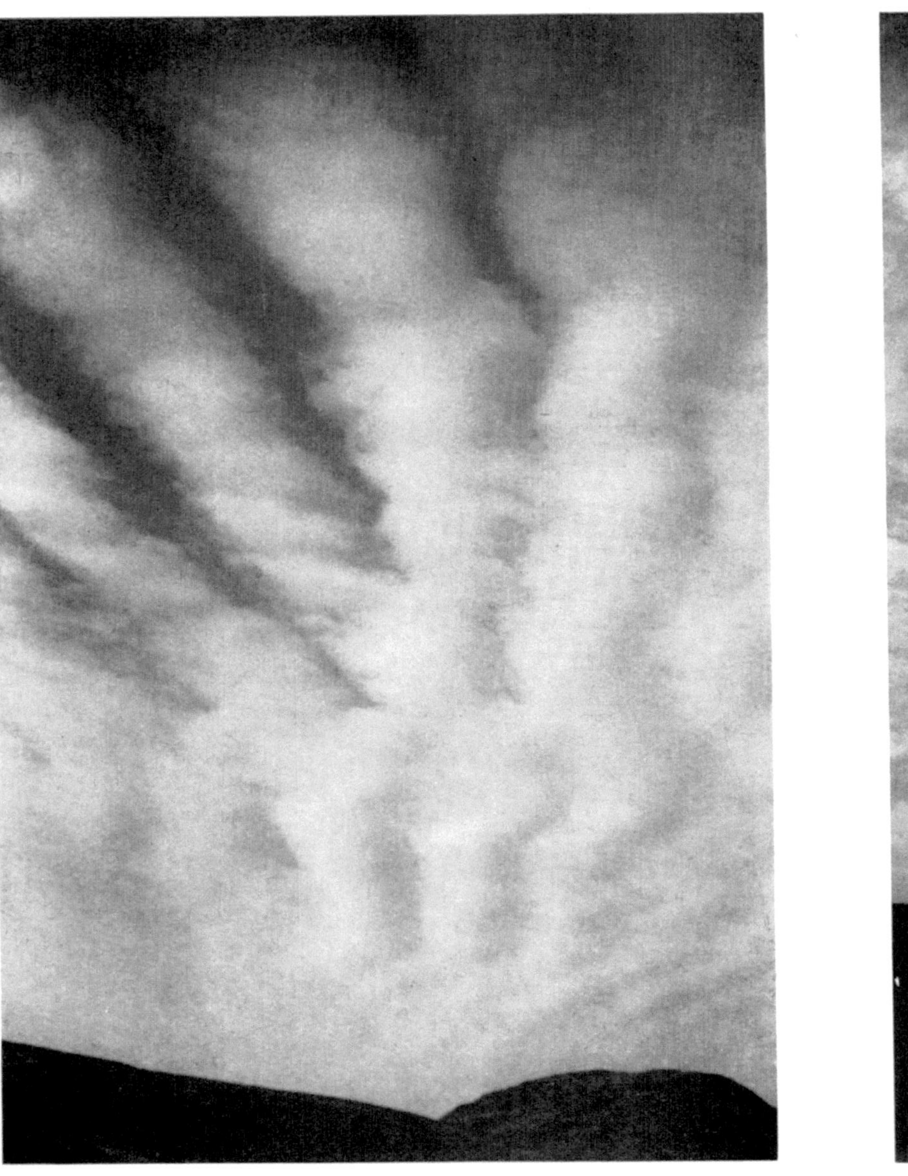

A. Large, low A K. Camp Lloyd, August 21, A.M., 1928 (photograph by David M. Potter)

B. C fibres above C K or A K. Camp Lloyd, August 5, P.M., 1928 (photograph by David M. Potter)

PLATE XIV

A. C S and C above A K; f K on horizon. Camp Lloyd looking south, August 5, P.M., 1928 (photograph by David M. Potter)

B. ^{v}C S, K and f K. Camp Lloyd, August 2 or 3, A.M., 1928 (photograph by David M. Potter)

PLATE XV

A. A K above small K. Camp Lloyd, August 27, A.M., 1928 (photograph by David M Potter)

B. A K and K, f K. Camp Lloyd, August 27, A.M., 1928 (photograph by David M. Potter)

PLATE XVI

A. ^{v}C S, K and f K. Camp Lloyd, August 27, A.M., 1928 (photograph by David M. Potter)

B. Small K — intense convection during calm. Camp Lloyd, July 16, A.M., 1928 (photograph by David M. Potter)

PLATE XVII

A. ${}^{v}C$ S, K, and f K. Camp Lloyd, August 2 or 3, A.M., 1928 (photograph by David M. Potter)

B. Typical foehn sky looking east from Sand Flat on Watson River, August 19, P.M., 1928 (photograph by David M. Potter)

PLATE XVIII

A. Foehn sky with dense blanket above and piled lenticular forms below. Sand Flat, August 19, P.M., 1928 (photograph by David M. Potter)

B. Foehn sky from Sand Flat, August 19, P.M., 1928 (photograph by David M. Potter)

PLATE XIX

Foehn sky with piles of lenticular S K. Sand Flat, August 19, 1928 (photograph by David M. Potter)

PLATE XX

A. Foehn clouds (A K) typical; lenticular A S just above horizon. From Mount Evans looking east, March 3, 10:46 A.M., 1929

B. Same as XX A, but taken at 10:47 A.M.

C. Foehn clouds (A K) multilayered, near horizon, Mount Evans, March 6, 3:18 P.M., 1929

D. A S at Mount Evans. Winter of 1928–29 (photographs by Leonard R. Schneider)

PLATE XXI

A

B

C

D

A. Foehn clouds (A K) multilayered, dissolving. A, B, C, and D were photographed successively at ten-minute intervals on March 8, 1929, beginning at 10:50 A.M. (photographs by Leonard R. Schneider)

PLATE XXII

A. C S veil, A K and A S. Foehn-like edges, large. View toward east from Mount Evans, March 9, 9:20 A.M., 1929

B. C changing to C S, both moving southwest. View southeast from Mount Evans, March 1, 2:14 P.M., 1929

C. K extending to and spreading out at A K level. View north-northwest from Mount Evans, June 18, 9:45 A.M., 1929

D. A K in vertical bands moving from southwest. View from Mount Evans towards zenith, March 16, 2:10 P.M., 1929 (photographs by Leonard R. Schneider)

PLATE XXIII

A. A K between S and C S. View toward south from Mount Evans, April 22, 11:00 A.M., 1929

B. Low C K or high A K. View southwest from Mount Evans, April 24, 3:54 P.M., 1929

C. Small, flat K or S K. View north from Mount Evans, May 11, 12:22 P.M., 1929

D. Small K and S K. View north-northwest from Mount Evans, May 11, 12:22 P.M., 1929 (photographs by Leonard R. Schneider)